# Application of Vision Technology and Artificial Intelligence in Smart Farming—2nd Edition

# Application of Vision Technology and Artificial Intelligence in Smart Farming—2nd Edition

Guest Editors

**Xiuguo Zou**
**Xiaochen Zhu**
**Wentian Zhang**
**Yan Qian**
**Yuhua Li**

MDPI

Basel • Beijing • Wuhan • Barcelona • Belgrade • Novi Sad • Cluj • Manchester

*Guest Editors*

Xiuguo Zou
College of Smart Agriculture
Nanjing Agricultural University
Nanjing
China

Xiaochen Zhu
School of Applied Meteorology
Nanjing University of Information Science and Technology
Nanjing
China

Wentian Zhang
Faculty of Engineering and Information Technology
University of Technology Sydney
Sydney
Australia

Yan Qian
College of Smart Agriculture
Nanjing Agricultural University
Nanjing
China

Yuhua Li
College of Smart Agriculture
Nanjing Agricultural University
Nanjing
China

*Editorial Office*
MDPI AG
Grosspeteranlage 5
4052 Basel, Switzerland

This is a reprint of the Special Issue, published open access by the journal *Agriculture* (ISSN 2077-0472), freely accessible at: https://www.mdpi.com/journal/agriculture/special_issues/V7B85K98W9.

For citation purposes, cite each article independently as indicated on the article page online and as indicated below:

Lastname, A.A.; Lastname, B.B. Article Title. *Journal Name* **Year**, *Volume Number*, Page Range.

**ISBN 978-3-7258-6208-5 (Hbk)**
**ISBN 978-3-7258-6209-2 (PDF)**
**https://doi.org/10.3390/books978-3-7258-6209-2**

# Contents

# About the Editors

**Xiuguo Zou**

Xiuguo Zou received his Ph.D. degree in Agricultural Bio-environment and Energy Engineering from Nanjing Agricultural University in 2013. Since 2013, he has been an associate professor at Nanjing Agricultural University. He is a senior member of the China Electronics Society, and also a member of the following: IEEE, United States; Rural Professional Technology Association of Jiangsu Province; Agricultural Automation Special Committee of Jiangsu Automation Society; the Facility Agricultural Equipment Branch of Jiangsu Agricultural Society; and the Artificial Intelligence Society of Jiangsu Province. His current research mainly focuses on agricultural electronics and information technology, machine vision and image processing, agricultural biological environment control and equipment, and big data agricultural technology.

**Xiaochen Zhu**

Xiaochen Zhu received his Ph.D. degree from the School of Applied Meteorology of Nanjing University of Information Science and Technology, Nanjing, China, in 2017. From October 2015 to October 2016 and from February to June 2018, he was a visiting scholar at the College of Agriculture, University of Kentucky. He is now an associate professor at Nanjing University of Information Science and Technology. His research interests include refined meteorological services, meteorological disaster prevention and reduction, and AI meteorology.

**Wentian Zhang**

Wentian Zhang received his Ph.D. degree from the University of Technology Sydney, Sydney, Australia, in 2020. He was an assistant research fellow with Prof. Steven Su at the University of Technology Sydney. His research interests include electronic nose systems, machine learning, signal processing, and medical devices.

**Yan Qian**

Yan Qian obtained her Ph.D. degree in Agricultural Electrification and Automation Engineering from Nanjing Agricultural University in 2014. She has been affiliated with Nanjing Agricultural University since 2006, first serving as a lecturer in the College of Engineering, and currently holding the position of an associate professor in the College of Smart Agriculture (College of Artificial Intelligence). Her main research interests include agricultural robots, computer vision technology, and virtual reality-based 3D reconstruction.

**Yuhua Li**

Yuhua Li received her Ph.D. degree in Information and Communication Engineering from Xi'an Jiaotong University, Xi'an, China, in 2015. Since 2015, she has been a University Lecturer in the Department of Electronic Information of Nanjing Algaculture University, Nanjing, China. Over the past decade, she has published over 20 papers spanning a range of theoretical and applied problems in machine learning and computer vision. Her research has been funded by the National Natural Science Foundation of China and the Fundamental Research Funds for the Central Universities. Her current research interests include image processing and pattern recognition, animal and plant phenotyping, and the applications of computer vision in algaculture.

*agriculture*

MDPI

*Editorial*

# Computer Vision and Artificial Intelligence Driving the Advancement of Agricultural Intelligence in Dynamic Environments

**Xiuguo Zou [1,*], Xiaochen Zhu [2], Wentian Zhang [3], Yan Qian [1] and Yuhua Li [1]**

[1] College of Artificial Intelligence, Nanjing Agricultural University, Nanjing 210031, China; qianyan@njau.edu.cn (Y.Q.); lyhresearch@njau.edu.cn (Y.L.)
[2] School of Applied Meteorology, Nanjing University of Information Science and Technology, Nanjing 210044, China; xiaochen.zhu@nuist.edu.cn
[3] Faculty of Engineering and Information Technology, University of Technology Sydney, Sydney, NSW 2007, Australia; wentian.zhang@alumni.uts.edu.au
* Correspondence: zouxiuguo@njau.edu.cn; Tel.: +86-25-58606585

The rise of agricultural digitalization is progressively reshaping the conventional extended management model through the profound integration of intelligent sensing technology and artificial intelligence. The efficient transfer and application of these technologies therefore remains a major challenge, since modern agricultural settings are dynamically complicated and mostly unstructured. In response, researchers have been refining these technologies with a focus on lightweight design and adaptability, aiming to facilitate the large-scale application of smart sensing in agriculture.

This Special Issue examines the application of and breakthroughs in computer vision and artificial intelligence in agriculture. It features fifteen articles reporting advances in areas such as crop growth monitoring, fruit grading detection, animal health and behavior recognition, and remote sensing image analysis. Table 1 illustrates the essential information contributed by these fifteen papers. These studies employ a range of advanced technologies, including multi-scenario real-time detection via YOLO-based frameworks, Transformer architecture-based approaches, methods employing generative adversarial networks (GANs), and other algorithmic strategies. Researchers have made significant strides in a variety of agricultural tasks and contexts. Field tests have successfully shown that these technologies are feasible in complex and dynamic agricultural settings, offering technical support for the expanded use of computer vision and artificial intelligence in agriculture.

The YOLO series of object detection algorithms is a typical framework in computer vision. Because of its high accuracy and real-time performance, it has quickly developed across a variety of fields and is now a robust baseline for many downstream tasks. In the agricultural domain, researchers have introduced several targeted improvements to the YOLO architecture in response to challenges such as complex background interference, small object detection, and the need for deployment on edge devices, focusing particularly on enhancing real-time detection capabilities and optimizing model efficiency. Xin et al. [1] developed a mobile system for the detection and retrieval of dead chickens utilizing an enhanced YOLOv6 model, with a success rate of 81.3% indicated in field trials. Sun et al. [2] created a lightweight YOLOv8-Pearpollen model for evaluating the germination vigor of pear tree pollen; through the application of knowledge distillation and model pruning, they achieved a detection speed of 147.1 FPS. Chen et al. [3] addressed the challenges posed by the small size and frequent occlusion of Chinese bayberry fruits with an improved YOLOv7-Tiny algorithm, which integrates partial convolutions, the SimAM attention mechanism, and a SIoU loss function. This optimized model retains a compact design and high efficiency, achieving a mere 4% miss rate and a 3% false detection rate in field tests. To tackle the issue of efficient citrus fruit detection in intricate orchard environments, Jing et al. [4] developed the YOLOv7-Tiny-BVP network by incorporating the BiFormer attention mechanism, the VoVGSCSP module, and partial convolutions into the YOLOv7-Tiny framework. This methodology produced a lightweight and highly precise citrus detection model, achieving enhancements of 0.9% in recognition accuracy, 2.02 FPS, and 1% in F1 score. Lin et al. [5] presented AG-YOLO, an enhanced algorithm based on the NextViT backbone and a Global Context Fusion Module. Specifically designed for occlusion-aware citrus detection, the research also involved the establishment of a diverse and comprehensive dataset, resulting in a testing performance of 83.2% mAP@0.5 and a detection speed of 34.22 FPS.

**Table 1.** Key information contributed by each paper.

| Authors | Objects | Models | Contributions |
|---|---|---|---|
| Xin et al. [1] | Dead yellow-feather broilers | YOLOv6 | Real-time detection and robotic grasping of dead yellow-feather broilers |
| Sun et al. [2] | Pollen germination vigor in pear trees | YOLOv8 | Real-time detection of pollen germination vigor in pear trees |
| Chen et al. [3] | Bayberry | YOLOv7-Tiny | Real-time detection and picking of Chinese bayberries |
| Jing et al. [4] | Citrus | YOLOv7-Tiny | Citrus fruit recognition under varying occlusion scenarios and lighting conditions |
| Lin et al. [5] | Citrus | YOLO, NextViT | Citrus fruit detection in complex environments |
| Jin et al. [6] | Wheat | Transformer, Symmetric Diffusion | Precise monitoring of wheat growth status and yield prediction in high-density agricultural environments |
| Huang et al. [7] | Pear | RT-DETR | Rapid detection of Xinli No. 7 fruit in natural environments |
| Huo et al. [8] | Tomato | StyleGAN3, Transformer | Recognition of growth stages in greenhouse tomato cultivation |
| Li et al. [9] | Weed | Semi-Supervised Diffusion Model | Weed detection in agricultural scenarios |
| Wang et al. [10] | Apple | Edge Attention Mechanism | Extraction of apple phenotypic features and recognition of growth abnormalities |
| Chen et al. [11] | Oudemansiella raphanipes | CNN, Three-Teacher Knowledge Distillation | Quality grading of Oudemansiella raphanipes |
| Cho et al. [12] | Dairy cows | Weakly Supervised Representation Learning | Classification and detection of mastitis in dairy cows |
| Yang et al. [13] | Yak | YOLOv7-pose | Detection and classification of behavior patterns in yaks |
| Wu et al. [14] | Zhuji Torreya | Machine Learning | Suitable habitats for Torreya in Zhuji City |
| Wu et al. [15] | Walnut | w-YOLO | Identification and counting of small walnut fruits in UAV remote sensing images |
| Pacioni et al. [16] | Vineyard shoots | Mask R-CNN (ResNet50 backbone), YOLOv8 | Direct detection of vine shoot cutting areas for intelligent pruning |

By leveraging self-attention mechanisms to model long-range dependencies in both sequential and spatial data, Transformer architectures enable high-precision crop phenotyping, yield forecasting, and pest-and-disease detection in agricultural settings, thereby markedly enhancing intelligent monitoring and decision-making. Jin et al. [6] presented SDS-Net, a Symmetric Diffusion Segmentation Network utilizing a Transformer backbone. By effectively combining Transformer blocks with symmetric diffusion modules and a symmetric attention mechanism, SDS-Net significantly enhances wheat instance segmentation accuracy in high-density planting environments, achieving an F1 score of 0.89 for segmentation and 0.92 for growth measurement. Huang et al. [7] addressed the challenge of identifying Xinli No. 7 pears in unstructured outdoor environments by proposing a lightweight Transformer-based detection framework. Their optimized model achieved a 48.47% reduction in total parameter count, a 56.2% decrease in computational overhead (FLOPs), and a 48.31% reduction in memory usage, all while maintaining robust detection accuracy.

However, the efficacy of deep learning models is fundamentally dependent on the availability of high-quality and diverse training data. In agricultural contexts, data collection often lacks standardized protocols, resulting in imbalanced datasets and challenges in achieving both efficiency and accuracy in recognition. Furthermore, traditional augmentation methods generate images of limited fidelity and diversity, rendering them insufficient for the rigorous requirements of effective model training. To address these challenges, Huo et al. [8] introduced an innovative framework that merges the capabilities of the StyleGAN3 generative adversarial network with Vision Transformer techniques, effectively addressing the shortcomings of traditional data collection and recognition approaches. Through the innovative use of StyleGAN3 to create high-fidelity images of tomato growth stages, they effectively tackled challenges related to data scarcity and class imbalance. This approach achieved a remarkable accuracy of 98.39% on the test set, with an average inference time of just 9.5 ms. Li et al. [9] utilized diffusion models for weed detection by creating a semi-supervised, diffusion-generated recurrent network that excels at generating high-quality synthetic data. By incorporating generation-aware attention modules and a hybrid diffusion loss function, alongside real-world images, this approach significantly lessens the dependence on extensive annotated datasets. It also boosts model adaptability and detection performance in challenging, data-scarce agricultural settings.

The studies discussed have mainly concentrated on object detection within conventional agricultural environments, investigating the use of YOLO-based models, Transformer architectures, and GANs in these contexts. With an increasing need for advanced agricultural practices, there is a notable trend towards collaborative efforts that encompass multiple tasks. In response, extensive work has been conducted to enhance algorithmic frameworks and introduce a range of innovations suited to diverse application scenarios.

In plant phenotyping and quality grading, Wang et al. [10] considered environmental variables like climate and soil by merging instance segmentation with natural language processing. This cross-modal fusion facilitated the integration of multidimensional data for the extraction of phenotypic traits in apples and the detection of growth anomalies, resulting in a more thorough framework for identifying anomalies and assessing quality. Similarly, Chen et al. [11] enhanced the automated grading of Oudemansiella raphanipes in resource-limited settings using a three-teacher knowledge distillation approach, showing that parallel, standard cascade, and residual-connected architectures each markedly improve the accuracy of compact models. Turning to livestock health monitoring, Cho et al. [12] introduced a weakly supervised classification approach for detecting mastitis utilizing representation learning techniques. This method utilizes a one-dimensional convolutional neural network autoencoder, incorporating a classifier branch within the latent space of the autoencoder. This facilitates efficient representation learning of milking data and the weakly supervised identification of mastitis symptoms. Current image-based techniques for detecting animal behavior face challenges in accurately differentiating between closely related behavioral patterns in yaks, including standing, walking, and excreting, as these actions exhibit similar leg postures. In response to this issue, Yang et al. [13] introduced a method for detecting yak behavior utilizing an enhanced YOLOv7-pose model. The integration of the Mish activation function, the SPPFCSPC feature extraction module, and a dynamic head module notably improved the model's capacity to identify yak behavioral patterns. The method demonstrated a precision of 89.9% and an mAP@0.5 of 90.4%. In the domain of remote sensing and habitat assessment, Wu et al. [14] employed various machine learning algorithms, such as Random Forest (RF), Support Vector Machine (SVM), and Naive Bayes (NB), to conduct suitability zoning for Torreya grandis cultivation in Zhuji City. The researchers integrated ecological factor raster data with the geographic coordinates of current Torreya grandis distribution points to pinpoint areas conducive to its growth. The results indicate that suitable locations for Torreya cultivation are primarily confined to mountainous and hilly terrains. The central basin and northern river plain zones are considered inappropriate. Among the variables examined, edaphic and topographic factors impose more significant constraints on Torreya distribution than climatic factors do. To address the challenge of detecting small walnut fruits in imagery obtained from UAVs, Wu et al. [15] developed an efficient detection network called w-YOLO. In direct comparisons with leading object detection frameworks, w-YOLO achieved a mAP@0.5 of 97% and an F1 score of 92%. The model demonstrated a 52.3% decrease in parameter count when compared to YOLOv8s. This demonstrates remarkable precision and effectiveness, even under the difficult lighting conditions commonly found in Yunnan Province. Finally, to resolve the issue of intelligent vineyard pruning via cutting area detection, Pacioni et al. [16] developed a labeled VidPrune Dataset and tested various segmentation models. The optimal model achieved mAP50 = 0.883 (shoot class) and a 55 ms inference time using Jetson AGX Orin, enabling efficient direct cutting area detection.

Intelligent detection in dynamic agricultural environments faces challenges such as varying lighting conditions, occlusion, limited computational resources, and the need for real-time performance, as well as the integration of multimodal data. The studies published in this Special Issue address these challenges from multiple technical perspectives, demonstrating breakthroughs in adapting to dynamic agricultural scenarios. These works validate the feasibility of applying next-generation artificial intelligence and computer vision techniques to smart agriculture. Researchers have utilized techniques such as pruning, quantization, knowledge distillation with cascade design, and Efficient Transformer architectures to reduce computational complexity in model lightweighting. The studies have integrated diffusion models, generative adversarial networks, and semi-supervised learning to address the challenges of limited data diversity and class imbalance in agricultural datasets. The integration of visual and historical information in application scenarios facilitates multimodal and comprehensive perception of environmental conditions. This Special Issue compiles various studies that align with the advanced techniques mentioned, aiming to inspire new directions in agricultural intelligence and promote ongoing research and advancements in related areas. Ongoing technological refinement and interdisciplinary collaboration are expected to drive the advancement of smart agriculture, resulting in concurrent improvements in productivity and sustainability.

**Acknowledgments:** We are thankful to Qiuhong Zhang, Hongfei Chen, and Jinlin Liu, who have contributed to helping with paper information sorting.

**Conflicts of Interest:** The authors declare no conflicts of interest.

## References


1. Xin, C.; Li, H.; Li, Y.; Wang, M.; Lin, W.; Wang, S.; Zhang, W.; Xiao, M.; Zou, X. Research on an Identification and Grasping Device for Dead Yellow-Feather Broilers in Flat Houses Based on Deep Learning. *Agriculture* **2024**, *14*, 1614. [CrossRef]
2. Sun, W.; Chen, C.; Liu, T.; Jiang, H.; Tian, L.; Fu, X.; Niu, M.; Huang, S.; Hu, F. YOLOv8-Pearpollen: Method for the Lightweight Identification of Pollen Germination Vigor in Pear Trees. *Agriculture* **2024**, *14*, 1348. [CrossRef]
3. Chen, Z.; Qian, M.; Zhang, X.; Zhu, J. Chinese Bayberry Detection in an Orchard Environment Based on an Improved YOLOv7-Tiny Model. *Agriculture* **2024**, *14*, 1725. [CrossRef]
4. Jing, J.; Zhai, M.; Dou, S.; Wang, L.; Lou, B.; Yan, J.; Yuan, S. Optimizing the YOLOv7-Tiny model with multiple strategies for citrus fruit yield estimation in complex scenarios. *Agriculture* **2024**, *14*, 303. [CrossRef]
5. Lin, Y.; Huang, Z.; Liang, Y.; Liu, Y.; Jiang, W. Ag-yolo: A rapid citrus fruit detection algorithm with global context fusion. *Agriculture* **2024**, *14*, 114. [CrossRef]
6. Jin, Z.; Hong, W.; Wang, Y.; Jiang, C.; Zhang, B.; Sun, Z.; Liu, S. A Transformer-Based Symmetric Diffusion Segmentation Network for Wheat Growth Monitoring and Yield Counting. *Agriculture* **2025**, *15*, 670. [CrossRef]
7. Huang, Z.; Zhang, X.; Wang, H.; Wei, H.; Zhang, Y.; Zhou, G. Pear Fruit Detection Model in Natural Environment Based on Lightweight Transformer Architecture. *Agriculture* **2024**, *15*, 24. [CrossRef]
8. Huo, Y.; Liu, Y.; He, P.; Hu, L.; Gao, W.; Gu, L. Identifying Tomato Growth Stages in Protected Agriculture with StyleGAN3–Synthetic Images and Vision Transformer. *Agriculture* **2025**, *15*, 120. [CrossRef]
9. Li, R.; Wang, X.; Cui, Y.; Xu, Y.; Zhou, Y.; Tang, X.; Jiang, C.; Song, Y.; Dong, H.; Yan, S. A Semi-Supervised Diffusion-Based Framework for Weed Detection in Precision Agricultural Scenarios Using a Generative Attention Mechanism. *Agriculture* **2025**, *15*, 434. [CrossRef]
10. Wang, Z.; Cui, W.; Huang, C.; Zhou, Y.; Zhao, Z.; Yue, Y.; Dong, X.; Lv, C. Framework for Apple Phenotype Feature Extraction Using Instance Segmentation and Edge Attention Mechanism. *Agriculture* **2025**, *15*, 305. [CrossRef]
11. Chen, H.; Huang, H.; Peng, Y.; Zhou, H.; Hu, H.; Liu, M. Quality Grading of Oudemansiella raphanipes Using Three-Teacher Knowledge Distillation with Cascaded Structure for LightWeight Neural Networks. *Agriculture* **2025**, *15*, 301. [CrossRef]
12. Cho, S.H.; Lee, M.; Lee, W.H.; Seo, S.; Lee, D.H. Mastitis Classification in Dairy Cows Using Weakly Supervised Representation Learning. *Agriculture* **2024**, *14*, 2084. [CrossRef]
13. Yang, Y.; Deng, Y.; Li, J.; Liu, M.; Yao, Y.; Peng, Z.; Gu, L.; Peng, Y. An Effective Yak Behavior Classification Model with Improved YOLO-Pose Network Using Yak Skeleton Key Points Images. *Agriculture* **2024**, *14*, 1796. [CrossRef]
14. Wu, L.; Yang, L.; Li, Y.; Shi, J.; Zhu, X.; Zeng, Y. Evaluation of the Habitat Suitability for Zhuji Torreya Based on Machine Learning Algorithms. *Agriculture* **2024**, *14*, 1077. [CrossRef]
15. Wu, M.; Yun, L.; Xue, C.; Chen, Z.; Xia, Y. Walnut Recognition Method for UAV Remote Sensing Images. *Agriculture* **2024**, *14*, 646. [CrossRef]
16. Pacioni, E.; Abengózar, E.; Macías Macías, M.; García-Orellana, C.J.; Gallardo, R.; González Velasco, H.M. Towards Intelligent Pruning of Vineyards by Direct Detection of Cutting Areas. *Agriculture* **2025**, *15*, 1154. [CrossRef]

agriculture

MDPI

Article

# Towards Intelligent Pruning of Vineyards by Direct Detection of Cutting Areas

**Elia Pacioni [1,2], Eugenio Abengózar [3], Miguel Macías Macías [1,4,*], Carlos J. García-Orellana [3,4], Ramón Gallardo [4,5] and Horacio M. González Velasco [4,5]**

1 Centro Universitario de Mérida, Universidad de Extremadura, Avda. Santa Teresa de Jornet, 38, 06800 Mérida, Spain; eliapacioni@unex.es
2 HES-SO Valais/Wallis, 3960 Sierre, Switzerland
3 Facultad de Ciencias, Universidad de Extremadura, Avda. de Elvas, s/n, 06006 Badajoz, Spain; eabengozar@unex.es (E.A.); cjgarcia@unex.es (C.J.G.-O.)
4 Instituto de Computación Científica Avanzada, Av. de la Investigación, s/n, 06006 Badajoz, Spain; rgallardo@unex.es (R.G.); hmgvelas@unex.es (H.M.G.V.)
5 Escuela Politécnica, Universidad de Extremadura, Avda. Universidad s/n, 10003 Cáceres, Spain
* Correspondence: mmacias@unex.es

**Abstract:** The development of robots for automatic pruning of vineyards using deep learning techniques seems feasible in the medium term. In this context, it is essential to propose and study solutions that can be deployed on portable hardware, with artificial intelligence capabilities but reduced computing power. In this paper, we propose a novel approach to vineyard pruning by direct detection of cutting areas in real time by comparing Mask R-CNN and YOLOv8 performances. The studied object segmentation architectures are able to segment the image by locating the trunk, and pruned and not pruned vine shoots. Our study analyzes the performance of both frameworks in terms of segmentation efficiency and inference times on a Jetson AGX Orin GPU. To compare segmentation efficiency, we used the mAP50 and AP50 per category metrics. Our results show that YOLOv8 is superior both in segmentation efficiency and inference time. Specifically, YOLOv8-S exhibits the best tradeoff between efficiency and inference time, showing an mAP50 of 0.883 and an AP50 of 0.748 for the shoot class, with an inference time of around 55 ms on a Jetson AGX Orin.

**Keywords:** computer vision; object segmentation; Mask R-CNN; YOLOv8; vineyard pruning

## 1. Introduction

The global wine industry plays a vital role in the economy, culture, and environment. Economically, it generates millions of jobs, drives significant exports from leading producers like France, Italy, Spain, or Chile, and adds high value to global markets. The wine industry is facing several technical challenges, such as precision viticulture, climate adaptation, fermentation control, and sustainability. Also, some operational requirements in wine production such as regulatory compliance and quality assurance need to be addressed [1].

The tasks that need to be performed by skilled labor in the agricultural sector are numerous, such as phytosanitary treatments, managing crops, harvesting, and pruning. All these activities together with the shortage of expert labor create a major problem [2]. At the same time, this issue favors investment in technology (around USD 106 billion [3]) and brings the agricultural sector closer to the technological world [4].

The present study addresses the issue of automatized vineyard pruning. This section offers an analysis of the manual techniques and technological processes that have been

 https://doi.org/10.3390/agriculture15111154

employed in this field to date. Subsequently, the technological advances achieved in pruning vineyards are presented, including the utilization of artificial intelligence (AI) in agricultural practices.

### 1.1. Vineyard Pruning

There are numerous reasons why a vineyard should be pruned. The most significant of these are growth control, grape quality, production improvement, disease prevention, and harvest facilitation [5]. There are multiple methods of growing vines outdoors (Royat hedge, Guyot hedge, vertical cordon, . . . ). This study focuses exclusively on the complete or partial removal of shoots. In the cases where the shoots are partially pruned, it is important to maintain a specific number of buds (typically two), as vine renewal process initiates at these points.

Numerous companies are investing in technology, with the objective of compensating for the labor shortage through the implementation of intelligent automated systems [6]. The use of Real-Time Kinematic (RTK) technology has facilitated the mechanization of many stages of agricultural activities, starting with planting, harvesting, and concluding with plant care. This technology provides the farmer with a Global Positioning System (GPS) coordinate map, which is useful to automate planned activities.

Nevertheless, although these procedures have achieved a high degree of acceptance, there are still activities that are performed manually by specialized personnel, which increases the problem of labor shortages. Although progress has been made in the area of mechanical prepruning (as illustrated in Figure 1), which serves to simplify the subsequent work of manual pruning, full automation of this process requires major developments.

**Figure 1.** Prepruning machine of a Royat vineyard.

In order to guide a robotic arm in the execution of optimal cuts during vine pruning, efficient and real-time identification of shoots is necessary, which is set around 30 FPS to roughly match the speed of human perception [7]. Although human perception is taken as a reference for the frame threshold, 30 FPS is not necessary to process in this study because the technical time required for the robot to perform the pruning should be considered. Therefore, even a significantly lower frame-rate can be considered suitable. In similar studies, as presented by Botterill et al. [8], they achieved a speed of 2.5 FPS, which was deemed suitable for the task.

This can be achieved by deploying AI algorithms and computer vision (CV) technologies on portable hardware, which typically has less computing power than a desktop Graphics Processing Unit (GPU).

### 1.2. Related Works

In this paper, we deal with the problem of vineyard pruning, which has been treated from different points of view in the literature. We analyzed several approaches to the problem such as reconstruction of the vine structure [8,9], volumetric identification of shoots and buds [10], or detection of branches and buds [11–13].

Within the first category, Botterill et al. [8] presented a complete robotic platform that specializes in creating a three-dimensional model of vines using a structure that completely covers the plant and prunes it in 2 min. Gentilhomme et al. [9] showed a method for automatic extraction of the grapevine structure from two-dimensional images using a Stacked Hourglass Network, allowing to address the problem of having an unknown number of branches in the vine plant. They achieved a 95% precision and 90% recall for node prediction. In the category of volumetric identification of shoots and buds, Lanza et al. [10] evaluated a portable vision-based acquisition device which merges depth, tracking and RGB information to estimate the wood volume of the shoots, and You Only Look Once (YOLO) version 8 to detect buds, with a 0.79 F1-score.

In the group of papers dealing with the direct detection of shoots and buds, Villegas et at. [11] implemented a MobileNet architecture to segment buds in the images reporting a 0.88 F1-score in this task. Chen et al. [12] dealt with the identification of the shoot cutting points by locating the buds with two networks, and they reported an F1-score of 0.80 for the YOLOv5 model. Finally, Oliveira et al. [13] used four versions of YOLO to identify nodes in two-dimensional grapevine images, concluding that YOLOv7 was the best option, with an F1-score reaching a value of 0.86.

In some of the works cited above, controlled backgrounds were used [8–10,14] to improve performance, but this option increases the cost of a final prototype, as in [8]. It would be desirable to avoid this element in order to develop a prototype, as in the approaches of [12,13].

Our approach attempts to test whether an algorithm is able to directly detect the section of the vine shoot where the cutting tool should act, even in vineyards where partial pruning was performed, which complicates the task. We presented preliminary results in [15] with a *Mask Region-based Convolutional Neural Network* (Mask R-CNN), and now we extend the work by comparing performance with one of the most reliable state-of-the-art network models (YOLOv8).

### 1.3. Object Detection Algorithms

We compared the performance in our problem of two of the most common instance segmentation models: Mask R-CNN and YOLOv8. The choice is due to the fact that Mask R-CNN represents a milestone in the field of mask-level segmentation, while YOLO is among the fastest and most suitable models for working in real-time projects. The Mask R-CNN algorithm was developed by *Facebook AI Research* [16] in 2017, born as an extension of Faster R-CNN [17]. We used the Mask R-CNN implementation available in Detectron2 [18] with a ResNet50 backbone. The YOLO family of algorithms [19] has evolved in time improving the quality of its predictions while preserving high inference speeds. We used the YOLOv8 implementation developed by Ultralytics [20].

The main differences between the Mask R-CNN and YOLO algorithms are as follows:

- Mask R-CNN uses two neural networks, one for boxes and classes and the other for masks, while YOLO uses only one;

- Mask R-CNN uses the anchor concept to propose regions, while YOLOv8 uses an anchor-free approach;
- Mask R-CNN uses the RoIAlign technique for mask alignment, while YOLOv8 does not provide anything analogous;
- Mask R-CNN is designed for segmentation mask generation, while YOLOv8 supports mask-level segmentation, but has historically been used for object detection tasks.

Our final aim is to run the object detector embedded in a robot, so we must use a compact and a high-performance computing module such as the Jetson family developed by NVIDIA. In this work, we want to check if the YOLOv8 algorithms can improve our previous results in terms of object segmentation and inference times using an NVIDIA Jetson AGX Orin Developer Kit 64 Gb (NVIDIA, Santa Clara, CA, USA).

The rest of the document is structured as follows. Section 2 defines the starting hypothesis and aims of this work. Section 3 describes used materials, applied methodology, configurations, and metrics to evaluate the results. Section 4 presents the experiments performed and compares the two algorithms chosen to approach the problem. Finally, Section 5 presents the conclusions of the study and outlines the trajectory of future work.

## 2. Hypothesis and Aims

Our starting hypothesis is that we can train an object detector to locate the vine shoots present in an image, but unlike other approaches, we only try to localize the initial portion of the shoots, including the first buds. The localization of these regions will allow us to estimate the possible cutting points of the shoots in a post-processing phase. Previous results [15] have indicated that to improve the performance in the pruning process, we should include three categories of objects that appear in a partially pruned vine plant: shoots, pruned shoots, and trunk. As our algorithms must run on board a robot, we need to use an embedded computing board. An NVIDIA Jetson AGX Orin was selected for this experiment, offering an AI performance of 275 TOPS, 2048 NVIDIA Ampere cores, 64 Tensor Cores, and a 64-bit 12-core Arm Cortex-A78AE CPU (https://www.nvidia.com/en-us/autonomous-machines/embedded-systems/jetson-orin/, accessed on 20 May 2025).

The aim of this work can be divided into two subobjectives:

1. To implement the use of the YOLOv8 algorithm in order to address the segmentation problem and compare the results with those previously obtained using Mask R-CNN;
2. To test the trained models on a Jetson AGX Orin system and determine under which conditions the obtained results allow a real-time operation of a robotic pruner.

## 3. Materials and Methods

This section outlines the dataset employed, the algorithms configurations, and the metrics utilized for evaluation.

### *3.1. Dataset*

In order to train and test the algorithms, we developed a labeled dataset of vineyard images [21]. The images were taken in the southwest of the province of Badajoz in Spain, where the pardina variety and the Royat hedge arrangement style are the most common. In this article, we worked with pardina vineyards pruned in the double Royat style, in which the shoots are vertically trellised on wires.

Images were taken using two different cameras: Canon EOS 2000D and Samsung SM-A715F (smartphone) (Suwon, Republic of Kore). The image collection began in 2021 and was repeated annually, though the 2023 images were not used in the current research. The prototype set and its ground truth (GT) were built using the VGG Image Annotator (VIA) software [22]. During the label creation process, the three classes *shoot*, *pruned shoot*, and *trunk* were registered. Each image always contains at least one *trunk* object, along with objects from the other two classes. The *pruned shoot* class was added to assist the system in developing generalizable solutions. However, it is not guaranteed that all three categories are present in every image.

The addition of the *pruned shoot* class to our work is the result of a previous study using Mask R-CNN on an initial set of images. Early models tended to segment pruned shoots as shoots to be pruned. Labeling examples of this new class improved the performance in detecting shoots that actually need to be pruned. Figure 2 illustrates an example of a partially pruned vine plant labeled through the VIA software. As we can see, for the *shoot class*, only the initial portion of the shoot is considered, as previously stated.

**Figure 2.** Examples of images labeled with VIA software for classes *shoot* (yellow), *pruned shoot* (blue), and *trunk* (green).

Table 1 presents the dataset containing 536 images with different resolutions. They were taken from different positions and distances, looking for the robustness of the system. After the annotation process, as presented in Table 2, we obtained 5329 objects of the class *shoot*, 887 for *pruned shoot*, and 752 for *trunk*. This dataset [21] is available to the research community as detailed in the Data Availability Statement section.

**Table 1.** Images resolution.

| Number of Images | Resolution |
| --- | --- |
| 151 | 6000 × 4000 |
| 261 | 4624 × 2604 |
| 23 | 2048 × 1532 |
| 22 | 1600 × 1197 |
| 79 | 800 × 600 |

**Table 2.** Structure of the dataset.

| Dataset | Shoots | Pruned Shoots | Trunk | Total | Images |
| --- | --- | --- | --- | --- | --- |
| Original | 5329 | 887 | 752 | 6968 | 536 |
| Balanced | 5329 | 5029 | 1741 | 12,099 | 1288 |

As we can see in Table 1, the dataset is clearly imbalanced, maintaining a 6-to-1 ratio between the number of prototypes of the shoot and pruned shoot classes. This is because the pruned shoot class was not included at the beginning of our work. It is well known that machine learning models struggle to differentiate minority classes and tend to treat them as noise with respect to majority classes [23]. In [23,24], the authors showed different strategies to mitigate the class imbalanced data. In this work, we balanced the dataset by applying data augmentation on the images where objects of the *pruned shoot* class are in the majority. In this way, when data augmentation generates a new image, the number of prototypes of the majority class is increased. However, it also increases, albeit to a lesser extent, those of the other classes present in the image. Figure 3 shows examples of the generated images.

Table 2 summarizes the new dataset distribution. Unfortunately, it is impossible to balance the instances of the trunk class, since each trunk instance is present once or, at most, twice in each image, making it impossible to augment its instances without further augmenting the other classes.

**Figure 3.** Example of the application of data augmentation to balance the dataset.

### *3.2. Algorithm Configurations*

In [15], we already studied the performance of a Mask R-CNN model on a preliminary dataset, and we obtained the optimized configuration parameters that we use in this work. We trained our Mask R-CNN-FPN networks by applying the transfer learning technique, using a ResNet50 backbone provided by the Detectron2 project as baseline model id 137849600. We trained our networks using data augmentation, enabling the algorithms specified in Table 3. The optimized *anchor sizes* were 16, 32, 64, 128, and 256 and the *anchor aspect ratios* were 0.2, 0.3, 0.5, 0.7, and 1. These values were chosen to favor the segmentation of unpruned shoots, which generally have a narrow and elongated shape. We performed a set of training experiments to empirically find the best learning rate for this model and found a value of 0.0001.

**Table 3.** Data augmentation operations used to train Mask R-CNN models.

| Augmentation Operation | Values | Description |
|---|---|---|
| Random Horizontal Flip | 0.5 | Flip the image horizontally with the given probability |
| Random Brightness | (0.8, 1.8) | Changes the image's brightness |
| Random Contrast | (0.6, 1.3) | Changes the image's contrast |
| Random Saturation | (0.8, 1.4) | Changes the image's saturation levels |
| Random Rotation | [−5, 5] | Rotates the image randomly in the interval |
| Random Crop | (0.8, 0.8) | Randomly crop a rectangle region out of an image |

In YOLOv8, the configuration was simplified by the Ultralytics framework, which provides an automatic optimizer that not only selects the type of optimizer to use from those available, but also automatically sets the learning rate, dynamics, and batch size.

To determine the appropriate batch size, YOLOv8 evaluates GPU memory availability. The framework executes a preliminary test to identify the largest batch size that utilizes approximately 60% of the available CUDA memory (https://docs.ultralytics.com/modes/train/#train-settings (accessed on 20 May 2025)). This strategy balances memory efficiency with training throughput and reduces the risk of out-of-memory errors.

Optimizer and learning rate selection are handled adaptively. In the initial training phase (first 10,000 iterations), YOLOv8 defaults to the AdamW optimizer, due to its ability to achieve faster convergence at early stages [25]. After surpassing the 10,000-iteration threshold, the optimizer automatically switches to Stochastic Gradient Descent (SGD), which generally exhibits better generalization performance in deep learning tasks.

In our specific training scenario, the dataset comprises 1288 images, partitioned into 10 folds for cross-validation (k = 10). With a batch size of 5, each fold processes approximately 1159 images, yielding around 232 iterations per epoch. Over the full training cycle, the total number of iterations exceeds 10,000, thus activating the optimizer transition mechanism. As a result, training begins with AdamW, using a learning rate of 0.001429 and momentum of 0.9, and subsequently, transitions to SGD, by YOLOv8's internal optimization logic (https://docs.ultralytics.com/reference/engine/trainer/#ultralytics.engine.trainer.BaseTrainer.build_optimizer (accessed on 20 May 2025).

Again, transfer learning was applied from the image segmentation model provided by Ultralytics pretrained on the Microsoft Common Objects in Context (COCO) dataset. Only the final layers of the network are trained, freezing the first 22 layers and training the head only. In this second group of experiments, data augmentation was also used with the operations listed in Table 4.

**Table 4.** Data augmentation operations used to train YOLO models.

| Augmentation Operation | Values | Description |
|---|---|---|
| Horizontal Flip | 0.5 | Flip the image horizontally with the given probability |
| Hue | 0.015 | Randomly adjust the hue gain with the this maximum |
| Saturation | 0.7 | Randomly adjust the saturation gain with this maximum |
| Value | 0.4 | Randomly adjust the value gain with this maximum |
| Scale | 0.5 | Scaling factor range from 0.5 to 1.5 |
| Translate | 0.1 | Maximum translation as a fraction of the image size |
| Erasing | 0.4 | Probability of randomly erasing a portion of the image |

Among the different versions of YOLO, we chose version 8 in the sizes nano (N), small (S), medium (M), large (L), and extra large (X). Table 5 reflects the number of trainable parameters of each of the models used, according to the specifications of their developers.

**Table 5.** Parameters of the neural models used in this work (in millions of parameters).

| Model | Mask R-CNN | YOLO-N | YOLO-S | YOLO-M | YOLO-L | YOLO-X |
|---|---|---|---|---|---|---|
| **Parameters** | 44 | 3.4 | 11.8 | 27.3 | 46.0 | 71.8 |
| **GFLOPs** | 289 | 12.6 | 42.6 | 110.2 | 220.5 | 334.1 |
| **Default Input Size** | 640–800 | 640 | 640 | 640 | 640 | 640 |
| **Employed Input Size** | 1024 | 1024 | 1024 | 1024 | 1024 | 1024 |

For both models, we used fixed size images of 1024 pixels width. Both models were trained on 1000 epochs using *early stopping* as the regularization technique with a patience of 100 epochs.

To guarantee the statistical reliability of the results, a 10-fold cross-validation scheme was employed. The 10-fold results are illustrated using box plots, which facilitate the visual evaluation of the segmentation performance for each experiment.

All training experiments were run on a dedicated machine with Ubuntu 20.04 operating system, 2 Intel Xeon Silver 4310 CPU, 512 GB RAM (Intel, Santa Clara, CA, USA), and four Nvidia A100 PCIe 40 GB (Nvidia, Santa Clara, CA, USA).

### 3.3. Metrics

COCOEvaluator [26] was employed to compute segmentation metrics. We considered Precision (P), Recall (R), mean Average Precision (mAP), and Average Precision (AP) by category to measure the system performance.

Precision, as illustrated by Equation (1), is the fraction of relevant objects among the ones that were predicted. Recall, as given by Equation (2), is the fraction of relevant objects that were predicted. Precision is crucial when the cost of false positives is high and Recall is vital when the cost of false negatives is high.

$$P = \frac{TP}{TP + FP} \quad (1)$$

$$R = \frac{TP}{TP + FN} \quad (2)$$

Finally, the area under the precision–recall curve (AP) can be used as a single metric to summarize the performance of the object segmentation model. The rate of TP, FP, and FN, and hence, the values of P, R, and AP, depend on the Intersection over the Union (IoU) which establishes when a detected object overlaps sufficiently with a real object. Thus, the AP50 measures the AP of the model by setting the IoU threshold at 0.5. In a multi-class object detection or segmentation task, the mAP is used, where individual AP is averaged over all classes.

The detection of the initial part of the shoots will be used to calculate the cut-off points at a later stage. These cut-off points are valid as long as they are at any position between the second and third buds. Hence, we use an IoU value of 0.5 which provides a good tradeoff between localization accuracy and robustness to small variations [27]. This ensures reliable detection of shoot segments for automated pruning.

## 4. Experiments and Results

This section presents the results of Mask R-CNN and YOLOv8 experiments, employing the N, S, M, L, and X versions of YOLOv8. The experiments analyze the performance of both frameworks in terms of segmentation efficiency and inference times on a Jetson AGX Orin GPU.

### 4.1. Segmentation

To compare segmentation performance, we used the mAP50 and AP50 per category metrics. The values of the medians of both parameters, resulting from the 10-fold cross-validation process, are shown in Table 6. To provide a clearer view of the cross-validation process, Figure 4a shows the cross-validation AP50 values for each network model, where each boxplot corresponds to the mAP50 or AP50 values obtained after the 10 runs in the K-fold cross-validation scheme.

To assess whether there exist statistically significant differences among the models performance, and considering that we have more than two groups of independent data and small sample sizes (equal to 10), we conducted a non-parametric Kruskal–Wallis test. The results yielded a $p$-value $< 0.001$, allowing us to conclude that significant differences exist between at least two groups. Subsequently, we performed pairwise comparisons using the Dwass-Steel-Critchlow-Fligner test [28] to identify which specific differences in model performance were statistically significant. The obtained $p$-values indicate that the performance of the Mask R-CNN model differs significantly from all the YOLO variants, both in terms of mAP50 and AP50 across all classes. Regarding comparisons among the YOLO models, we found significant differences between YOLO-N and the other YOLO models (aside from YOLO-S), but only for the mAP50 and AP50 of the *pruned shoot* class.

Therefore, we can conclude that Mask R-CNN has a lower segmentation performance than all YOLO8 variants when considering all classes. Specifically, Mask R-CNN has an mAP50 of 0.521, while the median value for all versions of YOLOv8 is greater than 0.87. Finally, although version M shows the best global performance with an mAP50 value of 0.897, the difference from the performance of version S is not significant, so we propose to use version S as it is simpler and its inference time is the lowest.

**Table 6.** mAP50 and AP50 values for the studied models and classes (ten-fold cross-validation median).

| Class | Metric | Mask R-CNN | YOLO-N | YOLO-S | YOLO-M | YOLO-L | YOLO-X |
|---|---|---|---|---|---|---|---|
| All | mAP50 | 0.521 | 0.875 | 0.883 | 0.897 | 0.891 | 0.891 |
| Trunk | AP50 | 0.682 | 0.971 | 0.974 | 0.980 | 0.980 | 0.979 |
| Pruned shoot | AP50 | 0.520 | 0.927 | 0.935 | 0.946 | 0.944 | 0.947 |
| Shoot | AP50 | 0.359 | 0.724 | 0.748 | 0.772 | 0.750 | 0.743 |

The use of a global mAP as the only metric for our study can hide segmentation errors in the most important class for the pruning task (*shoot*). The *trunk class* is only used in the post-processing phase to distinguish the lower and upper cutting points of the vine shoots, so errors in its correct segmentation are not critical. The *pruned shoot class* has been added to allow the use of the robotic pruner in vineyards where partial pruning is performed. Both an FP and FN in this class will have no effect on the management of the cutting tool. Hence, the most important errors are those that occur on the *shoot class*. An FP in this class would take the cutting tool to an area where there is nothing to cut and an FN would leave the shoot not pruned.

Therefore, it is necessary to study the AP50 values per category for each network model. Figure 4b shows AP50 behavior for the trunk class, and again, Mask R-CNN performs significantly worse than all YOLOv8 variants. In fact, the performance improvement achieved by YOLOv8 over Mask R-CNN is greater than 0.29 when comparing its median values. As there are statistically non-significant differences in the performance of the variants of YOLOv8 models in this class, we considered variant S, which achieves an AP50 of 0.974. The same considerations can also be extended to the AP50 for the *pruned shoot class* depicted in Figure 4c.

As the *shoot class* is the most important for the pruning task, a more comprehensive analysis should be conducted on this class in order to determine definitively which model is suitable for a production environment. Figure 4d shows a lower performance of all network models for this class, and Mask R-CNN also ranks last in performance with a median AP50 value of 0.359. Among the versions of YOLOv8, the differences in performance are very small (with a median AP50 value around 0.71–0.76) and are not statistically significant, so the use of the larger versions (L, M, or X), more complex and slow, is not justified. Again, YOLOv8-S seems to be the best option. In some contexts, where the training dataset is not large enough or even with correlation between the data, a smaller model such as S might offer better robustness to overfitting than larger models such as X or L, providing a better generalization ability in the data [29].

Analyzing the possible causes that lead to a lower performance of our system for the *shoot* class, compared to the other two classes, we found that the matching of the proposals to the ground-truth marks sometimes does not meet the established IoU. In our opinion, this is the most complex class, because we want the system to model a part of the object instead of the whole object, as is usual in segmentation.

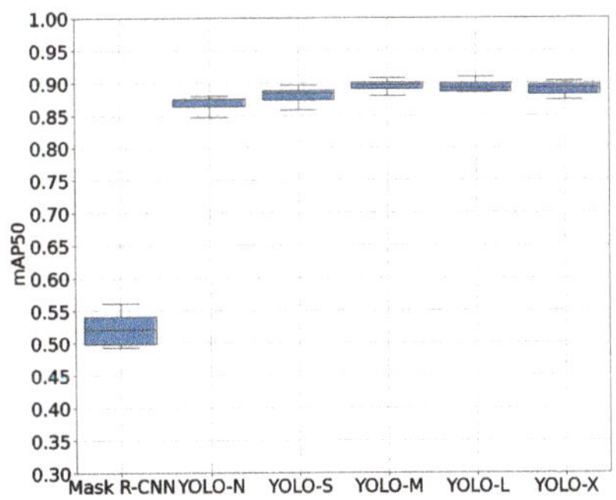


(**a**) All classes

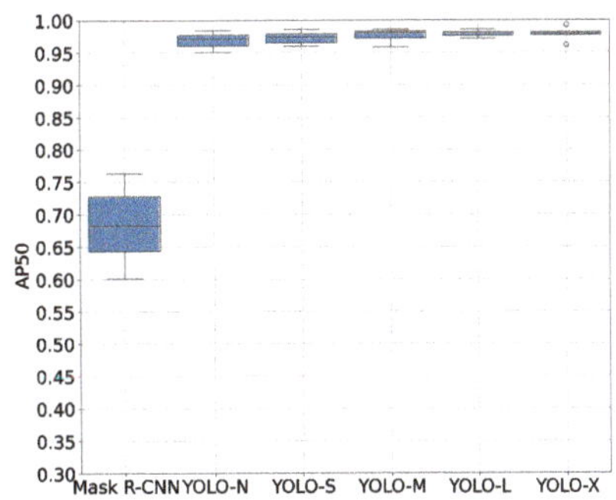


(**b**) Trunk class

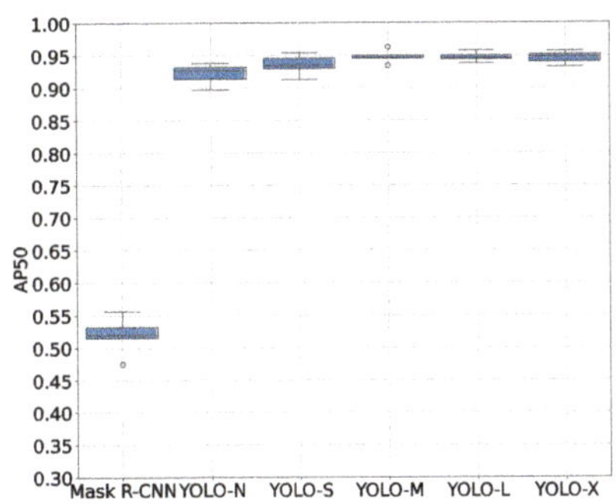


(c) Pruned shoot class

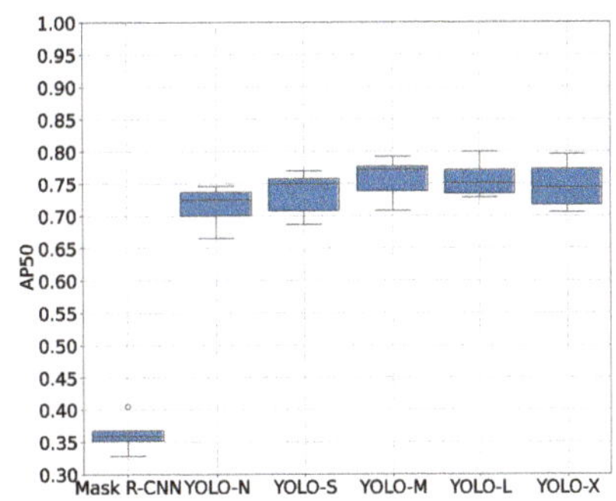


(**d**) Shoot class

**Figure 4.** mAP50 and AP50 values for the studied algorithms. (**a**) The results of the cross-validation procedure for all classes and (**b–d**) the AP50 results disaggregated by class.

### *4.2. Inference Time*

In the production environment, an intensive energy consumption increases the battery requirements of the robot. Therefore, it may be interesting to offload the inference process to near base station with stable power supply. The data link energy consumption will be significantly lower than on-board image processing. For this reason, two different inference tests were conducted: on the training server and on a Jetson AGX Orin.

Table 7 shows the average inference times of the studied models running on the two machines. The first test was performed on the same machine used in the training phase and Figure 5a shows the distribution of this parameter using a boxplot. On the one hand, we can observe that Mask R-CNN performs significantly worse than all YOLOv8 versions. On the other hand, YOLOv8 models have inference times ranging from 23.3 ms to 6.2 ms, which allows theoretic processing speeds from 38 to 161 frames per second (FPS) and significantly exceeds the minimum standards required for this application.

**Table 7.** Average inference times for the studied network models during the cross-validation process (milliseconds).

| Model | GPU | Jetson AGX Orin |
|---|---|---|
| Mask R-CNN | 43.3 | 171.3 |
| YOLOv8-X | 26.3 | 135.3 |
| YOLOv8-L | 16.0 | 108.6 |
| YOLOv8-M | 11.7 | 89.2 |
| YOLOv8-S | 6.7 | 55.0 |
| YOLOv8-N | 6.2 | 36.0 |

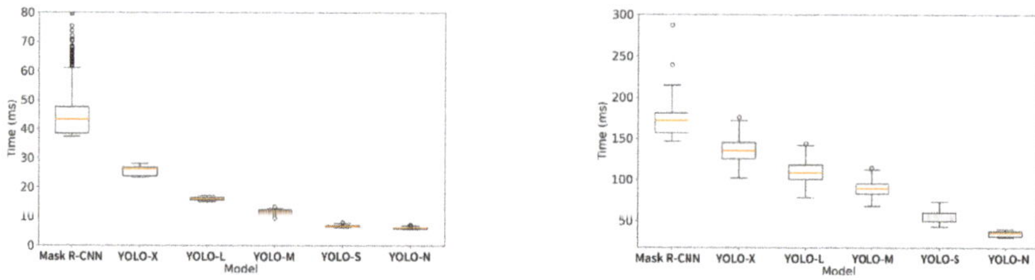


(**a**) Nvidia A100 GPU (**b**) Jetson AGX Orin

**Figure 5.** Inference time boxplots running on an Nvidia A100 GPU and Jetson AGX Orin.

Considering the integration and compactness objectives of the work, an integrated computing board was initially considered the optimal option for its development. For this scenario, Figure 5b shows the inference times distribution required by the network models under consideration running on a Jetson AGX Orin. First, we observe an increase in general inference times compared to those obtained on the training computer. Second, we see that the superiority of all YOLOv8 models over Mask R-CNN is repeated. Furthermore, finally, the inference time required for the YOLOv8-S network allows us to achieve a processing speed about 18 FPS.

In Figure 6, various YOLOv8 inference examples are presented, where ground truth marks are shown in red to facilitate comparison. Figure 6a shows the inference result with YOLOv8-L for the labeled image previously shown in Figure 2a. We can see that all the objects of the classes trunk and shoot are correctly detected but two FNs on the class pruned shoot are obtained. In spite of these FNs, all the objects of interest for our final purpose are correctly detected, and hence, the robotic arm would be guided to prune all the desired shoots in the image. Similarly, Figure 6b shows the results for a complex example (Figure 2b) correctly handled by one of the ten networks trained for YOLOv8-X, on an image not included in its training set. Finally, Figure 6c,d show the inference over YOLOv8-S and YOLOv8-N. Additional examples of how our system works can be found in Appendix A.

(**a**) YOLOv8-L (**b**) YOLOv8-X

(**c**) YOLOv8-S (**d**) YOLOv8-M

**Figure 6.** Inference examples obtained with various versions of YOLO. Ground truth marks are shown in red, inferred shoots are shown in yellow, pruned shoots are shown in blue, and inferred trunk are shown in green.

## 5. Conclusions

In this paper, we approached the automatic pruning of vineyards problem, using deep learning and computer vision algorithms, specifically Mask R-CNN and YOLOv8.

The results of the experiments demonstrate that YOLOv8 outperforms Mask R-CNN in the two test scenarios analyzed, both in inference time and segmentation success rate. Analyzing the results of the different YOLOv8 sizes, we can affirm that although version M shows the best global performance with an mAP50 value of 0.897, the difference from the performance of version S is not statistically significant, so we propose to use version S, as it is simpler and its inference time is the lowest.

The inference speed achieved on the Jetson AGX Orin leads us to believe that embedded deployment is feasible. In the following part of the project, we intend to drive the robot in front of the vine using RTK technology. Therefore, the inference time only affects the positioning of the cutting arm. Since the repositioning of the arm exceed 55 ms, we believe that the speed achieved is sufficient for the task.

As mentioned in Section 1.2, the purpose of several referenced papers [8,9,13] is the detection of full buds, and not to the estimation of the cutoff point. In addition, some of these works are based on other pruning techniques, such as Guyot. Hence, a direct comparison of our results with those achieved in other state-of-the-art proposals is somewhat complex.

The most important difference to highlight between our work and those analyzed in Section 1.2 lies on the attempt to directly estimate the safe pruning area in the shoot. By eliminating further post-processing, we believe that it simplifies the structure of a future robot and can increase the robustness of the system.

In conclusion, the results obtained represent a valuable advance over previous studies and open the door to future work:

- Establish a figure of merit to measure the accuracy in the cut-point identification process, test this functionality, and implement it;
- Estimate, according to the pruning density desired by the user, the shoots to be cut in the lower cutting point and those to be cut in the upper one;
- Expand the set of prototypes to ensure a correct response in any planting condition.

**Author Contributions:** Conceptualization, M.M.M.; methodology and software, E.P. and E.A.; validation, E.P. and M.M.M.; formal analysis, R.G. and H.M.G.V.; investigation, E.P. and M.M.M.; resources, M.M.M.; data curation, E.P.; writing—original draft preparation and writing—review and editing and visualization, R.G. and H.M.G.V.; supervision, C.J.G.-O.; project administration, M.M.M.; funding acquisition, M.M.M. and C.J.G.-O. All authors have read and agreed to the published version of the manuscript.

**Funding:** This research was funded by the Ministry of Science and Innovation through the project 'SISTEMA ROBOTIZADO PARA LA PODA INTELIGENTE DE VIÑEDOS (TED2021-131242B-I00)', corresponding to the 2021 Call for Projects oriented to the Ecological and Digital Transition, and has been co-financed 85% by the European Union, European Regional Development Fund (ERDF), and the Junta de Extremadura. Managing Authority. Ministerio de Hacienda, through project GR24086.

**Institutional Review Board Statement:** Not applicable.

**Data Availability Statement:** The set of images and their *ground truth* used in this work have been published in the open access journal *Data in Brief* under the Repository name: *VidPrune Dataset*.

**Acknowledgments:** The authors would like to thank Miguel González Velasco for their advice and help with the statistical analysis.

**Conflicts of Interest:** The authors declare no conflicts of interest.

## Abbreviations

The following abbreviations are used in this manuscript:

| | |
|---|---|
| AP | Average Precision |
| AP50 | Average Precision for an IoU threshold of 0.5 |
| CNN | Convolutional Neural Network |
| COCO | Microsoft Common Objects in Context |
| FN | False Negative |
| FP | False Positive |
| FPS | Frames Per Second |
| GPS | Global Positioning System |
| GPU | Graphics Processing Unit |
| IoU | Intersection over Union |
| mAP | Mean Average Precision |
| mAP50 | Mean Average Precision for an IoU threshold of 0.5 |
| R-CNN | Region-based Convolutional Neural Network |
| RTK | Real-Time Kinematic |
| TP | True Positive |
| TN | True Negative |
| VIA | VGG Image Annotator |
| YOLO | You Only Look Once |

## Appendix A. Outputs of the Inference Process

In the following figures inferred shoots are shown in yellow, pruned shoots are shown in blue and inferred trunks are shown in green.

## References

1. Jones, G.V.; Webb, L.B. Climate Change, Viticulture, and Wine: Challenges and Opportunities. *J. Wine Res.* **2010**, *21*, 103–106. [CrossRef]
2. Ryan, M. Labour and skills shortages in the agro-food sector. In *OECD Food, Agriculture and Fisheries Papers*; OECD Publishing: Paris, France, 2023. [CrossRef]
3. OECD. *Agricultural Policy Monitoring and Evaluation 2024*; OECD Publishing: Paris, France, 2024; p. 660. [CrossRef]
4. Ruzzante, S.; Labarta, R.; Bilton, A. Adoption of agricultural technology in the developing world: A meta-analysis of the empirical literature. *World Dev.* **2021**, *146*, 105599. [CrossRef]
5. Bruez, E.; Cholet, C.; Giudici, M.; Simonit, M.; Martignon, T.; Boisseau, M.; Weingartner, S.; Poitou, X.; Rey, P.; Geny-Denis, L. Pruning Quality Effects on Desiccation Cone Installation and Wood Necrotization in Three Grapevine Cultivars in France. *Horticulturae* **2022**, *8*, 681. [CrossRef]
6. Bogue, R. Robots addressing agricultural labour shortages and environmental issues. *Ind. Robot* **2024**, *51*, 1–6. [CrossRef]
7. Sze, S.; Kunze, L. Real-time 3D semantic occupancy prediction for autonomous vehicles using memory-efficient sparse convolution. In Proceedings of the 2024 IEEE Intelligent Vehicles Symposium (IV), Jeju Island, Republic of Korea, 2–5 June 2024; pp. 1286–1293. [CrossRef]
8. Botterill, T.; Paulin, S.; Green, R.; Williams, S.; Lin, J.; Saxton, V.; Mills, S.; Chen, X.; Corbett-Davies, S. A Robot System for Pruning Grape Vines. *J. Field Robot.* **2017**, *34*, 1100–1122. [CrossRef]
9. Gentilhomme, T.; Villamizar, M.; Corre, J.; Odobez, J.M. Towards smart pruning: ViNet, a deep-learning approach for grapevine structure estimation. *Comput. Electron. Agric.* **2023**, *207*, 107736. [CrossRef]

10. Lanza, B.; Nuzzi, C.; Botturi, D.; Pasinetti, S. First Step Towards Embedded Vision System for Pruning Wood Estimation. In Proceedings of the 2023 IEEE International Workshop on Metrology for Agriculture and Forestry (MetroAgriFor), Pisa, Ital, 6–8 November 2023; pp. 757–762. [CrossRef]
11. Villegas Marset, W.; Pérez, D.S.; Díaz, C.A.; Bromberg, F. Towards practical 2D grapevine bud detection with fully convolutional networks. *Comput. Electron. Agric.* **2021**, *182*, 105947. [CrossRef]
12. Chen, Z.; Wang, Y.; Tong, S.; Chen, C.; Kang, F. Grapevine Branch Recognition and Pruning Point Localization Technology Based on Image Processing. *Appl. Sci.* **2024**, *14*, 3327. [CrossRef]
13. Oliveira, F.; da Silva, D.Q.; Filipe, V.; Pinho, T.M.; Cunha, M.; Cunha, J.B.; dos Santos, F.N. Enhancing Grapevine Node Detection to Support Pruning Automation: Leveraging State-of-the-Art YOLO Detection Models for 2D Image Analysis. *Sensors* **2024**, *24*, 6774. [CrossRef] [PubMed]
14. Fernandes, M.; Scaldaferri, A.; Fiameni, G.; Teng, T.; Gatti, M.; Poni, S.; Semini, C.; Caldwell, D.G.; Chen, F. Grapevine Winter Pruning Automation: On Potential Pruning Points Detection through 2D Plant Modeling using Grapevine Segmentation. In Proceedings of the 2021 IEEE 11th Annual International Conference on CYBER Technology in Automation, Control, and Intelligent Systems (CYBER), Jiaxing, China, 27–31 July 2021; pp. 13–18. [CrossRef]
15. Pacioni, E.; Abengozar-García, E.; Macías Macías, M.; García Orellana, C.J.; Gallardo Caballero, R.; García Manso, A. Mask R-CNN aplicado a la poda de viñedos. In Proceedings of the XX Conferencia de la Asociación Española para la Inteligencia Artificial, A Coruña, Spain, 19–21 June 2024.
16. He, K.; Gkioxari, G.; Dollár, P.; Girshick, R. Mask R-CNN. In Proceedings of the IEEE International Conference on Computer Vision, Venice, Italy, 22–29 October 2017; pp. 2980–2988. [CrossRef]
17. Ren, S.; He, K.; Girshick, R.; Sun, J. Faster R-CNN: Towards Real-Time Object Detection with Region Proposal Networks. *IEEE Trans. Pattern Anal. Mach. Intell.* **2017**, *39*, 1137–1149. [CrossRef] [PubMed]
18. Wu, Y.; Kirillov, A.; Massa, F.; Lo, W.Y.; Girshick, R. Detectron2. 2019. Available online: https://github.com/facebookresearch/detectron2 (accessed on 25 November 2024).
19. Redmon, J.; Divvala, S.; Girshick, R.; Farhadi, A. You Only Look Once: Unified, Real-Time Object Detection. In Proceedings of the IEEE Conference on Computer Vision and Pattern Recognition, Las Vegas, NV, USA, 27–30 June 2016; pp. 779–788. [CrossRef]
20. Varghese, R.; Sambath, M. YOLOv8: A Novel Object Detection Algorithm with Enhanced Performance and Robustness. In Proceedings of the 2024 International Conference on Advances in Data Engineering and Intelligent Computing Systems (ADICS), Chennai, India, 18–19 April 2024; pp. 1–6. [CrossRef]
21. Pacioni, E.; Abengózar, E.; Macías Macías, M.; García Orellana, C.J.; González Velasco, H.M.; García Manso, A. Vineyard dataset for automatic pruning based on main parts localization. *Data Brief* **2025**, *59*, 111335. [CrossRef] [PubMed]
22. Dutta, A.; Zisserman, A. The VIA Annotation Software for Images, Audio and Video. In Proceedings of the 27th ACM International Conference on Multimedia, Nice, France, 21–25 October 2019; Association for Computing Machinery: New York, NY, USA, 2019. [CrossRef]
23. Dablain, D.A.; Bellinger, C.; Krawczyk, B.; Chawla, N.V. Efficient Augmentation for Imbalanced Deep Learning. In Proceedings of the 2023 IEEE 39th International Conference on Data Engineering (ICDE), Anaheim, CA, USA, 3–7 April 2023; pp. 1433–1446. [CrossRef]
24. Crasto, N. Class Imbalance in Object Detection: An Experimental Diagnosis and Study of Mitigation Strategies. *arXiv* **2024**, arXiv:2403.07113.
25. Keskar, N.S.; Socher, R. Improving Generalization Performance by Switching from Adam to SGD. *arXiv* **2017**, arXiv:1712.07628.
26. Lin, T.Y.; Maire, M.; Belongie, S.; Hays, J.; Perona, P.; Ramanan, D.; Dollár, P.; Zitnick, C.L. Microsoft COCO: Common Objects in Context. In *Proceedings of the Computer Vision–ECCV 2014*; Fleet, D., Pajdla, T., Schiele, B., Tuytelaars, T., Eds.; Springer: Cham, Switzerland, 2014; pp. 740–755.
27. Everingham, M.; Eslami, S.M.A.; Van Gool, L.; Williams, C.K.I.; Winn, J.; Zisserman, A. The Pascal Visual Object Classes Challenge: A Retrospective. *Int. J. Comput. Vis.* **2015**, *111*, 98–136. [CrossRef]
28. Douglas, C.E.; Michael, F.A. On distribution-free multiple comparisons in the one-way analysis of variance. *Commun. Stat.-Theory Methods* **1991**, *20*, 127–139. [CrossRef]
29. Zhang, C.; Bengio, S.; Hardt, M.; Recht, B.; Vinyals, O. Understanding deep learning requires rethinking generalization. In Proceedings of the International Conference on Learning Representations, Toulon, France, 24–26 April 2017.

agriculture

MDPI

Article

# A Transformer-Based Symmetric Diffusion Segmentation Network for Wheat Growth Monitoring and Yield Counting

**Ziyang Jin, Wenjie Hong, Yuru Wang, Chenlu Jiang, Boming Zhang, Zhengxi Sun, Shijie Liu and Chunli Lv ***

China Agricultural University, Beijing 100083, China
* Correspondence: lvcl@cau.edu.cn

**Abstract:** A wheat growth and counting analysis model based on instance segmentation is proposed in this study to address the challenges of wheat growth monitoring and yield prediction in high-density agricultural environments. The model integrates the transformer architecture with a symmetric attention mechanism and employs a symmetric diffusion module for precise segmentation and growth measurement of wheat instances. By introducing an aggregated loss function, the model effectively optimizes both segmentation accuracy and growth measurement performance. Experimental results show that the proposed model excels across several evaluation metrics. Specifically, in the segmentation accuracy task, the wheat instance segmentation model using the symmetric attention mechanism achieved a Precision of 0.91, Recall of 0.87, Accuracy of 0.89, mAP@75 of 0.88, and F1-score of 0.89, significantly outperforming other baseline methods. For the growth measurement task, the model's Precision reached 0.95, Recall was 0.90, Accuracy was 0.93, mAP@75 was 0.92, and F1-score was 0.92, demonstrating a marked advantage in wheat growth monitoring. Finally, this study provides a novel and effective method for precise growth monitoring and yield counting in high-density agricultural environments, offering substantial support for future intelligent agricultural decision-making systems.

**Keywords:** wheat growth monitoring; precision agriculture; yield prediction; instance segmentation; deep learning

## 1. Introduction

Wheat, as one of the most important food crops worldwide, plays a crucial role in agricultural production stability and food security [1–3]. Monitoring wheat growth and yield estimation are essential aspects of agricultural production management, influencing farmers' planting decisions, grain yield forecasting, market supply regulation, and agricultural policy making [4,5]. Traditional methods for wheat growth monitoring primarily rely on manual observation and remote sensing analysis based on vegetation indices (e.g., NDVI, EVI) [6,7]. While these methods can reflect vegetation coverage and growth health to some extent, their accuracy is often limited by data resolution and environmental factors. Although remote-sensing-based vegetation indices provide large-scale assessments of growth conditions, their spectral limitations make it difficult to accurately distinguish individual wheat plants in high-density vegetation environments. Moreover, insufficient spatial resolution constrains their application in precise wheat monitoring [8,9]. Accurate wheat yield estimation is a critical task in crop growth assessment. Traditional methods mainly rely on random sampling and statistical modeling for yield estimation [10]. However, these methods suffer from limited sampling density, high estimation errors, and difficulty adapting to variations in growth stages and field management practices. Therefore, leveraging computer vision techniques combined with deep learning to achieve efficient and precise wheat growth monitoring and yield estimation has become a significant research direction in precision agriculture.

In recent years, deep learning has achieved groundbreaking progress in computer vision, providing new technological means for agricultural image analysis [11]. Specifically, convolutional neural network (CNN)- and transformer-based methods have been widely applied in crop phenotyping, particularly in semantic and instance segmentation tasks [12,13]. Hong et al. [14] proposed a hybrid attention network (CTHNet) to count wheat spikes from RGB images, integrating local features and global contextual information. Their experimental results demonstrated mean absolute errors of 3.40 and 5.21, significantly outperforming previous studies. Guan et al. [15] proposed an improved YOLOv10 algorithm that substantially enhances the feature extraction and detection capabilities of wheat spike models by incorporating a bidirectional feature pyramid network (BiFPN), a separation and enhancement attention module (SEAM), and a global context network (GCNet). Their method achieved an Accuracy of 93.69%, a Recall of 91.70%, and a Mean Average Precision (mAP) of 95.10% in wheat spike detection, outperforming the baseline YOLOv10 model by 2.02%, 2.92%, and 1.56%, respectively. The improved YOLOv10 algorithm effectively addresses challenges in wheat spike detection under complex field conditions, providing robust support for agricultural production and research. Yao et al. [16] focused on addressing wheat spike adhesion issues and proposed a combined algorithm named "APW", integrating the alternating direction multiplier method (ADMM), Potts model, and watershed algorithm for rapid wheat spike identification and counting. Their results indicated optimal accuracy in low-density planting scenarios, with an $R^2$ value of 0.89 and a root mean square error (RMSE) of 3.72, demonstrating APW's capability to handle scenarios with significant background variation and low object adherence. These findings offer a novel approach for effective wheat spike counting in drone-acquired images. Sun et al. [17] developed a simulation strategy to replicate real wheat field conditions, enabling data collection in an indoor environment within a short timeframe. The results showed that YOLOv7 performed the best, achieving $R^2 = 0.963$ and RMSE = 2.463. While these methods have been preliminarily applied to agricultural object detection tasks, challenges remain, including segmentation errors in heavily overlapping objects and increased computational complexity leading to slower inference speed. Additionally, density-based counting methods have recently been integrated with deep learning approaches, such as CSRNet and Count-ception, which perform object counting through density estimation. However, these methods still suffer from significant background noise interference and cumulative counting errors.

To address the limitations of existing methods, this study proposes an instance segmentation-based wheat growth and yield analysis model, adopting a transformer-structured symmetric diffusion segmentation network (SDS-Net) to enhance wheat identification accuracy and counting precision. The key innovations of this model include the symmetric diffusion module, symmetric attention mechanism, and aggregated loss function. The symmetric diffusion module simulates wheat plant growth characteristics by enhancing boundary information through feature diffusion, improving segmentation accuracy in overlapping-object scenarios. The symmetric attention mechanism leverages multi-scale feature fusion and global context awareness to enhance wheat field object recognition, effectively reducing background noise interference. Additionally, the proposed aggregated loss function integrates instance segmentation loss, counting loss, and boundary loss, ensuring strong generalization performance in both growth analysis and precise counting tasks. The contributions of this study are as follows:

- Transformer-based symmetric diffusion segmentation network: Unlike traditional CNN-based architectures, this approach employs a transformer as the primary feature extraction module, combined with a symmetric diffusion module to enhance the boundary recognition of wheat plants, improving segmentation accuracy in densely planted environments.
- Symmetric diffusion module: A novel diffusion mechanism is introduced, enabling the model to effectively handle wheat plant overlap, reducing mis-segmentation and omission errors.
- Symmetric attention mechanism: Integrating self-attention mechanisms and multi-scale feature fusion enhances segmentation robustness in complex environments while optimizing target counting accuracy.

In summary, this study proposes a deep-learning-based approach for wheat growth monitoring and yield estimation, integrating instance segmentation and a symmetric diffusion mechanism to significantly improve detection accuracy and counting stability.

## 2. Related Work

### 2.1. Semantic Segmentation

Semantic segmentation is a core computer vision task that classifies each pixel for precise target region delineation [18–20]. In agriculture, it is widely used for crop recognition, disease detection, and growth monitoring [21–23]. Compared to object detection, it provides pixel-level masks for fine-grained analysis. Traditional methods, such as region growing, clustering, and threshold-based segmentation, struggle with complex backgrounds and occlusions. Recent deep-learning-based approaches, particularly CNN- and transformer-based models, have significantly improved segmentation accuracy. CNN models extract features via convolutional layers and use encoder–decoder architectures for segmentation [24–26]. U-Net employs skip connections for high-resolution feature retention, while DeepLab uses dilated convolutions to expand the receptive field [27,28]. However, CNNs suffer from limited receptive fields, leading to challenges in segmenting densely grown wheat fields, causing boundary blurring and omissions. Originally designed for NLP, transformers have demonstrated superior performance in semantic segmentation [29]. The self-attention mechanism in transformers effectively captures global dependencies, enhancing segmentation in complex backgrounds. ViT was the first fully transformer-based vision model, dividing images into patches for global feature modeling [30]. However, ViT's high computational cost makes it less efficient for large-scale agricultural applications. While transformer-based models excel in global feature extraction and dense wheat field segmentation, their computational complexity remains a challenge for real-world deployment.

### 2.2. Object Counting Based on Probability Density Estimation

Object counting is widely used in crowd counting, cell counting, and agricultural yield estimation [31,32]. In wheat growth analysis, accurate counting provides key insights into crop conditions and yield prediction [33]. However, dense wheat growth introduces challenges such as occlusions and varying lighting conditions, increasing the complexity of counting tasks. Traditional methods fall into detection-based, regression-based, and density-estimation-based approaches [34]. Among these, probability density estimation has gained attention for handling densely distributed targets and enabling efficient counting under weak supervision [35]. In density-estimation-based counting, the model learns a mapping from images to density maps, integrating density values to estimate the total count. This approach is particularly effective in dense wheat fields, where detection-based methods like Faster R-CNN and YOLO often fail due to overlapping objects [36]. For sparse wheat distributions, detection-based models may suffice, but for high-density scenarios, such as the heading stage, density estimation methods provide greater robustness. By modeling spatial distributions, these methods mitigate object omissions and enhance counting accuracy in agricultural environments.

## 3. Materials and Method

### *3.1. Dataset Collection*

The wheat image data used in this study were sourced from several locations, including field-collected images from the Science Park at the West Campus of China Agricultural University in Haidian District, Beijing, publicly available images from the internet, and the official Kaggle wheat detection dataset. The data encompass various wheat growth densities, including sparse wheat fields with larger plant spacing, moderate-density fields with medium plant spacing, and high-density fields with dense plant growth, as shown in Figure 1. Each category includes between 1600 and 1900 images, ensuring the balance and representativeness of the dataset, as shown in Table 1. Each category includes between 1600 and 1900 images, with data further divided based on acquisition methods: "UAV" represents images collected using aerial drone photography and "Camera" refers to images obtained using ground-based digital cameras. This differentiation allows the model to learn multi-scale features from both large-area wheat field perspectives and close-up views of individual plants.

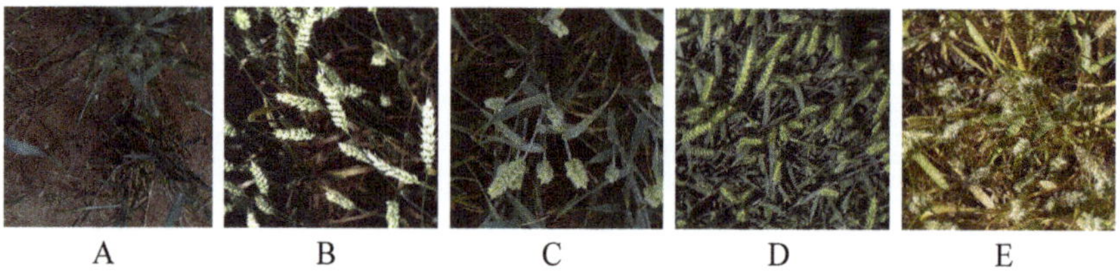


**Figure 1.** Dataset samples. (**A**) Sparse. (**B**,**C**) Moderate. (**D**,**E**) Dense.

**Table 1.** Data quantity for different types of wheat.

| Density | Data | UAV | Camera |
|---|---|---|---|
| Dense | 1682 | 693 | 989 |
| Moderate | 1804 | 962 | 842 |
| Sparse | 1752 | 837 | 915 |

Field collection was carried out between March 2023 and March 2024 at the Science Park of the West Campus of China Agricultural University in Haidian District, Beijing, where various wheat varieties were cultivated at different growth stages, including the jointing, booting, flowering, and grain filling stages. These stages fulfill the study's requirements for growth analysis across different wheat development phases. The collection equipment included a DJI Phantom 4 multispectral drone(DJI Technology Co., Ltd., Shenzhen, China) and a Canon EOS 5D Mark IV camera (Canon Inc., Tokyo, Japan). The multispectral camera mounted on the DJI Phantom 4 captures both visible and near-infrared bands, providing important vegetation growth parameters such as the NDVI (Normalized Difference Vegetation Index), while the Canon EOS 5D Mark IV, with its high-resolution imaging capabilities, accurately captures details of wheat ears and individual leaves. The combined use of these two devices ensures the diversity of the data while maintaining high recognition accuracy under various lighting conditions. During data collection, a combination of ground-based and drone-based imagery was employed. Ground-based imaging focused on capturing close-up images of wheat plants, leaves, and ears to obtain clear instance contours suitable for semantic segmentation tasks. Different angles, including frontal, lateral, and top-down views, were used to observe the growth condition of wheat from various perspectives, with close-up shots of critical areas like the leaves and ears to enhance feature learning by the model. Drone-based imagery was utilized to capture the large-scale field conditions, with the camera height set between 5 m and 20 m, balancing field-level growth analysis and the preservation of local details. Automatic and manual exposure modes were employed during shooting, adjusting ISO and shutter speed under cloudy or low-light conditions to ensure consistent image quality. Additionally, to mitigate the influence of lighting changes at different times of the day on image color and brightness, gray cards were placed on the ground for color calibration and post-processing lighting normalization was performed using the multispectral images. For wheat fields of varying density, stratified random sampling was applied to ensure a balanced distribution of data across the density categories. Sparse areas were mainly located in plots with uneven sowing or in the early stages of planting. Moderate-density areas represented typical wheat fields with standard sowing patterns, while high-density areas consisted of densely planted or vigorously growing wheat plots. In addition to the field-collected data, publicly available network data and the official Kaggle wheat detection dataset were also included. The network data primarily came from agricultural research institutions, crop disease databases, and high-resolution agricultural imaging databases, all of which were carefully selected for inclusion in this study. The Kaggle wheat detection dataset contained finely annotated wheat ear semantic segmentation data from various growing regions, providing valuable information for training the model's semantic segmentation capabilities. Both the network data and Kaggle dataset underwent rigorous screening and preprocessing, including noise filtering, color normalization, and geometric transformations, to ensure stylistic consistency with the field-collected data. Distinct feature differences were observed among the categories of wheat images with varying densities. In the sparse category, the contours of individual plants were clear, with a relatively simple background where soil or weeds were prominent, making it suitable for precise individual semantic segmentation. In the moderate-density category, plant spacing was smaller, with some overlap between plants, and the background was more complex than in the sparse category. In the high-density category, wheat plants grew densely and ears and leaves were heavily occluded, making it difficult to distinguish individual plants. Accurate segmentation in this category typically relied on advanced attention mechanisms and contextual information.

### *3.2. Data Annotation*

In computer vision tasks, dataset annotation serves as a crucial step in model training, particularly in semantic and semantic segmentation. Accurate annotations significantly enhance model generalization capabilities. In agricultural image analysis, wheat growth monitoring and counting tasks necessitate precise annotation to ensure that deep learning models can correctly learn the boundaries, morphology, and key structural features of wheat plants. The primary objective of data annotation is to generate semantic segmentation masks for each wheat plant within an image and to mark its key feature points, thereby enabling the model to effectively learn wheat structural information and ultimately improving growth assessment and yield estimation accuracy. In semantic segmentation tasks, each wheat plant is considered an independent target, requiring a unique mask for every individual plant during annotation. Given an input image $I \in \mathbb{R}^{H \times W \times C}$, where $H$ and $W$ represent the image height and width, respectively, and C denotes the number of channels, the goal of semantic segmentation is to learn a mapping function $f_\theta$ such that the model can predict the mask $M_i$ for the $i$-th wheat plant:

$$M_i = f_\theta(I), \tag{1}$$

where $M_i \in 0, 1^{H \times W}$ represents a binary mask, with a value of 1 indicating that a pixel belongs to the $i$-th wheat plant and a value of 0 denoting background. The final semantic segmentation result $M$ for an image containing $N$ wheat plants can be represented as the collection of all instance masks:

$$M = \sum_{i=1}^{N} M_i \tag{2}$$

A critical challenge in data annotation involves ensuring clear boundaries between instances to prevent mislabeling due to overlapping. Generally, plant contours are manually outlined using polygon-based (Polygon) or lasso-based (Lasso) methods to generate high-precision semantic segmentation masks. Mathematically, the closed boundary region $B_i$ of each wheat plant can be defined as follows:

$$B_i = (x, y) \mid F_i(x, y) = 0, \tag{3}$$

where $F_i(x, y)$ represents the implicit function of the plant boundary. When $F_i(x, y) = 0$, the pixel $(x, y)$ lies on the contour of the wheat plant. If $F_i(x, y) < 0$, the pixel is inside the plant, whereas $F_i(x, y) > 0$ indicates that the pixel is outside the plant. In addition to semantic segmentation masks, key feature point (Keypoint) annotation is also crucial for wheat growth analysis. Keypoint annotation typically includes structural information such as the wheat plant's root, stem, and spike. Each wheat plant $i$ is represented by $K$ keypoints $P_{i,j}j = 1^K$, where $Pi, j = (x_{i,j}, y_{i,j})$ denotes the coordinates of the $j$-th keypoint. The objective of keypoint detection is to learn a probability distribution $P(x, y)$, where $(x, y)$ represents the position of keypoints. The probability density function is defined as:

$$P(x, y) = \sum_{j=1}^{K} G_\sigma(x - x_{i,j}, y - y_{i,j}), \tag{4}$$

where $G_\sigma$ is a Gaussian kernel function used to generate a smoothed keypoint heatmap:

$$G_\sigma(x, y) = \frac{1}{2\pi\sigma^2} \exp\left(-\frac{x^2 + y^2}{2\sigma^2}\right) \tag{5}$$

During annotation, Gaussian distributions are commonly employed to model keypoint locations, thereby reducing annotation errors and mitigating the impact of label noise on model training. To improve annotation efficiency, modern annotation tools such as LabelMe and CVAT provide semi-automated annotation capabilities, including automatic segmentation using GrabCut, region clustering with K-means, and Mask R-CNN-based interactive segmentation. These approaches effectively reduce the time cost of manual annotation and enhance annotation consistency. For instance, GrabCut employs an energy minimization framework using a Bayesian model to estimate the probability distribution of foreground and background pixels. The energy function $E$ comprises a data term and a smoothness term:

$$E(M) = \sum_{x,y} (-\log P(I_{x,y} \mid M_{x,y}) + \lambda R(M)), \tag{6}$$

where $P(I_{x,y} \mid M_{x,y})$ represents the probability that pixel $(x, y)$ belongs to the foreground or background, $\lambda$ is a balancing parameter, and $R(M)$ is the smoothness regularization term. By minimizing $E(M)$, the optimized semantic segmentation mask can be obtained. In agricultural applications, wheat plants exhibit dynamic growth over time, it is necessary to consider the temporal information during data annotation to construct a more temporally consistent crop growth model. By continuously annotating wheat images from different growth stages, plant growth trends can be extracted, enabling further optimization of growth assessment algorithms. For example, the plant growth rate $G(t)$ can be defined as the change in projected area between two time points $t_1$ and $t_2$:

$$G(t) = \frac{A(t_2) - A(t_1)}{t_2 - t_1}, \quad (7)$$

where $A(t)$ represents the projected area of the wheat plant at time $t$, which can be computed using the semantic segmentation mask $M_i$:

$$A(t) = \sum_{x,y} M_i(x, y, t) \quad (8)$$

By continuously sampling wheat images at different time points and performing semantic segmentation and keypoint annotation, a more comprehensive crop growth monitoring system can be developed.

### 3.3. Data Preprocessing

During deep learning training, the distribution of input data significantly affects model convergence speed and stability. Since image pixel value distributions vary across different datasets, normalization is typically required to ensure that all input samples fall within a consistent range, thereby preventing instability in gradient updates. Given an input image $I$ with pixel values in the range $[0, 255]$, normalization is usually performed using either mean standardization or min–max normalization:

$$I' = \frac{I - \mu}{\sigma},$$

where $\mu$ represents the dataset mean and $\sigma$ denotes the standard deviation. Alternatively, min–max normalization is applied as follows:

$$I' = \frac{I - I_{\text{min}}}{I_{\text{max}} - I_{\text{min}}},$$

where $I_{\text{min}}$ and $I_{\text{max}}$ represent the minimum and maximum pixel values, respectively. The application of normalization ensures a more stable input distribution for neural networks, facilitating faster model convergence. Furthermore, in agricultural image processing, variations in camera equipment may lead to significant differences in image resolution, necessitating scale transformation. Given an original image size of $(H, W)$ and a target size of $(H', W')$, the linear interpolation formula for scale transformation is defined as:

$$I'(x', y') = I(x \cdot s_x, y \cdot s_y),$$

where $s_x = \frac{W'}{W}$ and $s_y = \frac{H'}{H}$, with $(x', y')$ representing the target image coordinates and $(x, y)$ denoting the original image coordinates. Common interpolation techniques include bilinear interpolation, nearest-neighbor interpolation, and cubic interpolation, with the choice of method depending on the specific task requirements. Data augmentation is employed as a strategy to generate additional training samples by applying various transformations to the original dataset. In wheat growth detection tasks, where field images exhibit significant variability, appropriate data augmentation techniques enhance model robustness, improving adaptability to changes in lighting, background, and occlusion.

Commonly used data augmentation methods include CutMix, Mosaic, GridMask, and synthetic data augmentation based on diffusion models, as shown in Figure 2.

**a** **b** **c**

**Figure 2.** Data annotation: (**a**) is CutMix; (**b**) is Mosaic; (**c**) is GridMask.

In this study, we utilized multiple data augmentation techniques to enhance model performance. CutMix increased the diversity of training data by randomly cropping and pasting regions from different images, effectively alleviating overfitting in our experiments. Mosaic combined four distinct images into a single composite image, which assisted the model in learning in multi-object environments and was beneficial for multi-object counting tasks in wheat growth monitoring. GridMask randomly masked certain regions in the input image, forcing the model to learn more discriminative features and improving the model's generalization ability. By integrating these preprocessing techniques, the model performance was significantly enhanced, providing reliable and efficient training for wheat growth monitoring and yield estimation tasks [37].

### *3.4. Proposed Method*

#### 3.4.1. Symmetric Diffusion Segmentation Network Based on Transformer

As shown in Figure 3, the symmetric diffusion segmentation network based on transformer proposed in this study consists of four key modules: the transformer encoder, the symmetric diffusion module, the symmetric attention mechanism, and the aggregation loss calculation module. This model adopts an end-to-end training approach, aiming to enhance the boundary segmentation capability of wheat targets and improve the stability of target counting. After the data preprocessing, the input is first passed through the transformer encoder, which uses the ViT structure to partition the input image $I \in \mathbb{R}^{H \times W \times C}$ into patches. Each patch is mapped to a feature vector through a linear projection, and the position encoding $E_p$ is added to form the input token.

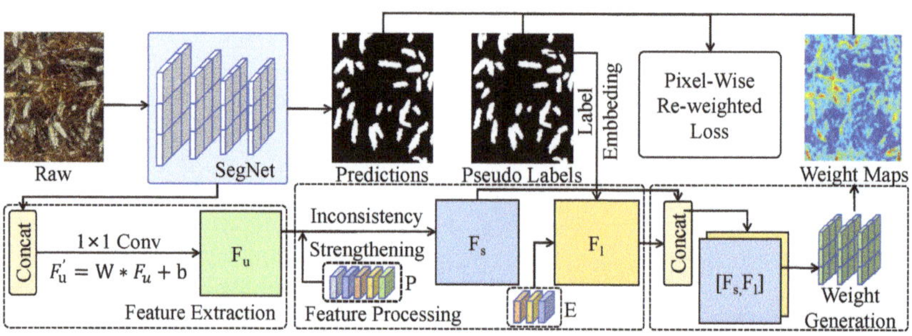


**Figure 3.** Overall framework of the wheat growth and counting analysis model based on semantic segmentation.

The framework begins with the transformer encoder, where the input image $I \in \mathbb{R}^{H\times W\times C}$ is first partitioned into patches consistent with the ViT structure. Each patch is mapped to a feature vector through a linear projection layer, capturing localized spatial relationships, and the position encoding $E_p$ is added to form the input token representation for subsequent transformer processing. This patch-based tokenization ensures that the model learns spatial dependencies effectively across different wheat growth regions.

$$X_0 = [X_p^1 E_p; X_p^2 E_p; \ldots; X_p^N E_p] \tag{9}$$

Once the initial token representation is generated, the transformer encoder extracts global features across the entire image, generating the feature representation $F_{global}$. The encoder's attention mechanism ensures that distant wheat plants and occlusions are effectively handled, improving robustness in segmentation. In the symmetric diffusion module, $F_{global}$ is further processed to simulate the growth diffusion process of wheat plants, enhancing the detailed expression of target boundaries. This module is inspired by biological growth diffusion models, where each feature point $(x, y)$ aggregates spatial information from neighboring pixels, simulating natural expansion and ensuring feature continuity across dense wheat regions. The diffused feature $F_{diff}$ is computed using the following equation:

$$F_{diff}(x,y) = \sum_{i=-k}^{k} \sum_{j=-k}^{k} W_{ij} \cdot F_{global}(x+i, y+j) \tag{10}$$

Here, $W_{ij}$ represents the adaptive diffusion weight, ensuring that the diffusion weight in high-density areas is smaller while diffusion is larger in low-density regions, thereby reducing background noise interference and enhancing the continuity of plant boundaries. The diffused feature $F_{diff}$ is then passed through the symmetric attention mechanism, which strengthens target information through multi-scale feature fusion. $F_{diff}$ is split into feature maps at different scales $F_s, F_l$, corresponding to local and global features, respectively:

$$F_s = \text{Conv}_s(F_{diff}), \quad F_l = \text{Conv}_l(F_{diff}) \tag{11}$$

These are then weighted and fused using the attention mechanism:

$$F_{attn} = \alpha F_s + (1-\alpha)F_l \tag{12}$$

where $\alpha$ is computed by the self-attention module to adaptively adjust the weight between local and global features. The final feature map $F_{attn}$ is passed through an semantic segmentation head to generate the final instance mask $M$, followed by loss calculation, which integrates segmentation loss, target counting error, and boundary optimization loss:

$$L_{total} = L_{seg} + \lambda_1 L_{count} + \lambda_2 L_{boundary} \tag{13}$$

where $L_{seg}$ optimizes segmentation accuracy, $L_{count}$ ensures the stability of target counting, and $L_{boundary}$ further strengthens boundary recognition capabilities. The inclusion of an aggregation loss function ensures that the model jointly optimizes multiple learning objectives, reducing inconsistencies between segmentation and counting tasks. The entire SDS-Net fully leverages the global information capture capability of the transformer structure, combining symmetric diffusion and attention mechanisms to improve wheat semantic segmentation accuracy while optimizing target counting stability in high-density environments. By explicitly integrating multi-scale feature fusion and detailing the ViT-based encoder structure, this framework ensures robust adaptability to different wheat growth stages and environmental conditions.

To explain the computation of diffusion weights in the symmetric diffusion module, let the input feature be $F \in \mathbb{R}^{H \times W \times C}$; the diffused feature $F_{diff}$ is obtained by a weighted sum of neighboring features, as shown in Algorithm 1:

$$F_{diff}(x, y) = \sum_{i=-k}^{k} \sum_{j=-k}^{k} W_{ij} \cdot F(x+i, y+j) \tag{14}$$

where $W_{ij}$ is the adaptive diffusion weight, defined as:

$$W_{ij} = \frac{\exp(-|F(x,y) - F(x+i, y+j)|^2)}{\sum_{m=-k}^{k} \sum_{n=-k}^{k} \exp(-|F(x,y) - F(x+m, y+n)|^2)} \tag{15}$$

This weight is controlled by feature similarity, ensuring that diffusion in high-density regions is reduced while diffusion in low-density regions is broader, thereby minimizing background noise interference with object boundaries.

**Algorithm 1** Symmetric Diffusion Algorithm

**Require:** Feature map $F \in \mathbb{R}^{H \times W \times C}$, diffusion range $k$
**Ensure:** Diffused feature map $F_{diff}$
Initialize $F_{diff} \leftarrow$ zeros of shape $(H, W, C)$
**for** $x \leftarrow 0$ **to** $H - 1$ **do**
  **for** $y \leftarrow 0$ **to** $W - 1$ **do**
    Extract local neighborhood:
    $F_{local} \leftarrow F[max(0, x-k) : min(H, x+k+1), max(0, y-k) : min(W, y+k+1)]$
    Compute similarity weights:
    $W_{ij} = \exp(-\|F(x,y) - F_{local}\|^2)$
    Normalize weights:
    $W_{ij} = \frac{W_{ij}}{\sum W_{ij}}$
    Compute diffused feature:
    $F_{diff}(x,y) = \sum W_{ij} \cdot F_{local}$
  **end for**
**end for**
**return** $F_{diff}$

### 3.4.2. Symmetric Diffusion Module

The core objective of the symmetric diffusion module is to simulate the growth diffusion process of wheat plants to enhance the boundary recognition capability of targets in semantic segmentation, especially in scenarios where wheat plants are densely grown and targets overlap.

As shown in Figure 4, this module effectively improves the model's robustness. The input features are first processed through channel reduction to decrease computational complexity, allowing the diffusion operation to be performed in a more compact feature representation space. Let the input feature be $F_{in} \in \mathbb{R}^{N \times C}$, where $N$ denotes the spatial dimension (such as $H \times W$) and $C$ represents the number of channels. Channel reduction is achieved through a linear mapping $W_r \in \mathbb{R}^{C \times C'}$:

$$F' = F_{in} W_r, \quad F' \in \mathbb{R}^{N \times C'} \tag{16}$$

where $C' \ll C$, ensuring that the main feature information is retained while reducing the computational cost. In the diffusion attention calculation stage, the input features are first projected into queries (Q), keys (K), and values (V). The query $Q \in \mathbb{R}^{N \times 1}$ is obtained by applying average pooling to the input features followed by a linear transformation, the key $K \in \mathbb{R}^{N' \times 1}$ is derived from the downsampled features, and the value $V \in \mathbb{R}^{N' \times C'}$ is obtained from the dimensionality-reduced features:

$$Q = W_q \cdot \text{AvgPool}(F') \tag{17}$$

$$K = W_k \cdot \text{DownSample}(F') \tag{18}$$

$$V = W_v \cdot F' \tag{19}$$

where $W_q$, $W_k$, $W_v$ are the respective linear transformation matrices. The attention weights are then computed to obtain the diffusion attention map $A$:

$$A = \text{softmax}\left(\frac{QK^T}{\sqrt{d_k}}\right) \tag{20}$$

where $d_k$ is the scaling factor for the key dimension, ensuring gradient stability. Subsequently, the attention weights $A$ are applied to the value $V$ to generate the diffused feature:

$$F_{diff} = AV \tag{21}$$

This feature $F_{diff}$ represents the diffusion result for the local region, which is further fused with the original feature $F'$ to retain detailed information about the target region while enhancing global coherence:

$$F_{out} = F' + W_d F_{diff} \tag{22}$$

where $W_d$ is the fusion weight parameter. The final output feature $F_{out} \in \mathbb{R}^{N \times C'}$ is mapped back to the original channel dimension using a channel recovery mapping $W_o$, forming the final diffusion-enhanced feature:

$$F_{final} = F_{out} W_o \tag{23}$$

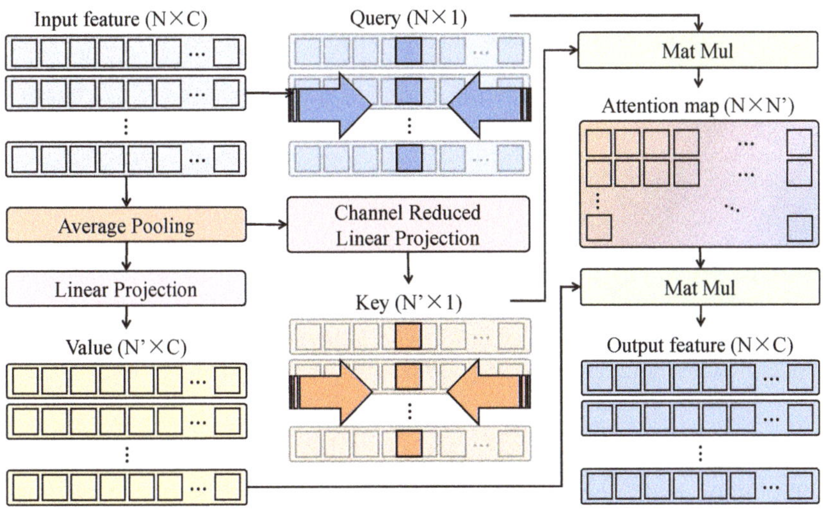


**Figure 4.** Diagram of the symmetric diffusion module.

The design advantage of the symmetric diffusion module lies in its ability to enhance boundary information. Through the diffusion operation, the semantic information of the target region is propagated to the neighboring areas, thus improving boundary accuracy during semantic segmentation and reducing target adhesion issues. In contrast to CNNs, which rely on small receptive fields for convolution operations, this module uses attention mechanisms to obtain a larger receptive field, enabling the model to capture long-distance feature dependencies. This is particularly suitable for target recognition tasks in densely planted wheat environments. Furthermore, by adopting dimensionality reduction and local attention mapping, the computational complexity is reduced from $\mathcal{O}(N^2)$ to $\mathcal{O}(Nd_k)$, making it more suitable for processing high-resolution agricultural images. In dense target environments, traditional object detection methods are prone to target omission, but this module, through adaptive diffusion calculations, enables the segmentation model to more accurately distinguish dense plants and improve the stability of target counting.

3.4.3. Symmetric Attention Mechanism

As shown in Figure 5, the symmetric attention mechanism presents a significant structural optimization compared to the traditional self-attention mechanism, particularly excelling in detection tasks involving densely packed objects.

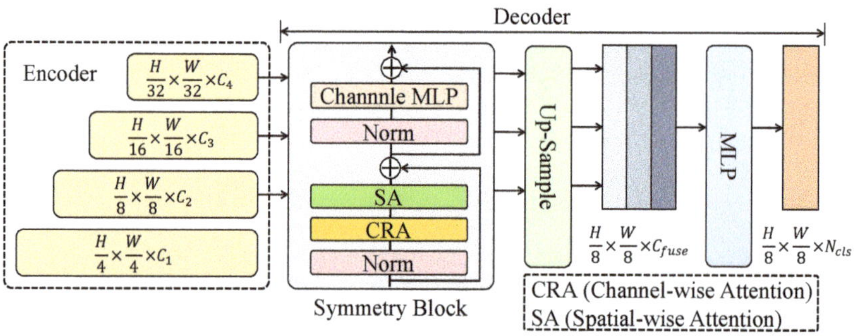


**Figure 5.** Schematic of the symmetric attention mechanism.

However, in agricultural image analysis tasks, the computational complexity of the standard transformer is $\mathcal{O}(N^2)$, resulting in a high computational overhead for high-resolution image processing. Additionally, while the self-attention mechanism primarily focuses on global information, there is significant local correlation between objects in dense target scenarios. As a result, directly using global attention may cause blurred boundaries and reduced object distinguishability. To address this, the symmetric attention mechanism combines channel attention and spatial attention, establishing a balance between feature maps at different scales and ensuring sufficient interaction of information at various scales, thereby improving the segmentation accuracy of local targets. The core of the symmetric attention mechanism consists of several symmetric attention blocks, including the channel attention module (CRA) and spatial attention module (SA). The input to the mechanism is a feature map of size $H \times W \times C$, where $H$ and $W$ represent the image height and width, respectively, and $C$ represents the number of channels. To effectively integrate multi-scale feature information, the network structure employs hierarchical feature extraction, where multiple encoder layers progressively refine spatial and semantic representations. The first layer has feature dimensions of $H/4 \times W/4 \times C_1$, the second layer $H/8 \times W/8 \times C_2$, the third layer $H/16 \times W/16 \times C_3$, and the fourth layer $H/32 \times W/32 \times C_4$. This hierarchical structure allows the model to capture both fine-grained local details and high-level global information, ensuring robust segmentation across different spatial scales. The network structure is composed of feature maps at multiple scales, with the first layer having feature dimensions of $H/4 \times W/4 \times C_1$, the second layer $H/8 \times W/8 \times C_2$, the third layer $H/16 \times W/16 \times C_3$, and the fourth layer $H/32 \times W/32 \times C_4$. In the encoder part, each layer contains normalization layers, symmetric attention computation, channel MLP, and multi-scale feature fusion, while the decoder part restores the resolution through upsampling and performs feature remapping using MLP. The computation of the channel attention module is as follows:

$$F_{avg} = \frac{1}{H \times W} \sum_{i=1}^{H} \sum_{j=1}^{W} F(i,j) \tag{24}$$

$$F_{max} = \max_{i,j} F(i,j) \tag{25}$$

$$W_c = \sigma(W_1(F_{avg}) + W_2(F_{max})) \tag{26}$$

Here, $W_1$ and $W_2$ are fully connected layers and $\sigma$ represents the Sigmoid activation function. Additionally, the spatial attention computation is as follows:

$$F'_c = \text{Conv}(F, k = 7) \tag{27}$$

$$W_s = \sigma(F'_c) \tag{28}$$

The final symmetric attention computation is expressed as:

$$F_{SAM} = W_c \cdot F + W_s \cdot F \tag{29}$$

In the SDS-Net structure, the symmetric attention mechanism mainly acts on the connection between the encoder and decoder, ensuring that during the feature extraction phase, not only is global information considered, but local features are also retained. This compensates for the lack of local information in standard transformers for semantic segmentation tasks. On the encoder side, after the transformer extracts global features $F_{global}$, the symmetric attention mechanism enhances the local feature information through multi-scale spatial attention, as shown below:

$$F' * global = F * global + F_{SAM} \tag{30}$$

On the decoder side, during the process of progressively restoring image resolution, the symmetric attention mechanism ensures that features at different scales are fully fused, making boundary information clearer, thus optimizing the semantic segmentation accuracy:

$$F' * decoder = \text{UpSample}(F * decoder) + F_{SAM} \tag{31}$$

Furthermore, to enhance feature representation across different spatial scales, the multi-scale feature fusion mechanism integrates outputs from multiple encoder layers before passing them to the decoder. This ensures that the network learns both fine-grained and high-level contextual features, improving segmentation robustness in dense wheat fields. The inclusion of multi-scale feature fusion mitigates the issue of feature degradation in deep networks, allowing the model to maintain consistent performance across varying wheat growth stages and occlusion levels. In addition to multi-scale feature fusion, the Channel MLP module plays a crucial role in enhancing inter-channel dependencies. Located within the Symmetric Block, the Channel MLP module processes the extracted features by refining channel-wise representations, ensuring effective feature recalibration before they are passed to the decoder. Unlike conventional MLP layers, the Channel MLP module is optimized for segmentation tasks by dynamically adjusting the weights of different channels, thereby improving feature discrimination. The integration of the Channel MLP with normalization layers, symmetric attention computation, and multi-scale feature fusion provides a balanced trade-off between computational efficiency and segmentation accuracy.

By introducing the symmetric attention mechanism, SDS-Net effectively combines the ability to model global information and extract local features, allowing the network to better adapt to dense target scenarios, improving the segmentation accuracy of object boundaries and reducing interference between instances. This optimization mechanism enhances the model's robustness and accuracy in semantic segmentation tasks in agricultural environments, especially for precise detection and counting tasks in dense wheat plant growth environments.

#### 3.4.4. Aggregated Loss Function

In semantic segmentation tasks, traditional loss functions typically employ binary cross-entropy loss (BCE) or Dice loss for optimization. However, each of these methods has limitations. BCE is primarily used for pixel-level classification but lacks the ability to capture boundary information, leading to target adhesion or mis-segmentation. Dice loss improves segmentation performance but struggles with background noise in densely distributed agricultural images. To address these limitations, this study proposes an aggregated loss function that jointly optimizes semantic segmentation, target counting, and boundary refinement. The weight selection process for these loss components was determined through extensive experiments and cross-validation to ensure balanced optimization.The aggregated loss function consists of three components. The first is the semantic segmentation loss $L_{seg}$, which optimizes the match between the segmented region and the ground truth, defined by the Dice loss as:

$$L_{seg} = 1 - \frac{2\sum PG}{\sum P + \sum G}, \tag{32}$$

where $P$ is the predicted segmentation mask and $G$ is the ground truth mask. To determine the optimal weight for $L_{seg}$, a grid search was conducted over a range of values, considering its impact on segmentation accuracy across different wheat densities. The second component is the target counting loss $L_{count}$, which ensures the stability of the predicted target count, defined using mean squared error (MSE):

$$L_{count} = \sum_{x=1}^{H} \sum_{y=1}^{W} \left(\hat{D}(x,y) - D(x,y)\right)^2, \tag{33}$$

where $\hat{D}(x,y)$ is the predicted density map and $D(x,y)$ is the ground truth density distribution. Since wheat distribution varies across fields, the weight of $L_{count}$ was optimized using 5-fold cross-validation, ensuring the loss function remains effective under different target densities. To further enhance boundary precision, the boundary loss $L_{boundary}$ is introduced:

$$L_{boundary} = \sum ||\nabla P - \nabla G||^2, \tag{34}$$

where $\nabla P$ and $\nabla G$ represent the gradient information of the predicted segmentation mask and ground truth mask, respectively. This loss component was weighted lower in early training phases to prevent overemphasis on boundaries before segmentation features were sufficiently refined. To determine the optimal loss weight combination, an extensive hyperparameter search was conducted over different weight values for $\lambda_1$, $\lambda_2$, and $\lambda_3$ in the total loss function:

$$L_{total} = \lambda_1 L_{seg} + \lambda_2 L_{count} + \lambda_3 L_{boundary}. \quad (35)$$

A grid search was performed over $\lambda_1, \lambda_2, \lambda_3 \in [0.1, 1.0]$ with a step size of 0.1, evaluating segmentation accuracy (mAP@75), target counting stability (MSE), and boundary precision (Boundary IoU) on a validation set. Five-fold cross-validation was used to ensure generalization across different field conditions. The results indicated that $\lambda_1 = 0.6$, $\lambda_2 = 0.3$, and $\lambda_3 = 0.1$ provided the best balance between segmentation accuracy, counting stability, and boundary clarity. The final optimization ensures that SDS-Net effectively captures global target distributions while maintaining detailed local feature refinement. This adaptive weighting approach enhances performance across varying wheat densities, making SDS-Net more robust in real-world agricultural applications.

### 3.5. Experimental Configuration

#### 3.5.1. Hardware and Software Configuration

Experiments were carried out on a high-performance computing platform. The hardware consisted of an NVIDIA A100 GPU(NVIDIA Corporation, Santa Clara, CA, USA) (Ampere architecture, 6912 CUDA cores, 432 Tensor cores, 40 GB HBM2 memory), an AMD EPYC 7742 64-core/128-thread processor, 512 GB of RAM, and NVMe SSDs for storage. The software used PyTorch 2.6.0 as the deep learning framework, running on Ubuntu 22.04 LTS with CUDA 12.1 and cuDNN 8.8. PyTorch Lightning managed the training workflow, while OpenCV and Albumentations were used for data preprocessing. NVIDIA Apex was utilized for mixed-precision training.

#### 3.5.2. Dataset Partitioning and Hyperparameters

The dataset was partitioned into training, validation, and test sets at an 80%:10%:10% ratio. A 5-fold cross-validation was also employed for stability. To determine the optimal optimizer, we conducted comparative experiments using Adam, SGD, RMSprop, and AdamW. The results indicated that AdamW outperformed the other optimizers in terms of convergence stability and final segmentation accuracy, particularly in dense wheat growth scenarios. The effectiveness of AdamW is attributed to its decoupled weight decay, which mitigates overfitting and enhances generalization. Based on empirical evaluation, we set the hyperparameters to $\beta_1 = 0.9$ and $\beta_2 = 0.999$ and weight decay to $\lambda = 0.01$. The learning rate was selected through a grid search, exploring initial values from $10^{-3}$ to $10^{-5}$, with different decay strategies, including exponential decay, cosine annealing, and linear decay. The experimental results demonstrated that a linear decay strategy provided the best balance between training stability and final performance. Thus, we adopted a learning rate schedule starting from $\eta_0 = 10^{-4}$ over $T = 100$ epochs, incorporating a warm-up phase in the first 10 epochs to prevent sudden weight updates at the beginning of training. The batch size was set to 16 with a gradient accumulation step of 4, ensuring efficient memory utilization without compromising training performance. To further enhance robustness, an early stopping mechanism was implemented, halting training if the validation loss did not decrease for 10 consecutive epochs. This helped prevent unnecessary computation and reduced the risk of overfitting. The combination of these hyperparameters, determined through extensive experimental evaluation, ensured the proposed model achieved optimal performance in wheat segmentation and growth measurement tasks.

3.5.3. Evaluation Metrics

To comprehensively evaluate model performance, we used Precision ($P$), Recall ($R$), Accuracy ($Acc$), F1-score ($F1$), and Mean Average Precision (mAP@75). These metrics, common in object detection and semantic segmentation, help assess classification accuracy, object coverage, and segmentation fidelity. Precision is the ratio of correctly predicted positive samples to all predicted positive ones. Recall is the proportion of correctly detected objects among all ground-truth objects. Accuracy gauges overall classification correctness, and the F1-score balances precision and recall. mAP@75 measures performance across different IoU thresholds, especially at 0.75. These metrics rely on True Positives (TPs), False Positives (FPs), False Negatives (FNs), and True Negatives (TNs). In semantic segmentation, TPs are correctly detected objects, FPs are incorrect detections, and FNs are missed ones. Predictions with IoU > 0.5 or 0.75 are considered correct, i.e., they are TPs. Their formulas are:

$$P = \frac{TP}{TP + FP} \tag{36}$$

$$R = \frac{TP}{TP + FN} \tag{37}$$

$$Acc = \frac{TP + TN}{TP + TN + FP + FN} \tag{38}$$

$$F1 = \frac{2 \times P \times R}{P + R} \tag{39}$$

$$AP = \int_0^1 P(R)dR \tag{40}$$

For multi-class tasks, mAP is the mean of all class-wise AP values ($mAP = \frac{1}{N}\sum_{i=1}^{N} AP_i$, where $N$ is the number of classes). In agricultural applications, mAP@75 uses a strict IoU threshold of 0.75 for accurate semantic segmentation. Wheat growth monitoring requires precise classification of plant regions, accurate target counting, and robust boundary delineation, making these metrics crucial for assessing segmentation quality and counting stability. Precision reflects the ability to minimize false positives, ensuring that misclassified background regions are not falsely detected as wheat plants, which is essential for avoiding overestimation in yield prediction. Recall measures the proportion of correctly detected wheat targets among all ground-truth instances, ensuring the model captures all wheat plants, even in dense growth scenarios. Accuracy provides a broad evaluation of classification correctness but is less informative in highly imbalanced scenarios where background pixels dominate. The F1-score balances precision and recall, making it useful for dense wheat fields where accurate plant detection is necessary without excessive false positives or omissions. mAP@75 is particularly important for wheat growth monitoring because it evaluates segmentation performance under a strict IoU threshold of 0.75, ensuring precise plant boundary detection. In dense wheat growth environments, achieving a high mAP@75 value indicates that the model can accurately segment individual plants without merging adjacent crops, which is critical for monitoring wheat development and estimating biomass. The calculations of these metrics rely on True Positives (TPs), False Positives (FPs), False Negatives (FNs), and True Negatives (TNs), defined as follows in semantic segmentation: TPs are correctly detected wheat regions, FPs are misclassified background areas, and FNs are missed wheat plants. Predictions with IoU > 0.5 or 0.75 are considered as correct detections (TPs), reinforcing the need for precise segmentation.

3.5.4. Baseline

To evaluate our proposed wheat growth detection and counting model, we selected several state-of-the-art semantic segmentation models as baselines. These include Tiny-Segformer [38], U-Net [39], Mask R-CNN [40], Segment Anything Model (SAM) [41], and

InternImage [42]. Tiny-Segformer is a lightweight transformer-based semantic segmentation network suitable for edge devices. U-Net is a classic encoder–decoder with skip connections. Mask R-CNN combines object detection and semantic segmentation. SAM uses prompt-based segmentation, and InternImage is a CNN-transformer hybrid. Comparing with these baselines, we comprehensively analyze the advantages of our proposed method for wheat growth detection and counting tasks.

## 4. Results and Discussion

### 4.1. Segmentation Accuracy Results

The objective of this experiment is to evaluate the segmentation accuracy of various semantic segmentation models in wheat growth monitoring and target counting tasks, as well as to validate the effectiveness of the proposed method in high-density target detection scenarios. By comparing Tiny-Segformer, UNet, Mask R-CNN, Segment Anything Model (SAM), InternImage, Swin Transformer, DETR, ViT, and the proposed method, the experiment primarily focuses on the models' performance in Precision, Recall, Accuracy, mAP@75, and F1-score, aiming to comprehensively assess the adaptability of different methods in the wheat semantic segmentation task. Additionally, we include a comparison of model parameters (Parameters (M)) and computational complexity (FLOPS (G)) to evaluate the efficiency and feasibility of the proposed method for real-world deployment.

As shown in Table 2 and Figure 6, the proposed method achieved the best performance across all metrics, with Precision and Recall reaching 0.91 and 0.87, respectively, and the F1-score being 0.89, significantly outperforming other baseline models. Compared to InternImage, the proposed method's mAP@75 was further improved to 0.88, indicating a stronger ability to identify target boundaries under high IoU thresholds. The relatively lower Recall suggests that some targets were missed, but the improvement in Precision indicates that the method reduced the likelihood of misclassifications during segmentation. This high-precision segmentation capability is critical for precise agricultural monitoring, particularly in densely growing wheat environments, where the accurate identification and counting of each plant directly impact yield predictions. Beyond segmentation accuracy, we further analyze the efficiency of the proposed method by reporting Parameters (M) and FLOPS (G). Parameters (M) represent the number of trainable parameters in the model, which directly affects memory consumption and model size. A lower parameter count is desirable for deployment on edge devices, while larger models often lead to improved performance at the cost of increased storage requirements. The proposed method achieves a balance between these factors, containing 15.2 million parameters, which is significantly smaller than Mask R-CNN (44.5 M) and SAM (641 M) but larger than Tiny-Segformer (5.2 M), ensuring high segmentation accuracy with a manageable model size.

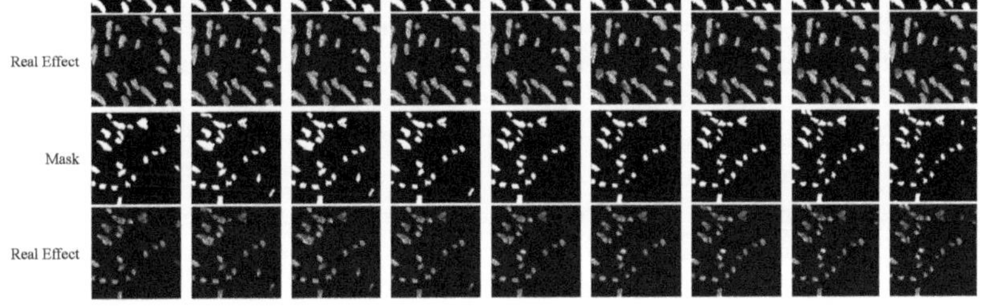


**Figure 6.** The visual results of the method on our dataset.

**Table 2.** Experimental results of wheat growth and counting models.

| Model | P | R | Acc | mAP@75 | F1-Score | Parameters (M) | FLOPS (G) |
|---|---|---|---|---|---|---|---|
| Tiny-Segformer | 0.80 | 0.77 | 0.79 | 0.79 | 0.79 | 5.2 | 4.4 |
| UNet | 0.82 | 0.79 | 0.81 | 0.81 | 0.80 | 7.76 | 65 |
| Mask R-CNN | 0.84 | 0.82 | 0.83 | 0.83 | 0.82 | 44.5 | 180 |
| SAM | 0.86 | 0.84 | 0.85 | 0.85 | 0.85 | 641 | 2963 |
| Swin Transformer | 0.83 | 0.80 | 0.82 | 0.81 | 0.81 | 88 | 15.4 |
| DETR | 0.85 | 0.81 | 0.82 | 0.83 | 0.82 | 41 | 86 |
| ViT | 0.87 | 0.84 | 0.85 | 0.86 | 0.85 | 307 | 190.7 |
| InternImage | 0.89 | 0.85 | 0.87 | 0.86 | 0.86 | 30 | 5 |
| Proposed Method | 0.91 | 0.87 | 0.89 | 0.88 | 0.89 | 15.2 | 30.1 |

FLOPS (G), on the other hand, quantifies the number of floating-point operations required for a single forward pass of the model. This metric is crucial for evaluating computational efficiency, as a lower FLOPS value indicates faster inference and lower power consumption. The proposed method requires 30.1 G FLOPS, which is significantly lower than SAM (2963 G) and Mask R-CNN (180 G), demonstrating that it is well-suited for deployment in agricultural monitoring applications where computational resources may be limited. Compared to InternImage, which has a FLOPS of 5G, the proposed method requires more computation but offers significantly better segmentation accuracy and robustness, making it a viable trade-off between efficiency and performance. It is important to note that Parameters (M) and FLOPS (G) are reported specifically in Table 2 because this table focuses on overall model evaluation. Other tables, such as those analyzing ablation studies or dataset-specific tests, primarily focus on comparative accuracy and robustness rather than computational efficiency. Including these metrics in Table 2 allows for a comprehensive assessment of both segmentation accuracy and computational feasibility in a single comparison. From a mathematical perspective, the performance of different models is greatly influenced by their structural design. As a lightweight transformer architecture, Tiny-Segformer benefits from computational efficiency; however, its local attention window limits its ability to model global information, resulting in slightly lower segmentation performance, with an F1-score of 0.79. UNet, utilizing a symmetric encoder–decoder structure and skip connections, preserves some high-resolution features, leading to improvements in both Precision and Recall, with an F1-score of 0.80. However, due to its reliance on CNNs, the model's receptive field is relatively small, and boundary blurring may occur in dense target segmentation tasks. Mask R-CNN combines object detection and semantic segmentation capabilities, enhancing target region extraction with RPN, which results in an mAP@75 of 0.83. However, due to the fixed window size of RoIAlign, this method still faces information loss issues in dense target detection. SAM utilizes a prompt-based segmentation mechanism that effectively adapts to different target categories, improving Recall and achieving an F1-score of 0.85. InternImage, as a CNN-transformer hybrid model, enhances the fusion of local and global features through the Large Kernel Attention (LKA) mechanism, further improving segmentation performance. The proposed method integrates transformer-based global feature extraction with a symmetric diffusion mechanism and symmetric attention mechanism, reducing background noise interference while enhancing the segmentation accuracy of dense targets. Mathematically, the method adjusts the weight of local and global features through adaptive attention, ensuring accurate boundary segmentation even in high-density target scenarios, thereby improving Precision and mAP@75. Additionally, the proposed aggregation loss function effectively optimizes semantic segmentation loss, target counting loss, and boundary loss, increasing Accuracy to 0.89 and ensuring the model's stability across different target densities. Computational complexity is a key factor in evaluating the deployability of deep learning models. Although lightweight models such as MobileNet and EfficientNet offer advantages in computational cost, their limited receptive fields and feature extraction capabilities often constrain segmentation accuracy in high-density object scenarios. The proposed method enhances feature extraction effectiveness by optimizing the attention mechanism and diffusion computation, enabling high inference efficiency despite an increase in computational cost. Additionally, mixed-precision computation (AMP) and tensor optimization strategies are employed during inference to reduce computational redundancy, further improving efficiency.

### 4.2. Growth Measurement Results

This experiment evaluates the accuracy of various segmentation models in wheat growth measurement and assesses the proposed method's adaptability in complex field environments. Accurate growth measurement enhances plant counting, leaf coverage estimation, and spatial distribution analysis across growth stages.

As shown in Table 3, the proposed method achieves the highest scores in all metrics, with Precision at 0.95, Recall at 0.90, and mAP@75 at 0.92, significantly outperforming the other models. This suggests fewer false positives and improved stability in wheat segmentation. Compared to InternImage, which achieves an Accuracy of 0.91, the proposed method shows better boundary detection and feature retention. While SAM and Mask R-CNN perform well globally, they suffer from mis-segmentation in dense target environments. Traditional CNN models such as UNet and Tiny-Segformer struggle with boundary details, leading to lower F1-scores (0.84 and 0.82, respectively). From a computational perspective, Tiny-Segformer employs local window attention ($\mathcal{O}(N \cdot W)$ complexity), limiting global feature modeling and reducing Recall in dense regions. UNet's encoder–decoder structure preserves spatial information but results in boundary blurring due to localized convolutions. Mask R-CNN, with its loss function $L = L_{cls} + L_{box} + L_{mask}$, accurately segments objects but suffers from fixed RoIAlign window limitations. SAM leverages attention-based segmentation ($Y = \text{Attention}(WX) + X$) for global feature extraction, enhancing boundary segmentation. InternImage, integrating Large Kernel Attention (LKA) and Self-Attention, improves feature extraction but at a higher computational cost. The proposed method enhances boundary segmentation through symmetric diffusion ($F_{diff}(x,y) = \sum_{i=-k}^{k} \sum_{j=-k}^{k} A_{i,j} F(x+i, y+j)$), combining transformer-based global extraction with adaptive diffusion computation. The symmetric attention mechanism further enhances feature stability while reducing background noise. The loss function optimizes segmentation, counting, and boundary accuracy, ensuring stable performance across varying densities.

**Table 3.** Experimental results of wheat growth measurement models.

| Model | Precision | Recall | Accuracy | mAP@75 | F1-Score |
|---|---|---|---|---|---|
| Tiny-Segformer | 0.83 | 0.80 | 0.82 | 0.81 | 0.82 |
| UNet | 0.85 | 0.82 | 0.84 | 0.83 | 0.84 |
| Mask R-CNN | 0.88 | 0.85 | 0.86 | 0.85 | 0.87 |
| SAM | 0.89 | 0.87 | 0.88 | 0.87 | 0.88 |
| InternImage | 0.92 | 0.89 | 0.91 | 0.90 | 0.91 |
| Proposed Method | 0.95 | 0.90 | 0.93 | 0.92 | 0.92 |

### 4.3. Ablation Study of Different Attention Mechanisms

This experiment evaluates how different attention mechanisms impact wheat segmentation and growth measurement, comparing Standard Self-Attention, CBAM, and the proposed symmetric attention mechanism across five evaluation metrics.

As shown in Table 4, the symmetric attention mechanism consistently outperforms the other methods, with Precision reaching 0.95, Recall at 0.90, and mAP@75 at 0.92 in the growth measurement task. CBAM achieves an F1-score of 0.85, while Standard Self-Attention lags at 0.74, highlighting its limitations in dense wheat segmentation. For semantic segmentation, the symmetric attention mechanism scores 0.89 in F1-score, outperforming CBAM (0.83) and Standard Self-Attention (0.72). Mathematically, these performance differences stem from their computational approaches. Standard Self-Attention uses global feature interactions with $\mathcal{O}(N^2)$ complexity, making it computationally expensive and less stable for high-resolution agricultural images. CBAM enhances feature selection via channel and spatial attention: $W_c = \sigma(W_1(F_{avg}) + W_2(F_{max}))$, $W_s = \sigma(\text{Conv}(F, k = 7))$, with fusion $F_{CBAM} = W_c \cdot F + W_s \cdot F$. While CBAM efficiently enhances local features, its separate processing of channel and spatial information limits global feature modeling. In contrast, the symmetric attention mechanism integrates global modeling with optimized local attention using $F_{SAM} = W_c \cdot F + W_s \cdot F$, further stabilizing local feature interactions via multi-scale fusion. This enables adaptive weighting between global and local features, ensuring precise segmentation even in high-density wheat fields. The resulting performance—Accuracy of 0.93 and Precision of 0.95 in growth measurement—significantly outperforms CBAM and Standard Self-Attention. Figure 7 illustrates heatmaps that visualize model attention distribution during segmentation, highlighting how the symmetric attention mechanism improves boundary detection and feature clarity.

**Table 4.** Ablation study of different attention mechanisms.

| Model | Precision | Recall | Accuracy | mAP@75 | F1-Score |
|---|---|---|---|---|---|
| Segmentation-Standard Self | 0.73 | 0.70 | 0.72 | 0.71 | 0.72 |
| Segmentation-CBAM | 0.85 | 0.82 | 0.83 | 0.82 | 0.83 |
| Segmentation-Symmetric | 0.91 | 0.87 | 0.89 | 0.88 | 0.89 |
| Measurement-Standard Self | 0.77 | 0.72 | 0.74 | 0.73 | 0.74 |
| Measurement-CBAM | 0.88 | 0.84 | 0.86 | 0.85 | 0.85 |
| Measurement-Symmetric | 0.95 | 0.90 | 0.93 | 0.92 | 0.92 |

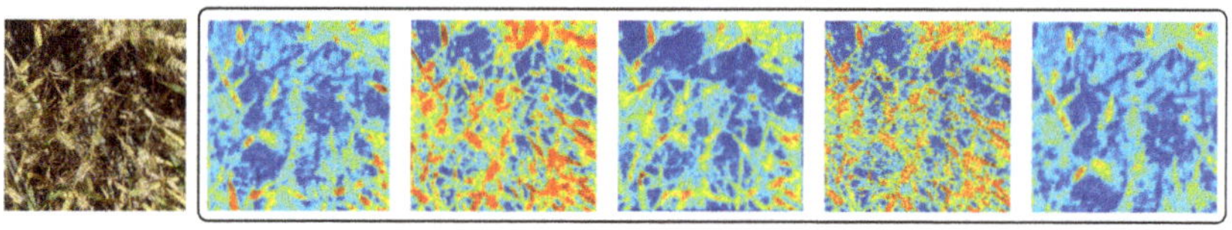


**Figure 7.** Heatmap demonstrating the effectiveness of the symmetric attention mechanism.

### *4.4. Ablation Study of Different Data Augmentation Strategies*

The objective of this experiment is to evaluate the impact of different data augmentation strategies on model performance and to validate the effectiveness of the proposed method in this regard. Data augmentation is a commonly employed strategy, particularly in agricultural scenarios where wheat plants grow densely under complex environmental conditions. Challenges such as insufficient training data, occlusions, and variations in lighting conditions may affect model performance. Therefore, an appropriate data augmentation approach can effectively enhance the generalization ability of the model and improve its adaptability to complex environments. This experiment compares three classical data augmentation methods, namely CutMix, Mosaic, and GridMask, with the proposed method, analyzing their effects on Precision, Recall, Accuracy, Mean Average Precision at 75% IoU threshold (mAP@75), and F1-score.

As shown in Table 5, different data augmentation techniques have varying degrees of impact on model performance. CutMix exhibited relatively low Precision (0.68) and Recall (0.62). Mosaic showed improved Precision (0.72) and Recall (0.69) compared to CutMix, whereas GridMask performed slightly worse, with Precision and Recall values of 0.64 and 0.60, respectively. In contrast, the proposed method achieved the best performance across all metrics, with a Precision of 0.91, Recall of 0.87, mAP@75 of 0.88, and an F1-score of 0.89. These results demonstrate that the proposed data augmentation strategy more effectively enhances the segmentation and counting capabilities of the model, thereby improving its generalization ability in agricultural environments. From a mathematical perspective, different data augmentation techniques influence the feature space in distinct ways, leading to variations in model performance. CutMix augments training diversity by randomly blending regions from different images. However, due to the similarity in color and texture features among wheat plants in agricultural scenarios, random blending may cause boundary confusion, thereby degrading segmentation accuracy. Mosaic stitches four different images together, enhancing the model's adaptability to diverse backgrounds. However, this method may compromise object integrity, increasing the risk of misclassification. GridMask applies structured occlusion to the images, compelling the model to focus on local features. However, in high-density object environments, this augmentation approach may obscure critical features, leading to lower recall performance. In contrast, the proposed method is designed with agricultural scene characteristics in mind. By employing adaptive regional augmentation, it enables the model to effectively learn key features at varying target densities, thereby improving segmentation accuracy and stability.

**Table 5.** Ablation study of different data augmentation strategies.

| Model | Precision | Recall | Accuracy | mAP@75 | F1-Score |
|---|---|---|---|---|---|
| CutMix | 0.68 | 0.62 | 0.65 | 0.65 | 0.64 |
| Mosaic | 0.72 | 0.69 | 0.71 | 0.70 | 0.70 |
| GridMask | 0.64 | 0.60 | 0.62 | 0.61 | 0.62 |
| Proposed Method | 0.91 | 0.87 | 0.89 | 0.88 | 0.89 |

### *4.5. Ablation Study on the Symmetric Diffusion Module*

The objective of this experiment is to investigate the role of the symmetric diffusion module (SDM) in wheat instance segmentation and to analyze the impact of different diffusion strategies on model performance. Diffusion mechanisms are widely used in computer vision tasks to enhance feature representations, particularly in dense object detection, where an appropriate diffusion strategy can effectively improve the model's ability to recognize object boundaries and reduce noise interference. This experiment compares three different settings: no diffusion, Gaussian diffusion, and the proposed symmetric diffusion module. The evaluation is conducted based on five key metrics.

As shown in Table 6, the model's segmentation performance is relatively poor when no diffusion is applied, with a Precision of only 0.70, a Recall of 0.63, and an mAP@75 of merely 0.67. This suggests that in the absence of diffusion, the model struggles to capture boundary information effectively, resulting in incomplete object segmentation. The Gaussian diffusion approach partially mitigates this issue, improving Precision and Recall to 0.78 and 0.84, respectively, while increasing mAP@75 to 0.80. These results suggest that Gaussian diffusion enhances feature propagation to some extent, improving boundary continuity. However, compared to the proposed symmetric diffusion module, Gaussian diffusion still exhibits certain limitations. The proposed method achieves the best performance across all metrics, with a Precision of 0.91, Recall of 0.87, mAP@75 of 0.88, and F1-score of 0.89. These findings indicate that, compared to traditional diffusion strategies, the symmetric diffusion module more effectively optimizes object boundaries, enhancing segmentation stability and accuracy. From a mathematical perspective, the performance differences among various diffusion strategies stem primarily from differences in diffusion computation methods. In the absence of diffusion, the model relies on self-attention mechanisms for feature extraction. However, in dense object environments, self-attention struggles to establish stable object boundary representations within a limited receptive field, leading to suboptimal Recall and mAP@75 performance. Gaussian diffusion smooths features using convolution kernels, leveraging neighborhood information to enhance local consistency. While this method improves local consistency, its diffusion weights are determined solely by spatial position and do not adapt to feature properties. As a result, it can introduce boundary blurring, limiting segmentation effectiveness in complex environments. The proposed symmetric diffusion module incorporates an adaptive feature diffusion strategy, where diffusion weights are determined based on feature similarity. This ensures reduced feature propagation in high-density regions and broader diffusion in low-density regions, thereby minimizing boundary blurring and improving segmentation accuracy. This property enables the model to accurately capture object contours even in dense target environments, leading to superior mAP@75 and F1-score performance.

**Table 6.** Ablation study on the symmetric diffusion module.

| Model | Precision | Recall | Accuracy | mAP@75 | F1-Score |
|---|---|---|---|---|---|
| no diffusion | 0.70 | 0.63 | 0.68 | 0.67 | 0.64 |
| gaussian diffusion | 0.78 | 0.84 | 0.82 | 0.80 | 0.81 |
| Proposed Method | 0.91 | 0.87 | 0.89 | 0.88 | 0.89 |

### *4.6. Test on Other Dataset*

Table 7 and Figure 8 present the performance of different models on the Kaggle dataset, including metrics such as Precision, Recall, Accuracy, mAP@75, and F1-score. From these experimental results, it is evident that the proposed method outperforms other baseline models across all evaluation metrics, particularly in Precision and Recall, where it achieves 0.89 and 0.85, respectively, significantly higher than other models like Tiny-Segformer and UNet.

**Table 7.** Experimental results of wheat growth and counting models in Kaggle.

| Model | Precision | Recall | Accuracy | mAP@75 | F1-Score |
|---|---|---|---|---|---|
| Tiny-Segformer | 0.78 | 0.75 | 0.76 | 0.77 | 0.76 |
| UNet | 0.81 | 0.76 | 0.79 | 0.78 | 0.78 |
| Mask R-CNN | 0.82 | 0.80 | 0.81 | 0.80 | 0.81 |
| SAM | 0.84 | 0.81 | 0.82 | 0.83 | 0.82 |
| Swin Transformer | 0.83 | 0.81 | 0.82 | 0.82 | 0.81 |
| DETR | 0.85 | 0.82 | 0.83 | 0.84 | 0.82 |
| ViT | 0.86 | 0.83 | 0.84 | 0.84 | 0.85 |
| InternImage | 0.86 | 0.82 | 0.84 | 0.83 | 0.82 |
| Proposed Method | 0.89 | 0.85 | 0.87 | 0.86 | 0.85 |

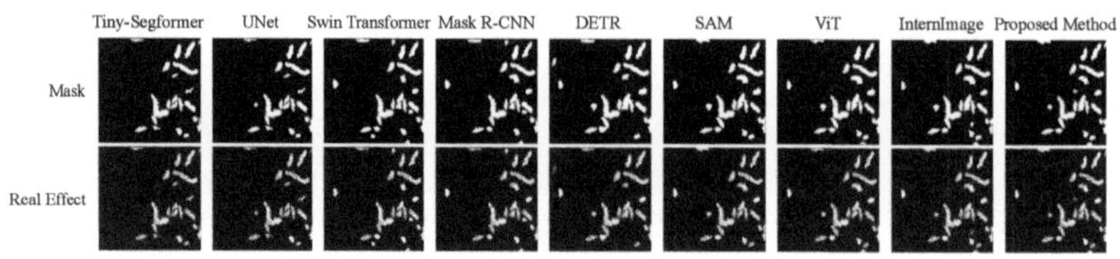


**Figure 8.** The visual results of the method on the Kaggle dataset.

### 4.7. Application Deployment

Table 8 presents the frames per second (FPS) of different models on different platforms, which indirectly reflects the computational efficiency and runtime speed of the models. By comparing the FPS of the "Proposed Method" with other baseline models (such as Tiny-Segformer, UNet, Mask R-CNN, SAM, and InternImage) on the Jetson and iPhone14 Pro Max platforms, it is evident that the proposed method outperforms the other models on both platforms. Particularly on the Jetson platform, the FPS of the proposed method is 50, significantly higher than other models, such as Tiny-Segformer (24 FPS) and UNet (20 FPS), indicating that the proposed method has higher computational efficiency on edge computing platforms, potentially leading to reduced operational costs. On the iPhone14 Pro Max platform, the proposed method achieves an FPS of 39, also outperforming other baseline models. These experimental results support the claim of cost reduction mentioned previously, as a higher FPS value means more images can be processed in the same amount of time, thus improving efficiency and lowering operational costs.

**Table 8.** FPS of different models on different platform.

| Platform | Tiny-Segformer | UNet | Mask R-CNN | SAM | InternImage | Proposed Method |
|---|---|---|---|---|---|---|
| Jetson | 24 | 20 | — | 3 | 31 | 50 |
| IPhone14 Pro Max | 18 | 13 | — | — | 27 | 39 |

## 5. Conclusions

A wheat growth and counting analysis model based on instance segmentation is proposed in this study to address the challenges of wheat growth detection in agricultural production. With the advancement of digitalization and automation in agriculture, traditional manual monitoring methods are no longer able to meet the demands for efficiency and precision, particularly in large-scale farmland applications. The innovations in this work are primarily reflected in the following aspects: First, a transformer-based symmetric diffusion segmentation network is proposed, which, by incorporating the symmetric diffusion module and symmetric attention mechanism, effectively solves the problem of high-density target detection in wheat instance segmentation. Second, an aggregated loss function is introduced, which further optimizes the accuracy of the task by combining the results of instance segmentation. The experimental results demonstrate that the proposed model achieves significant advantages across multiple evaluation metrics. In wheat instance segmentation tasks, the model incorporating the symmetric attention mechanism performs excellently on all indicators, achieving a Precision of 0.91, Recall of 0.87, Accuracy of 0.89, mAP@75 of 0.88, and an F1-score of 0.89, significantly outperforming other baseline methods. Moreover, the model also performs outstandingly in growth measurement tasks, with Precision reaching 0.95, Recall at 0.90, and mAP@75 improving to 0.92, fully validating the effectiveness of the symmetric attention mechanism and symmetric diffusion module in high-density target detection. Ablation studies on different attention mechanisms further confirm the advantages of the symmetric attention mechanism in optimizing instance segmentation accuracy and boundary recognition. Compared to the standard self-attention mechanism and CBAM, the symmetric attention mechanism proposed in this work improves the interaction between local and global features, effectively enhancing the model's performance in complex scenarios. The contributions of this study are not only in proposing a new instance segmentation method but also in providing a novel intelligent solution for the agricultural field, particularly in wheat growth detection tasks. This work demonstrates the potential application of deep learning techniques in precision agriculture.

Although the wheat growth and counting analysis model based on semantic segmentation proposed in this study demonstrates significant advantages in several aspects, there remain certain limitations that future research can address to further enhance the model's performance and application scope. First, the experiments in this study were primarily based on a single wheat dataset for training and testing. Although good results were achieved on this dataset, the diversity and coverage of the dataset remain limiting factors. Therefore, future research should consider using wheat image data from different regions, varying climatic conditions, and different growth stages to expand the diversity of the dataset, thereby enhancing the model's generalization ability and robustness. Second, although the proposed symmetric attention mechanism and symmetric diffusion module show excellent performance in high-density target detection, there is still room for improvement in the model's computational efficiency due to limitations in computational resources and training data. Future studies could focus on optimizing the model architecture by employing more efficient network designs or lightweight models (such as MobileNet, EfficientNet, etc.) in order to improve inference speed and computational efficiency. This would enable the model to be deployed on resource-constrained edge devices, facilitating its practical application. Furthermore, although the aggregated loss function proposed in this study optimizes the accuracy of growth measurement, the model may still be influenced by factors such as noise, occlusion, and lighting variations in complex environments, leading to suboptimal performance in certain scenarios. Therefore, future work could incorporate more sophisticated noise processing methods and data cleaning techniques during data preprocessing and model training stages to enhance the model's stability in complex environments. Also, we plan to integrate additional biological and non-biological variables, such as climate data, soil characteristics, and crop varieties, to improve the model's robustness and adaptability to different agricultural environments. By combining these factors, we aim to enhance the model's predictive accuracy and generalization across diverse agricultural settings. In future research, we aim to integrate segmentation and counting results with plant physiological models, exploring how morphological features impact crop growth and yield. This will enable more accurate analysis of crop dynamics and improve yield prediction, further enhancing the practical application of the model in precision agriculture.

**Author Contributions:** Conceptualization, Z.J., W.H., Y.W. and C.L.; Data curation, B.Z. and Z.S.; Formal analysis, C.J., Z.S. and S.L.; Funding acquisition, C.L.; Investigation, C.J.; Methodology, Z.J., W.H., Y.W. and C.L.; Project administration, C.L.; Resources, B.Z.; Software, Z.J., W.H. and Y.W.; Supervision, S.L.; Validation, C.J. and S.L.; Visualization, B.Z. and Z.S.; Writing—original draft, Z.J., W.H., Y.W., C.J., B.Z., Z.S., S.L. and C.L. Z.J. and W.H. contributed equally to this work. All authors have read and agreed to the published version of the manuscript.

**Funding:** The authors would like to express their sincere gratitude to the Computer Association of China Agricultural University (ECC) for their valuable technical support. Upon the acceptance of this paper, the project code and the dataset will be made publicly available to facilitate further research and development in this field.

**Data Availability Statement:** The data presented in this study are available on request from the corresponding author.

**Conflicts of Interest:** The authors declare no conflicts of interest.

## References

1. Mottaleb, K.A.; Kruseman, G.; Snapp, S. Potential impacts of Ukraine-Russia armed conflict on global wheat food security: A quantitative exploration. *Glob. Food Secur.* **2022**, *35*, 100659.
2. Zhang, Y.; Wa, S.; Zhang, L.; Lv, C. Automatic plant disease detection based on tranvolution detection network with GAN modules using leaf images. *Front. Plant Sci.* **2022**, *13*, 875693.

3. Zhang, Y.; Wa, S.; Liu, Y.; Zhou, X.; Sun, P.; Ma, Q. High-accuracy detection of maize leaf diseases CNN based on multi-pathway activation function module. *Remote Sens.* **2021**, *13*, 4218. [CrossRef]
4. Alkhudaydi, T.; De la Lglesia, B. Counting spikelets from infield wheat crop images using fully convolutional networks. *Neural Comput. Appl.* **2022**, *34*, 17539–17560.
5. Ye, J.; Yu, Z.; Wang, Y.; Lu, D.; Zhou, H. WheatLFANet: In-field detection and counting of wheat heads with high-real-time global regression network. *Plant Methods* **2023**, *19*, 103. [PubMed]
6. Shammi, S.A.; Meng, Q. Use time series NDVI and EVI to develop dynamic crop growth metrics for yield modeling. *Ecol. Indic.* **2021**, *121*, 107124. [CrossRef]
7. Yu, J.; Zhang, S.; Zhang, Y.; Hu, R.; Lawi, A.S. Construction of a winter wheat comprehensive growth monitoring index based on a fuzzy degree comprehensive evaluation model of multispectral UAV data. *Sensors* **2023**, *23*, 8089. [CrossRef]
8. Li, S.; Jin, G.; Lou, Z. Research on the extraction of winter wheat planting area in Henan Province based on time-series EVI. In Proceedings of the Second International Conference on Geographic Information and Remote Sensing Technology (GIRST 2023), Qingdao, China, 21–23 July 2023; Volume 12797, pp. 237–242.
9. Maheswaran, S.; Arunkumar, T.; Gomathi, R.; Rithika, P.; Shafiq, I.M.; Nandita, S. Blockchain-Based Digital Twin for Agricultural Field Management. In *Blockchain-Based Digital Twins: Research Trends and Challenges*; CRC Press: Boca Raton, FL, USA, 2025; p. 279.
10. Wang, D.; Zhang, D.; Yang, G.; Xu, B.; Luo, Y.; Yang, X. SSRNet: In-field counting wheat ears using multi-stage convolutional neural network. *IEEE Trans. Geosci. Remote Sens.* **2021**, *60*, 4403311.
11. Sun, J.; Yang, K.; Chen, C.; Shen, J.; Yang, Y.; Wu, X.; Norton, T. Wheat head counting in the wild by an augmented feature pyramid networks-based convolutional neural network. *Comput. Electron. Agric.* **2022**, *193*, 106705.
12. Li, L.; Hassan, M.A.; Yang, S.; Jing, F.; Yang, M.; Rasheed, A.; Wang, J.; Xia, X.; He, Z.; Xiao, Y. Development of image-based wheat spike counter through a Faster R-CNN algorithm and application for genetic studies. *Crop J.* **2022**, *10*, 1303–1311.
13. Yousafzai, S.N.; Nasir, I.M.; Tehsin, S.; Fitriyani, N.L.; Syafrudin, M. FLTrans-Net: Transformer-based feature learning network for wheat head detection. *Comput. Electron. Agric.* **2025**, *229*, 109706.
14. Hong, Q.; Liu, W.; Zhu, Y.; Ren, T.; Shi, C.; Lu, Z.; Yang, Y.; Deng, R.; Qian, J.; Tan, C. CTHNet: A network for wheat ear counting with local-global features fusion based on hybrid architecture. *Front. Plant Sci.* **2024**, *15*, 1425131.
15. Guan, S.; Lin, Y.; Lin, G.; Su, P.; Huang, S.; Meng, X.; Liu, P.; Yan, J. Real-time detection and counting of wheat spikes based on improved YOLOv10. *Agronomy* **2024**, *14*, 1936. [CrossRef]
16. Yao, Z.; Zhang, D.; Tian, T.; Zain, M.; Zhang, W.; Yang, T.; Song, X.; Zhu, S.; Liu, T.; Ma, H.; et al. APW: An ensemble model for efficient wheat spike counting in unmanned aerial vehicle images. *Comput. Electron. Agric.* **2024**, *224*, 109204. [CrossRef]
17. Sun, X.; Jiang, T.; Hu, J.; Song, Z.; Ge, Y.; Wang, Y.; Liu, X.; Bing, J.; Li, J.; Zhou, Z.; et al. Counting wheat heads using a simulation model. *Comput. Electron. Agric.* **2025**, *228*, 109633.
18. Mo, Y.; Wu, Y.; Yang, X.; Liu, F.; Liao, Y. Review the state-of-the-art technologies of semantic segmentation based on deep learning. *Neurocomputing* **2022**, *493*, 626–646. [CrossRef]
19. Li, Q.; Zhang, Y. Confidential Federated Learning for Heterogeneous Platforms against Client-Side Privacy Leakages. In Proceedings of the ACM Turing Award Celebration Conference 2024, Changsha, China, 5–7 July 2024; pp. 239–241.
20. Li, Q.; Zhang, Y.; Ren, J.; Li, Q.; Zhang, Y. You Can Use But Cannot Recognize: Preserving Visual Privacy in Deep Neural Networks. *arXiv* **2024**, arXiv:2404.04098.
21. Luo, Z.; Yang, W.; Yuan, Y.; Gou, R.; Li, X. Semantic segmentation of agricultural images: A survey. *Inf. Process. Agric.* **2023**, *11*, 172–186.
22. Anand, T.; Sinha, S.; Mandal, M.; Chamola, V.; Yu, F.R. AgriSegNet: Deep aerial semantic segmentation framework for IoT-assisted precision agriculture. *IEEE Sens. J.* **2021**, *21*, 17581–17590.
23. Kitzler, F.; Barta, N.; Neugschwandtner, R.W.; Gronauer, A.; Motsch, V. WE3DS: An RGB-D image dataset for semantic segmentation in agriculture. *Sensors* **2023**, *23*, 2713. [CrossRef]
24. Peng, M.; Liu, Y.; Qadri, I.A.; Bhatti, U.A.; Ahmed, B.; Sarhan, N.M.; Awwad, E. Advanced image segmentation for precision agriculture using CNN-GAT fusion and fuzzy C-means clustering. *Comput. Electron. Agric.* **2024**, *226*, 109431.
25. Li, Q.; Ren, J.; Zhang, Y.; Song, C.; Liao, Y.; Zhang, Y. Privacy-Preserving DNN Training with Prefetched Meta-Keys on Heterogeneous Neural Network Accelerators. In Proceedings of the 2023 60th ACM/IEEE Design Automation Conference (DAC), San Francisco, CA, USA, 9–13 July 2023; pp. 1–6.
26. Zhang, Y.; Yang, X.; Liu, Y.; Zhou, J.; Huang, Y.; Li, J.; Zhang, L.; Ma, Q. A time-series neural network for pig feeding behavior recognition and dangerous detection from videos. *Comput. Electron. Agric.* **2024**, *218*, 108710.
27. Cui, J.; Tan, F.; Bai, N.; Fu, Y. Improving U-net network for semantic segmentation of corns and weeds during corn seedling stage in field. *Front. Plant Sci.* **2024**, *15*, 1344958.
28. Mahmud, M.N.; Osman, M.K.; Ismail, A.P.; Ahmad, F.; Ahmad, K.A.; Ibrahim, A. Road image segmentation using unmanned aerial vehicle images and DeepLab V3+ semantic segmentation model. In Proceedings of the 2021 11th IEEE International Conference on Control System, Computing and Engineering (ICCSCE), Penang, Malaysia, 27–28 August 2021; pp. 176–181.

29. Shen, Y.; Wang, L.; Jin, Y. AAFormer: A multi-modal transformer network for aerial agricultural images. In Proceedings of the IEEE/CVF Conference on Computer Vision and Pattern Recognition, New Orleans, LA, USA, 18–24 June 2022; pp. 1705–1711.
30. Fu, X.; Ma, Q.; Yang, F.; Zhang, C.; Zhao, X.; Chang, F.; Han, L. Crop pest image recognition based on the improved ViT method. *Inf. Process. Agric.* **2024**, *11*, 249–259.
31. Cheng, Z.Q.; Dai, Q.; Li, H.; Song, J.; Wu, X.; Hauptmann, A.G. Rethinking spatial invariance of convolutional networks for object counting. In Proceedings of the IEEE/CVF Conference on Computer Vision and Pattern Recognition, New Orleans, LA, USA, 18–24 June 2022; pp. 19638–19648.
32. You, Z.; Yang, K.; Luo, W.; Lu, X.; Cui, L.; Le, X. Few-shot object counting with similarity-aware feature enhancement. In Proceedings of the IEEE/CVF Winter Conference on Applications of Computer Vision, Waikoloa, HI, USA, 2–7 January 2023; pp. 6315–6324.
33. Khaki, S.; Safaei, N.; Pham, H.; Wang, L. WheatNet: A lightweight convolutional neural network for high-throughput image-based wheat head detection and counting. *Neurocomputing* **2022**, *489*, 78–89.
34. Wang, Y.; Hou, J.; Hou, X.; Chau, L.P. A self-training approach for point-supervised object detection and counting in crowds. *IEEE Trans. Image Process.* **2021**, *30*, 2876–2887.
35. Hallin, A.; Isaacson, J.; Kasieczka, G.; Krause, C.; Nachman, B.; Quadfasel, T.; Schlaffer, M.; Shih, D.; Sommerhalder, M. Classifying anomalies through outer density estimation. *Phys. Rev. D* **2022**, *106*, 055006.
36. Jiang, L.; Wang, Y.; Wu, C.; Wu, H. Fruit Distribution Density Estimation in YOLO-Detected Strawberry Images: A Kernel Density and Nearest Neighbor Analysis Approach. *Agriculture* **2024**, *14*, 1848. [CrossRef]
37. Wang, Y.; Wang, P.; Tansey, K.; Liu, J.; Delaney, B.; Quan, W. An interpretable approach combining Shapley additive explanations and LightGBM based on data augmentation for improving wheat yield estimates. *Comput. Electron. Agric.* **2025**, *229*, 109758.
38. Zhang, Y.; Lv, C. TinySegformer: A lightweight visual segmentation model for real-time agricultural pest detection. *Comput. Electron. Agric.* **2024**, *218*, 108740.
39. Ronneberger, O.; Fischer, P.; Brox, T. U-Net: Convolutional Networks for Biomedical Image Segmentation. In Proceedings of the International Conference on Medical Image Computing and Computer-Assisted Intervention, Munich, Germany, 5–9 October 2015.
40. He, K.; Gkioxari, G.; Dollár, P.; Girshick, R. Mask r-cnn. In Proceedings of the IEEE International Conference on Computer Vision, Venice, Italy, 22–29 October 2017; pp. 2961–2969.
41. Kirillov, A.; Mintun, E.; Ravi, N.; Mao, H.; Rolland, C.; Gustafson, L.; Xiao, T.; Whitehead, S.; Berg, A.C.; Lo, W.Y.; et al. Segment Anything. *arXiv* **2023**, arXiv:2304.02643.
42. Wang, W.; Dai, J.; Chen, Z.; Huang, Z.; Li, Z.; Zhu, X.; Hu, X.; Lu, T.; Lu, L.; Li, H.; et al. Internimage: Exploring large-scale vision foundation models with deformable convolutions. In Proceedings of the IEEE/CVF Conference on Computer Vision and Pattern Recognition, Vancouver, BC, Canada, 17–24 June 2023; pp. 14408–14419.

*agriculture*

MDPI

*Article*

# A Semi-Supervised Diffusion-Based Framework for Weed Detection in Precision Agricultural Scenarios Using a Generative Attention Mechanism

**Ruiheng Li [1,†], Xuaner Wang [1,†], Yuzhuo Cui [1,†], Yifei Xu [1], Yuhao Zhou [1], Xuechun Tang [1,2], Chenlu Jiang [1], Yihong Song [1], Hegan Dong [3,4,*] and Shuo Yan [1,*]**

1 China Agricultural University, Beijing 100083, China
2 Beijing Foreign Studies University, Beijing 100089, China
3 College of Life Sciences, Shihezi University, Shihezi 832003, China
4 Xinjiang Production and Construction Corps Key Laboratory of Oasis Town and Mountain-basin System Ecology, Shihezi 832003, China
* Correspondence: donghegan@shzu.edu.cn (H.D.); yanshuo@cau.edu.cn (S.Y.)
† These authors contributed equally to this work.

**Abstract:** The development of smart agriculture has created an urgent demand for efficient and accurate weed recognition and detection technologies. However, the diverse and complex morphology of weeds, coupled with the scarcity of labeled data in agricultural scenarios, poses significant challenges to traditional supervised learning methods. To address these issues, a weed detection model based on a semi-supervised diffusion generative network is proposed. This model integrates a generative attention mechanism and semi-diffusion loss to enable the efficient utilization of both labeled and unlabeled data. Experimental results demonstrate that the proposed method outperforms existing approaches across multiple evaluation metrics, achieving a precision of 0.94, recall of 0.90, accuracy of 0.92, and mAP@50 and mAP@75 of 0.92 and 0.91, respectively. Compared to traditional methods such as DETR, precision and recall are improved by approximately 10% and 8%, respectively. Additionally, compared to the enhanced YOLOv10, mAP@50 and mAP@75 are increased by 1% and 2%, respectively. The proposed semi-supervised diffusion weed detection model provides an efficient and reliable solution for weed recognition and introduces new research perspectives for the application of semi-supervised learning in smart agriculture. This framework establishes both theoretical and practical foundations for addressing complex target detection challenges in the agricultural domain.

**Keywords:** artificial intelligence in agriculture; weed detection; multi-scale feature extraction; deep learning; precision agriculture

## 1. Introduction

Agriculture serves as the foundation for human survival and development, and crop yield and quality directly determine the economic benefits of agriculture and food security [1–3]. However, weeds, as one of the most common biological interferences in agricultural production, can severely reduce crop yield and quality by competing with crops for water, nutrients, sunlight, and growing space. It is estimated that weeds cause crop yield losses of up to 10–15% globally each year [4,5]. The weed problem is particularly prominent in the cultivation of field crops, affecting not only yield but also potentially leading to imbalances in the agricultural ecosystem. Effective weed control is a key element in ensuring efficient and sustainable agricultural development.

Traditional weed management methods mainly rely on the use of chemical herbicides or manual weeding [6,7]. While chemical herbicides work quickly and cover large areas, prolonged and excessive use can lead to soil and water pollution, damage to the ecological environment, and an increase in herbicide resistance. On the other hand, manual weeding is effective in reducing environmental pollution but is labor-intensive, inefficient, and not feasible in large-scale agricultural fields due to high cost and labor constraints [8]. With the increasing attention to environmental protection and sustainable agriculture, finding ways to effectively control weeds while minimizing negative impacts on the ecosystem has become a critical challenge.

In recent years, the development of smart agriculture technologies has provided new solutions for the precise identification and control of weeds [9]. Deep learning techniques have enabled image-based weed detection and classification tasks, laying the foundation for efficient and environmentally friendly weed management systems. Compared to traditional methods, deep learning-based weed recognition offers advantages such as higher automation, better detection accuracy, and stronger adaptability [10,11]. However, many challenges remain in current research and applications. Supervised learning is the dominant technology in weed detection, but it heavily depends on large-scale labeled datasets. To address this issue, Saleh Alzayat et al. [12] proposed a new semi-supervised weed detection method that includes two main components: first, multi-scale feature representation techniques are used to capture unique weed features at different scales; second, an adaptive pseudo-labeling strategy is introduced that dynamically assigns confidence scores to pseudo-labels generated from unlabeled data during training. Experiments on the COCO dataset and five well-known weed datasets demonstrated that this method achieved state-of-the-art performance in weed detection while requiring significantly fewer labeled data compared to existing methods. Muvva Vijaya Bhaskar Reddy et al. [13] proposed a simple and efficient deep learning-based weed detection model to address the main drawbacks of feature extraction, such as loss of originality and image quality issues. To address real-time monitoring issues, Narayana Ch Lakshmi et al. [14] proposed a weed detection system based on YOLOv7. Experimental results showed that the YOLOv7 model achieved 99.8% accuracy, but further modeling with machine vision systems deployed on onboard computers requires real-world setting evaluations and updates.

Li et al. [15] evaluated the effectiveness of a semi-supervised learning framework for multi-class weed detection, using two object detection frameworks: FCOS (Fully Convolutional One-Stage Object Detection) and Faster-RCNN (Faster Region-based Convolutional Neural Networks). Experimental results showed that the proposed method achieved detection accuracies of approximately 76% and 96% with only 10% of the labeled data in the CottonWeedDet3 and CottonWeedDet12 datasets, respectively. Zhang et al. [16] proposed an optimized Faster R-CNN-based soybean seedling weed recognition method. They used the trained Faster R-CNN algorithm to detect soybeans and weeds in natural field environments, addressing the issue of low attention focus during model training and comparing the method with two classic object detection algorithms, SSD and YOLOv4. The experimental results indicated that the Faster R-CNN algorithm using VGG19-CBAM as the backbone feature extraction network effectively identified soybeans and weeds in complex backgrounds. The average recognition speed per image was 336 ms, with an average recognition accuracy of 99.16%, which improved by 5.61% compared to the unoptimized version and by 2.24% compared to SSD and 1.24% compared to YOLOv4. However, the agricultural environment is highly dynamic, and field conditions are greatly influenced by weather, lighting, and weed growth stages, which poses challenges for model generalization.

To address the challenges mentioned above, a semi-supervised weed detection and recognition framework is proposed in this paper, combining high-quality synthetic data generated by a diffusion model to enhance the model's performance with limited labeled data. Specifically, the main contributions of this paper include the following:

- Introduction of diffusion model-based generative cyclic network: This paper is the first to apply a diffusion model to weed detection tasks, designing a semi-supervised diffusion-based generative cyclic network capable of generating high-quality synthetic data. The model improves the realism and diversity of the generated data through cyclic optimization strategies. This approach significantly reduces the reliance on large-scale labeled data.
- Design of a combined generation and detection attention mechanism: A generation attention mechanism is proposed in this paper, which integrates the features generated by the diffusion model into the detection network. By dynamically adjusting the weight distribution, this mechanism enhances the detection model's ability to express fine-grained features in complex scenes, especially in scenarios where weeds and crops have highly similar appearances.
- Proposal of a new semi-supervised loss function: To optimize the model's performance with limited labeled data, a new loss function, semi-diffusion loss, is introduced. This function combines the characteristics of supervised and generative learning, effectively balancing the training weights between labeled and generated data, thereby improving the model's overall robustness and detection accuracy.

In conclusion, the semi-supervised weed detection framework proposed in this paper not only provides a new approach for accurate weed identification but also opens up new directions for the application of deep learning in smart agriculture.

## 2. Related Work

### 2.1. Supervised Learning

Supervised learning is one of the core techniques in computer vision. Its principle involves using a training dataset with explicit labels to guide the model in learning and performing feature extraction and task decision-making [17–22]. The mathematical modeling of supervised learning is generally based on optimizing the objective function to minimize the gap between the model's predictions and the true labels. Let the training dataset $\mathcal{D} = (x_i, y_i)_{i=1}^N$, where $x_i$ represents the $i$-th image, and $y_i$ represents its corresponding label, which could be the class label in a classification task or a bounding box and its associated class in an object detection task. The basic objective of supervised learning is to optimize the model's parameters $\theta$ by learning the prediction function $f(x;\theta)$ so that it approximates the true label $y_i$. In classification tasks, the most commonly used optimization objective is the cross-entropy loss function, defined as:

$$L_{CE} = -\frac{1}{N}\sum_{i=1}^{N}\sum_{c=1}^{C} y_i^c \log \hat{y}_i^c, \tag{1}$$

where $C$ represents the total number of classes, $y_i^c$ is the true distribution of the sample $i$ for class $c$ (typically one-hot encoded), and $\hat{y}_i^c$ is the predicted probability for class $c$. By minimizing $L_{CE}$, the model adjusts the parameters $\theta$ so that the predicted distribution $\hat{y}_i^c$ is as close as possible to the true distribution $y_i^c$. In object detection tasks, in addition to classification loss, the bounding box regression loss must also be considered. Bounding box regression typically uses the smooth $L_1$ loss function, defined as:

$$L_{bbox} = \frac{1}{N} \sum_{i=1}^{N} \sum_{j \in \{x,y,w,h\}} \text{Smooth}_{L_1}(t_{ij} - \hat{t}_{ij}), \tag{2}$$

where $t_{ij}$ and $\hat{t}_{ij}$ represent the true and predicted coordinates for the bounding box in the $x$, $y$ positions, as well as the width $w$ and height $h$. The smooth $L_1$ function is defined as:

$$\text{Smooth}_{L_1}(x) = \begin{cases} 0.5x^2, & \text{if } |x| < 1, \\ |x| - 0.5, & \text{otherwise.} \end{cases} \tag{3}$$

The total loss for object detection is typically the weighted sum of classification and bounding box regression losses:

$$L_{total} = L_{CE} + \lambda L_{bbox}, \tag{4}$$

where $\lambda$ is the balancing coefficient used to adjust the relative weight of classification and bounding box regression losses. In weed detection, YOLO series are the most common supervised learning frameworks. YOLO models, using a single-stage detection framework, simplify the object detection task to an end-to-end regression problem, significantly improving detection speed [23–25]. Specifically, YOLO divides the input image into an $S \times S$ grid, with each grid responsible for predicting several bounding boxes and their corresponding class probabilities. Its loss function includes classification loss, localization loss, and confidence loss:

$$L_{YOLO} = L_{class} + L_{bbox} + L_{conf}. \tag{5}$$

Although supervised learning performs excellently in weed detection tasks, its limitations are quite apparent. Firstly, acquiring high-quality annotated data is costly, especially in complex agricultural scenarios, where annotations for different weed species and growth stages require expert knowledge, greatly increasing the difficulty of data labeling. Secondly, the complexity of agricultural environments (e.g., lighting changes, soil background interference) poses a higher demand on the model's generalization ability [26]. Furthermore, the diversity of weed species and their similarity to crop morphology lead to significant class confusion, which limits the model's detection accuracy [18,27,28].

### 2.2. Semi-Supervised Learning

In the field of image recognition, especially in agricultural scenarios like weed detection, semi-supervised learning has become an important research direction [17,29–32]. Semi-supervised learning lies between supervised and unsupervised learning, utilizing a large amount of unlabeled data along with a small amount of labeled data for training, thus reducing the dependency on labeled data. The advantages of semi-supervised learning in agricultural image analysis are particularly significant because agricultural data often faces challenges like high labeling costs and scarce samples, particularly in weed detection tasks where labeling different weed species and complex environments is even more difficult [33]. In pseudo-label generation, the model first predicts the unlabeled data and assigns the high-confidence predictions as pseudo-labels to the unlabeled samples [34]. The goal of the pseudo-label generation process is to expand the training dataset by assigning labels to unlabeled data, thereby improving the model's generalization ability. Its loss function can be expressed as:

$$L_u = \frac{1}{M}\sum_{j=1}^{M}\mathbb{I}(\hat{y}_j \geq \tau)\cdot \mathrm{CE}(\hat{y}_j, f(x_j;\theta)), \quad (6)$$

where $\mathbb{I}(\hat{y}_j \geq \tau)$ is the indicator function, which indicates that the pseudo-label $\hat{y}_j$ is used to calculate the loss when the predicted probability exceeds the threshold $\tau$. $\mathrm{CE}(\hat{y}_j, f(x_j;\theta))$ is the cross-entropy loss, which measures the gap between the pseudo-label and the model's prediction. Semi-supervised learning has made significant progress in agricultural applications, especially in tasks such as weed detection, crop recognition, and disease detection. In weed detection, semi-supervised learning is especially crucial, as the diversity of weed species and their morphological similarities to crops pose challenges to traditional supervised learning methods, which require large amounts of labeled samples to achieve high detection accuracy [35–37]. Semi-supervised learning can effectively utilize unlabeled image data for enhanced training, overcoming the limitations of insufficient labeled data.

## 3. Materials and Methods

### 3.1. Dataset Collection

The collection of the dataset forms the foundational work for implementing the semi-supervised weed recognition model and directly impacts the accuracy of model training and validation. In this study, image data were primarily collected from experimental fields in Zhuozhou, Hebei province, Changchun, Jilin Province, and publicly available online sources, as shown in Figure 1, from January 2023 to October 2024.

A B C D E

**Figure 1.** Dateset samples: **(A)** is Xanthium spinosum; **(B)** is Xanthium italicum; **(C)** is Amaranthus rudis; **(D)** is Setaria viridis; **(E)** is Cyclachaena xanthiifolia.

The selected weed types include Setaria viridis, Xanthium spinosum, Cyclachaena xanthiifolia, Xanthium italicum, and Amaranthus rudis. The number of images per weed type ranged from 800 to 2200, as summarized in Table 1, ensuring the dataset's diversity and sufficiency.

**Table 1.** Number of images for different weed types.

| Weed Type | Number of Images |
|---|---|
| Setaria viridis | 1591 |
| Xanthium spinosum | 842 |
| Cyclachaena xanthiifolia | 2094 |
| Xanthium italicum | 1796 |
| Amaranthus rudis | 1339 |

During image acquisition, high-resolution digital cameras and drone-mounted imaging devices were employed to ensure the capture of high-quality images from various angles and heights. The digital cameras were primarily used for ground-level close-up images, while drones were utilized to obtain large-scale field images from aerial perspectives. This multi-angle approach aids subsequent model recognition by adapting to the various growth stages of weeds and lighting conditions. Particular attention was given to capturing the distinctive features of different growth cycles for each weed type. Additionally, differences between similar weed variants, such as Xanthium spinosum and Cyclachaena xanthiifolia, were meticulously observed. Although these weeds share similar appearances, differences in leaf shapes and spine lengths were noted through careful on-site observation and consultation with agricultural experts, ensuring these subtle distinctions were accurately reflected in the collected images. To further enhance the dataset's representativeness and adaptability, additional images were sourced from online platforms. These images primarily originated from agricultural fields in various regions, covering diverse soil types, lighting conditions, and climatic environments, thereby improving the dataset's generalization capability across both national and global contexts. During the incorporation of online images, only high-quality and accurately labeled resources were selected to maintain the dataset's authenticity and reliability. Through this comprehensive data collection process, a high-quality, diverse, and information-rich weed image dataset was established, providing a solid foundation for training the semi-supervised learning model.

### *3.2. Data Annotation*

In weed detection tasks, due to the complexity of agricultural environments and the diversity of weed species, accurate annotation is not only crucial for improving model performance but also the foundation for ensuring the reliability of experimental results. The weed dataset in this study includes various types of weeds, each exhibiting unique characteristics under different growth stages and environmental conditions. To enable effective training, it is essential to accurately annotate the weeds in each image. During the data annotation process, we employed expert-assisted manual annotation. Agricultural experts carefully examined each image and precisely annotated every weed region in the image. Specifically, the experts first identified the weeds in the image and marked their corresponding bounding boxes. These bounding boxes not only indicated the position of the weeds but also included category information about the weed types. Additionally, the experts annotated the weeds based on their growth stages. For example, the leaves of Setaria viridis are narrow and light green in the early growth stage, while they become broader and darker as the plant matures. These subtle differences were carefully recorded during the annotation process, providing more contextual information for subsequent model training.

We paid special attention to the consistency and comprehensiveness of the annotations. Consistency in annotation refers to ensuring that the same type of weed follows the same annotation standards across different images, avoiding data noise caused by inconsistent annotations. Comprehensiveness means ensuring that the performance of each weed category is fully covered under all environmental conditions, lighting situations, and growth stages. In the construction of this study's dataset, the accuracy and quality of annotations are critical factors that influence the model's training effectiveness. To ensure high-quality annotations, we implemented a rigorous annotation process, combining expert guidance and automated tools to assist with the annotation, ensuring that the data annotation meets high standards.

### *3.3. Data Augmentation*

Data augmentation techniques are widely used in deep learning to improve the robustness and generalization ability of models, especially when training data are insufficient. By applying various transformations to the training data, data augmentation can generate additional training samples, enabling the model to learn richer features. In the agricultural field, especially in tasks such as weed detection and crop disease recognition, data augmentation plays a crucial role in enhancing model accuracy, as shown in Figure 2.

**Figure 2.** Data augmentation methods used in this paper.

CutOut is a simple and effective image data augmentation technique that enhances the model's robustness by occluding a rectangular region ("cutout") in the input image. The principle of this method is to randomly select a region in the image and set the pixel values of that region to zero, preventing the model from overly relying on a particular region or local feature. Given an input image $x$, the CutOut operation randomly generates a rectangular region with size $w \times h$ at a random location and sets the pixel values in that region to zero. This operation can be represented by the following formula:

$$\tilde{x} = x \odot (1 - M), \tag{7}$$

where $\odot$ represents element-wise multiplication, and $M$ is a binary matrix with the same dimensions as the input image. Mixup is another augmentation method that generates new samples through linear interpolation. It combines two images and their corresponding labels in a weighted manner to create new training samples. Given two input images $x_1$ and $x_2$ and their corresponding labels $y_1$ and $y_2$, the Mixup operation generates a new image $\tilde{x}$ and label $\tilde{y}$ as follows:

$$\tilde{x} = \lambda x_1 + (1 - \lambda) x_2, \tag{8}$$

$$\tilde{y} = \lambda y_1 + (1 - \lambda) y_2, \tag{9}$$

where $\lambda$ is a randomly generated weight, usually sampled from a beta distribution, i.e., $\lambda \sim \text{beta}(\alpha, \alpha)$, where $\alpha$ is a hyperparameter that determines the shape of the weight distribution. In this way, mixup generates smoother decision boundaries, improving the model's generalization ability. GridMask is a grid-based occlusion augmentation method designed to enhance the model's ability to learn local features. Unlike cutout, gridmask does not occlude the entire rectangular region but instead divides the image into several grids and randomly occludes some of the grids to achieve data augmentation. Given that the input image $x$ has size $H \times W$, gridmask divides the image into $G \times G$ grids and randomly selects several grids for occlusion. The occlusion operation can be expressed as the following formula:

$$\tilde{x} = x \odot (1 - M), \tag{10}$$

where $M$ is a binary matrix of size $H \times W$, representing the location of the grids. In each grid, some grids are randomly selected for occlusion. The size and shape of the occlusion area are determined by the grid size $G$ and the occlusion probability. By this method, GridMask generates different types of local occlusion, increases the diversity of training data, and forces the model to extract useful features from multiple local regions in the image.

### 3.4. Proposed Method

To address the challenges of limited annotated agricultural image data, the complexity of weed species, and the high morphological similarity between crops and weeds, a weed detection network framework based on a semi-supervised diffusion model is proposed. The overall framework integrates the powerful generative capabilities of diffusion models with the feature extraction and classification capacities of detection networks, leading to the design of an innovative semi-supervised learning mechanism. This mechanism enables efficient and accurate weed recognition and detection under conditions of limited annotated data. The framework comprises three primary modules: the semi-supervised diffusion weed detection network, the semi-supervised diffusion data generation loop, and the detection network based on the generative attention mechanism. To further enhance training effectiveness, a novel semi-supervised loss function, referred to as semi-diffusion loss, is introduced. This loss function enables the collaborative optimization of supervised and generative signals by combining the learning processes of annotated and generated data. The network structure is illustrated in Figure 3.

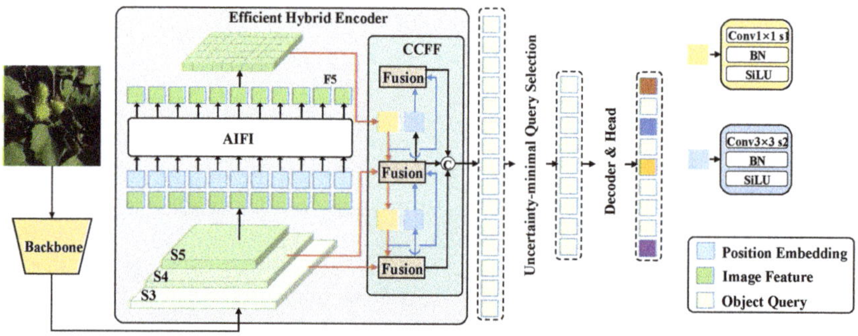


**Figure 3.** The figure illustrates the overall framework of the proposed semi-supervised diffusion weed detection network (SSDWDN).

#### 3.4.1. Semi-Supervised Diffusion Weed Detection Network

The proposed semi-supervised diffusion weed detection network (SSDWDN) consists of four core modules: the efficient hybrid encoder, the uncertainty-minimal query selection module, the cross-scale compact feature fusion module (CCFF), the decoder, and detection head. The overall design effectively combines the generative capabilities of the diffusion model with the feature extraction and fusion capacities of the detection network.

The efficient hybrid encoder plays a crucial role in extracting image features. After passing through the backbone, the input image is processed by the encoder's multi-scale feature extraction layers. The encoder consists of three key feature layers (S3, S4, S5) that extract low-level, mid-level, and high-level features, respectively. Each feature layer applies the SiLU activation function and Batch Normalization (BN) to maintain gradient flow stability. The outputs from these feature extraction layers are aggregated into the CCFF through hierarchical feature fusion, with the resolution of each feature progressively decreasing to adapt to multi-scale target detection tasks. Within the encoder, an adaptive integrated feature interaction (AIFI) module is employed to enhance information interaction across different scales using a dynamic attention mechanism. The computation of the AIFI module is expressed as follows:

$$F_{\text{AIFI}} = \text{softmax}(Q \cdot K^{\top}) \cdot V, \tag{11}$$

where $Q$, $K$, and $V$ represent the query, key, and value matrices, respectively. The softmax function is applied to normalize the attention weights, ensuring effective integration of information across scales.

The query ($Q$), key ($K$), and value ($V$) matrices represent the feature representations, and $F_{\text{AIFI}}$ denotes the output features. The softmax function is applied to normalize the attention weights. This dynamic interaction effectively enhances the network's capability to capture critical features of weed targets. The CCFF module integrates the features generated by the diffusion model with the multi-scale features extracted by the encoder. Features generated by the diffusion model are encoded with position embeddings before being weighted and fused with image features at each layer. The fusion weights are computed as follows:

$$W_{\text{fusion}} = \text{softmax}(F_{\text{gen}} \cdot F_{\text{img}}^{\top}), \tag{12}$$

where $F_{\text{gen}}$ and $F_{\text{img}}$ represent the generated features and image features, respectively, and $W_{\text{fusion}}$ is the fusion weight. The fused features are added to the original image features through residual connections, and the results are passed to the decoder. The uncertainty-minimal query selection module optimizes pseudo-label generation by selecting high-confidence target regions during detection, ensuring higher-quality pseudo-data generated by the diffusion model. This is achieved by calculating the prediction uncertainty for each target region and selecting the targets with the lowest uncertainty as pseudo-label training samples. The uncertainty is computed based on the standard deviation ($\sigma$):

$$\sigma = \sqrt{\frac{1}{n}\sum_{i=1}^{n}(\hat{y}_i - \bar{y})^2}, \tag{13}$$

where $\hat{y}_i$ represents the predicted class probabilities for each iteration, and $\bar{y}$ is the mean of the predictions. By filtering out targets with lower uncertainty, the credibility of the generated data during training is improved.

The decoder employs a multi-layer convolutional neural network (Conv1×1, Conv3×3) to gradually restore feature resolution and uses an object query detection head to predict target regions. The object query is generated with position-encoded features using the following formula:

$$Q_{\text{obj}} = \sin\left(\frac{pos}{10000^{2i/d}}\right), \tag{14}$$

where $pos$ denotes the positional encoding, and $d$ represents the embedding dimension. This formula incorporates positional information to enhance the network's target localization capabilities. The final predictions include category classification and bounding box regression, optimized with an IoU-based loss function.

The mathematical core of this design lies in the integration of generative model features with multi-scale features of the detection network. Positional attention and cross-feature fusion significantly enhance the expressive capacity of the detection network. First, the efficient hybrid encoder optimizes information flow across scales through the AIFI module, enabling the network to capture detailed features of weed targets across different scales. Second, the CCFF module improves the network's adaptability to generated data by fusing generative and image features, alleviating the challenges posed by insufficient labeled data. Compared to traditional detection models, the SSDWDN demonstrates superior performance by extending the utilization of unlabeled data and further optimizing pseudo-label quality through the uncertainty-minimal query selection mechanism. Moreover, the attention mechanism based on generative features reduces the network's sensitivity to complex backgrounds, improving precise localization of target regions. This design proves highly effective in addressing complex agricultural scenarios and sparse labeled data, offering a robust and efficient solution for weed detection and agricultural intelligence.

#### 3.4.2. Semi-Supervised Diffusion Data Generation Loop

The semi-supervised diffusion data generation loop (SSDDGL) is designed to iteratively optimize the quality and diversity of generated data, thereby enhancing the performance of the detection network. The framework consists of two core components: the generative module and the discriminative module. Additionally, a teacher–student architecture and diffusion-based generation mechanism are incorporated. The overall structure of the network is illustrated in Figure 4 and includes key components such as the encoder, decoder, generative diffusion loss module, and discriminative loss module.

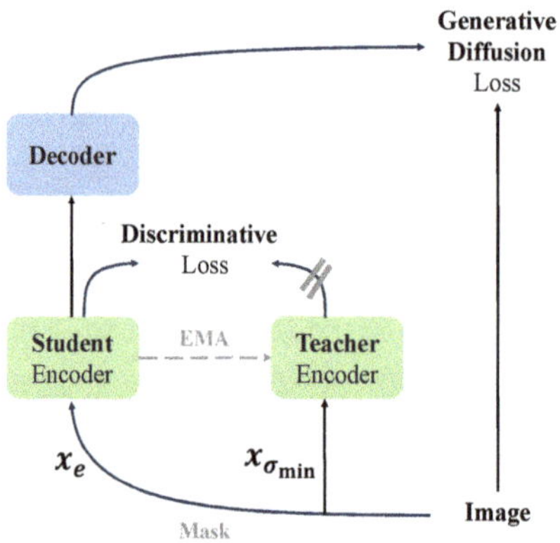


**Figure 4.** The figure illustrates the overall structure of the proposed SSDDGL. The network consists of a teacher encoder, a student encoder, a generative diffusion loss module, and a discriminative loss module.

The encoder and decoder are implemented with a symmetric structure. The encoder consists of five convolutional layers, with input dimensions progressively reduced as $(256, 256, 3)$, $(128, 128, 64)$, $(64, 64, 128)$, $(32, 32, 256)$, and $(16, 16, 512)$. The convolutional kernel size is $3 \times 3$ with a stride of 2, and the channel count increases from 3 to 512. The decoder restores the feature dimensions through transposed convolutions, with output dimensions gradually increased as $(16, 16, 512)$, $(32, 32, 256)$, $(64, 64, 128)$, $(128, 128, 64)$, and $(256, 256, 3)$. The final layer of the decoder utilizes the SiLU activation function to ensure smoothness and nonlinearity in the output features.

A teacher–student architecture is introduced to improve the quality of the generated data. The parameters of the teacher network are updated via the exponential moving average (EMA) of the student network's parameters, expressed as:

$$\theta_T = \alpha\theta_T + (1 - \alpha)\theta_S, \tag{15}$$

where $\theta_T$ and $\theta_S$ denote the parameters of the teacher and student networks, respectively, and $\alpha$ is the momentum coefficient, typically set to 0.999. This design ensures the stability of the teacher network during training and provides high-quality supervisory signals to the student network.

The generative module is based on the reverse diffusion process, which generates synthetic data from pure noise. The generation process is described as

$$p_\theta(x_{t-1}|x_t) = \mathcal{N}(x_{t-1}; \mu_\theta(x_t, t), \sigma_t^2 I), \tag{16}$$

where $\mu_\theta(x_t, t)$ is the predicted mean, $\sigma_t^2$ is the fixed noise variance, $x_t$ represents the image state at the current timestep, and $x_{t-1}$ is the state at the previous timestep.

The generative diffusion loss function ($L_{\text{gen}}$) optimizes the model by predicting the noise during the generation process, defined as

$$L_{\text{gen}} = \mathbb{E}_{t,x_t,\epsilon}\left[\|\epsilon - \epsilon_\theta(x_t, t)\|^2\right], \tag{17}$$

where $\epsilon$ represents the true noise, and $\epsilon_\theta$ denotes the predicted noise.

The discriminative module is responsible for optimizing the quality of pseudo-labels by supervising the consistency between generated and real data distributions. The discriminative loss function ($L_{\text{disc}}$) is given as

$$L_{\text{disc}} = -\mathbb{E}_{x\sim p_{\text{data}}}[\log D(x)] - \mathbb{E}_{x\sim p_{\text{gen}}}[\log(1 - D(x))], \tag{18}$$

where $D(x)$ represents the discriminator's predicted probability for input data $x$, and $p_{\text{data}}$ and $p_{\text{gen}}$ are the distributions of real and generated data, respectively.

SSDDGL iteratively generates and optimizes pseudo-labels to improve the model's performance. Specifically, generated data are first input into the detection network, where features are extracted, and pseudo-labels are generated. These pseudo-labels are then used as supervisory signals to optimize the next iteration of the diffusion generation process. The cycle loss is defined as

$$L_{\text{cycle}} = \mathbb{E}_{x\sim p_{\text{gen}}}\left[\|y_{\text{pred}} - y_{\text{true}}\|^2\right], \tag{19}$$

where $y_{\text{pred}}$ represents the pseudo-label predicted by the detection network, and $y_{\text{true}}$ is the approximate ground truth label generated by the teacher network. By minimizing $L_{\text{cycle}}$, the network progressively enhances the realism and diversity of the generated data.

The diffusion model generates high-quality images through the reverse diffusion process, grounded in the learning of probability density functions. In SSDDGL, the generative module optimizes the mean squared error of noise prediction to approximate the real data distribution $p_{\text{data}}$. The teacher–student architecture ensures stable supervisory signals, guiding the student network's optimization and enabling high generalization performance with limited labeled data. Compared to traditional generative models such as GANs, diffusion models exhibit significant advantages in stability and diversity. The stepwise generation mechanism of diffusion introduces greater randomness at each step, enhancing data diversity. Furthermore, the iterative optimization strategy of SSDDGL progressively improves the quality of generated data, and the supervision signals from the discriminative module ensure the accuracy of pseudo-labels. This design effectively addresses the issue of scarce annotations in agricultural data and significantly enhances the performance of detection networks, particularly in complex scenarios.

### 3.4.3. Generative Attention Mechanism

The generative attention mechanism significantly differs from standard self-attention mechanisms in both its design objectives and sources of information, as shown in Figure 5. While standard self-attention mechanisms focus on capturing global dependencies within the features of the input data by calculating the similarity between queries ($Q$), keys ($K$), and values ($V$) of the input features, the generative attention mechanism introduces additional features generated by diffusion models. By fusing generated features with input features, this mechanism provides richer semantic information, enhancing the network's ability to recognize target regions. This design is particularly suitable for semi-supervised learning scenarios, where generative models are employed to augment the feature representation of real data. Specifically, the generative attention mechanism dynamically integrates the generated features $F_g$ from the diffusion model and the real image features $F_r$, capturing a more comprehensive spatial and semantic context. In contrast, traditional self-attention mechanisms rely solely on local correlations within the input data.

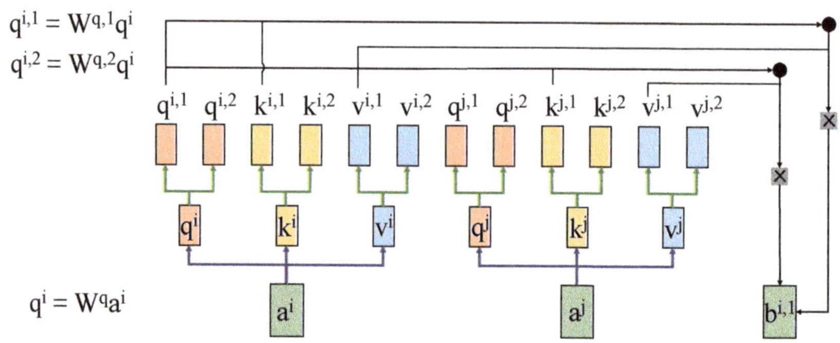


**Figure 5.** Generative attention mechanism architecture.

The generative attention mechanism is implemented through a multi-layer feature fusion network comprising three modules: the generative feature embedding module, the feature fusion module, and the weighted output module. The input dimensions for the real features are $F_r \in \mathbb{R}^{H \times W \times C_r}$, and the generated features are $F_g \in \mathbb{R}^{H \times W \times C_g}$, where the height $H = 32$, the width $W = 32$, the number of channels for real features $C_r = 256$, and for generated features $C_g = 128$.

1. Generative feature embedding module: The generated features $F_g$ are first passed through a $1 \times 1$ convolutional layer to reduce the number of channels to 128 while preserving the spatial dimensions, thereby reducing computational overhead. The embedding process is expressed as

$$F'_g = \text{Conv}_{1\times1}(F_g), \tag{20}$$

where $F'_g \in \mathbb{R}^{H \times W \times 128}$ represents the embedded features.

2. Feature fusion module: Complementary information between the generated features $F'_g$ and the real features $F_r$ is captured by computing the attention weight matrix $W$. The attention weights are computed as

$$W = \text{softmax}(F'_g \cdot F_r^{\top}), \tag{21}$$

where $W \in \mathbb{R}^{H \times H}$ denotes the normalized attention weights. The fused features $F_a$ are obtained by weighting the real features $F_r$ and adding the generated features $F'_g$:

$$F_a = W \cdot F_r + F'_g. \tag{22}$$

3. Weighted output module: The fused features $F_a$ are further processed through a $3 \times 3$ convolutional layer to enhance local information and adjust the number of channels to the target dimension $C_o = 256$:

$$F_o = \text{Conv}_{3\times3}(F_a), \tag{23}$$

where $F_o \in \mathbb{R}^{H\times W\times C_o}$ represents the final fused feature output.

The generative attention mechanism is embedded into the feature extraction stage of the semi-supervised diffusion weed detection network (SSDDN). It dynamically fuses the generated features $F_g$ from the diffusion model with the real image features $F_r$ at different scales, thereby enhancing the representation of multi-scale features. This integration provides several key advantages. First, the mechanism incorporates semantic information from the generative model, which is particularly beneficial for recognizing complex targets in scenarios with limited labeled data. The generated features $F_g$, enriched with contextual information through the diffusion model, significantly improve the model's focus on target regions when fused with the real features $F_r$. Second, the dynamic attention weights $W$ enable the generative attention mechanism to adaptively adjust the contribution of generated and real features, allowing the model to flexibly adjust feature representations based on the characteristics of different scenes. Additionally, the use of local convolution operations ensures computational efficiency while refining the detail of the feature fusion.

Theoretically, the generative attention mechanism enhances the alignment and integration of different information sources within the feature space, enabling the SSDDN to more effectively adapt to complex backgrounds and diverse targets. This leads to higher precision and robustness in weed detection tasks. By addressing the challenges of background interference and target diversity in agricultural image scenarios, the proposed mechanism significantly improves the applicability of semi-supervised learning in intelligent agriculture.

#### 3.4.4. Semi-Diffusion Loss

Traditional supervised learning loss functions, such as cross-entropy loss and mean squared error loss, primarily rely on labeled data by directly optimizing the discrepancy between model outputs and true labels to update parameters. However, in semi-supervised learning, relying solely on labeled data is insufficient to fully exploit the potential of unlabeled data. The semi-diffusion loss combines supervised loss from labeled data and pseudo-supervised loss from unlabeled data, establishing a comprehensive optimization objective that maximizes the utility of unlabeled data under conditions of limited labeled samples.

The semi-diffusion loss consists of two components: the supervised loss $L_s$ and the pseudo-supervised loss $L_u$. The pseudo-supervised loss optimizes generated pseudo-labels for unlabeled data. The core idea is to use high-confidence pseudo-labels as optimization targets. Specifically, pseudo-labels are included in the loss computation when the model's prediction confidence for unlabeled samples exceeds a predefined threshold $\tau$. The pseudo-supervised loss is formulated as

$$L_u = \frac{1}{M}\sum_{j=1}^{M}\mathbb{I}(\max(\hat{y}_j) \geq \tau)\|\hat{y}_j - y_j^{\text{pseudo}}\|^2, \tag{24}$$

where $\hat{y}_j$ denotes the predictions for the unlabeled sample, $y_j^{\text{pseudo}}$ represents the generated pseudo-labels, $\mathbb{I}(\cdot)$ is an indicator function that evaluates to 1 if the prediction confidence exceeds $\tau$ and 0 otherwise, and $M$ is the number of unlabeled samples.

The final semi-diffusion loss is defined as the weighted sum of the supervised loss and pseudo-supervised loss:

$$L_{\mathrm{SDL}} = \lambda_s L_s + \lambda_u L_u, \tag{25}$$

where $\lambda_s$ and $\lambda_u$ are hyperparameters controlling the relative contributions of supervised loss and pseudo-supervised loss during the training process. By adjusting these weights, the impact of different data sources on model optimization can be flexibly controlled.

In the SSDDN, the semi-diffusion loss is closely integrated with the diffusion model's generative capability and the detection network's feature learning process. The diffusion model incrementally generates pseudo-labels, providing supervisory signals for unlabeled data, while the detection network jointly optimizes features from both labeled and generated data. During each training iteration, the computation of the semi-diffusion loss follows these steps:

1. Pseudo-labels $y_j^{\mathrm{pseudo}}$ for unlabeled data are generated by the diffusion model.
2. The detection network computes the supervised loss $L_s$ for labeled data and the pseudo-supervised loss $L_u$ for unlabeled data.
3. The losses $L_{\mathrm{SDL}}$ are combined to update the network parameters.

Through this approach, the semi-diffusion loss enables the network to fully utilize the potential information contained in unlabeled data and enhances the robustness of the detection network with high-quality pseudo-labels generated by the diffusion model. The design of the semi-diffusion loss is grounded in the following theoretical principles: minimizing the supervised loss $L_s$ ensures accurate feature representation learned from labeled data, while minimizing the pseudo-supervised loss $L_u$ effectively leverages unlabeled data, thereby expanding the distributional coverage of the training samples. The ultimate goal is to optimize the network's performance across the entire data distribution $p(x)$.

The reliability of the pseudo-supervised loss $L_u$ is ensured by the introduction of a confidence threshold $\tau$, which ensures that only high-confidence pseudo-labels are used for optimization. According to semi-supervised learning theory, high-confidence pseudo-labels are more likely to approximate the true distribution, thereby guaranteeing the effectiveness of unlabeled data. Furthermore, the flexible adjustment of the weights $\lambda_s$ and $\lambda_u$ mitigates the risk of noise introduced by inaccurate pseudo-labels, further stabilizing the training process. This design demonstrates significant potential for improving the performance of detection networks in scenarios with limited labeled data.

### 3.5. Evaluation Metrics

To comprehensively assess the model's performance, various common evaluation metrics were selected, including precision ($p$), recall ($r$), accuracy ($acc$), mean average precision (mAP@50), mAP@75, and frames per second (FPS). Precision is defined as the proportion of true positive samples among those predicted as positive by the model. It reflects the accuracy of the model's predictions, with higher values indicating that more of the predicted positive samples are correct. Recall is the proportion of true positive samples correctly predicted by the model among all actual positive samples. Accuracy is the proportion of correctly predicted samples among all samples. To evaluate the performance of the model in object detection tasks, particularly in multi-object detection, mean average precision (mAP) was adopted. mAP is a commonly used evaluation metric in object detection tasks, mainly measuring the model's precision performance at different recall rates. The calculation of mAP is based on the precision–recall curve and measures the average precision at different recall rates. mAP@50 and mAP@75 represent specific evaluation standards of mAP, corresponding to an intersection over union (IoU) threshold of 50% and 75%, respectively. Specifically, mAP@50 calculates the average precision when

IoU is greater than or equal to 50%, while mAP@75 calculates it for IoU greater than or equal to 75%. FPS measures the model's processing speed, indicating how many images the model can process per second, providing an insight into the computational efficiency of the model. These metrics are defined as follows:

$$p = \frac{TP}{TP + FP} \tag{26}$$

$$r = \frac{TP}{TP + FN} \tag{27}$$

$$acc = \frac{TP + TN}{TP + TN + FP + FN} \tag{28}$$

$$\text{mAP@50} = \frac{1}{N}\sum_{i=1}^{N} AP_{i,IoU\geq 0.5} \tag{29}$$

$$\text{mAP@75} = \frac{1}{N}\sum_{i=1}^{N} AP_{i,IoU\geq 0.75} \tag{30}$$

$$\text{FPS} = \frac{\text{Number of images processed}}{\text{Total time taken to process these images}} \tag{31}$$

where $TP$ represents the number of true positives, $FP$ represents false positives, $FN$ represents false negatives, $TN$ represents true negatives, and $N$ is the number of classes. $AP_{i,IoU\geq 0.5}$ represents the average precision for class $i$ when IoU is greater than or equal to 50%, and similarly for $AP_{i,IoU\geq 0.75}$. Through these evaluation metrics, the performance impact of different models and configurations on the weed recognition task can be accurately measured, providing a basis for further optimization and improvement.

## 4. Results and Discussion

### 4.1. Results

#### 4.1.1. Baseline

In this study, multiple well-known baseline models were used to comprehensively evaluate the performance of the proposed model. These include YOLOv9l [38], YOLOv10l [39], leafdetection [40], TinySegFormer [41], and DETR [42]. The YOLO series models, as classic object detection algorithms, are particularly adept at handling real-time detection tasks. They predict object bounding boxes and class labels directly using regression methods, achieving a good balance between speed and accuracy. YOLOv9l and YOLOv10l are iterative improvements based on the YOLO series. YOLOv10l introduces more complex network architectures and an improved loss function, further enhancing the model's detection accuracy and robustness. Leafdetection is a deep learning model specifically designed for plant leaf disease detection. It utilizes CNNs to extract detailed features of plant leaves, accurately identifying disease regions. TinySegFormer is a small-scale target segmentation model based on transformers. By combining traditional CNNs with transformer architectures, it effectively improves image segmentation tasks, particularly in fine-grained segmentation in complex backgrounds. DETR is an end-to-end object detection model based on transformers. It uses the self-attention mechanism to globally model features, effectively handling detection tasks for various objects in complex scenes. The same dataset was used for all models, ensuring a fair and consistent comparison between the proposed model and the baseline models. This dataset includes various weed types under different growth stages and environmental conditions. By comparing these baseline models, the advantages and areas for improvement of the proposed method in the weed recognition task can be better understood, further verifying its effectiveness and feasibility in practical applications.

4.1.2. Hardware and Software Platform

In this experiment, high-performance computational hardware was utilized to support the training and inference processes of complex models in deep learning tasks. Specifically, the hardware platform used an NVIDIA A100 GPU, which possesses powerful computational capabilities, making it suitable for large-scale deep learning training tasks. To ensure the efficiency of the training process, the hardware was also equipped with 64 GB of memory and high-frequency processors (such as AMD Ryzen 9 5950X or Intel i9-series processors), enabling the quick processing of large volumes of training data and multi-task parallel processing. Additionally, the hardware was integrated with a high-bandwidth data storage system to support the rapid reading and storage of large image datasets, ensuring that training efficiency was not hindered by data bottlenecks.

In terms of the software platform, Python 3.8 was used as the programming language, paired with the popular deep learning framework PyTorch. These frameworks provide flexible and efficient model training and inference capabilities, supporting various deep learning algorithms, such as CNNs and generative models (e.g., diffusion models). For data processing and augmentation, libraries such as OpenCV and PIL were employed to preprocess and enhance the images, ensuring the quality and diversity of the input data. Model training was accelerated using NVIDIA's CUDA and cuDNN libraries to maximize GPU performance and improve training speed.

4.1.3. Optimizer and Hyperparameters

For this experiment, the dataset was divided into training, validation, and test sets. Specifically, 80% of the dataset was used for training, and 20% was used for testing. This ratio ensures that the training set is sufficiently large for the model to learn effectively while maintaining the independence of the test set for model evaluation. During training, five-fold cross-validation was adopted to further optimize the model's performance. The basic idea of cross-validation is to divide the dataset into five equal subsets, using one subset as the validation set and the remaining four as the training set. This process is repeated for each subset, helping to reduce the model's dependence on a single data partition and providing more reliable evaluation results.

Regarding hyperparameters, the learning rate was initially set to 0.001 and then adjusted based on the changes in the loss function during training using a learning rate decay strategy. A larger learning rate was used in the early stages to accelerate convergence and gradually reduced in later stages to fine-tune model parameters. The Adam optimizer was employed, with its update rule given as

$$\theta_t = \theta_{t-1} - \alpha \cdot \frac{m_t}{\sqrt{v_t} + \epsilon}, \tag{32}$$

where $\theta_t$ is the model parameter, $\alpha$ is the learning rate, $m_t$ and $v_t$ are the first and second moment estimates of the gradient, and $\epsilon$ is a smoothing term to prevent division by zero. The Adam optimizer has been shown to perform well during deep neural network training, especially when dealing with sparse and noisy gradients, with faster convergence. Additionally, the batch size was set to 32, which is the number of samples used to update the model parameters in each training iteration. This batch size was found to perform stably in the experiment, balancing memory usage and training efficiency. Dropout, with a rate of 0.5, was also used to prevent overfitting. This means that during training, each neuron has a 50% chance of being temporarily discarded, forcing the model to learn more robust features. Hyperparameters such as the regularization coefficient and gradient clipping thresholds were also fine-tuned. The regularization coefficient $\lambda$ was set to 0.001 to avoid overfitting, and gradient clipping was applied to prevent gradient explosion, with the clipping threshold set to 5.0, meaning that when the gradient magnitude exceeds 5.0, it is clipped to that threshold.

#### 4.1.4. Weed Detection Results

This experiment was designed to compare the performance of different models on weed recognition tasks, evaluate the superiority of the proposed model for this task, and analyze the applicability of various models in real-world applications. The evaluation metrics include precision, recall, accuracy, mAP@50, and mAP@75, which collectively assess the models' performance in terms of precision, coverage, and overall capability. As shown in Table 2 and Figure 6, the proposed method outperforms all other compared models across all metrics, demonstrating its effectiveness in handling weed detection in complex agricultural scenarios.

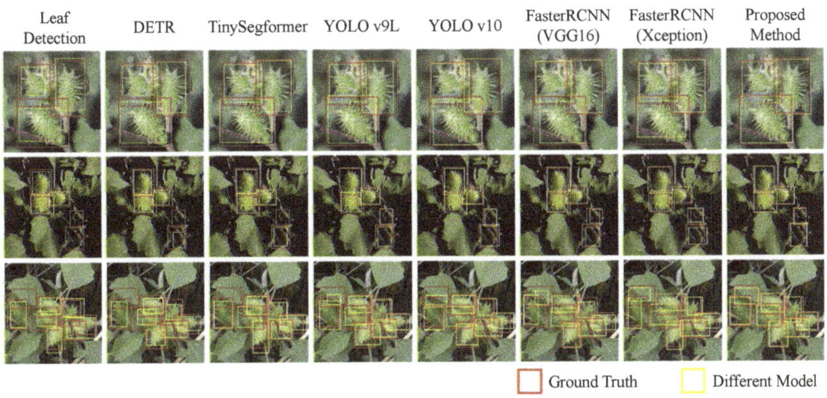


**Figure 6.** Visualization of weed detection results on different methods.

Compared to the baseline model, leafdetection, and the classic object detection model DETR, the proposed method improves precision and recall by approximately 10% and 8%, respectively, indicating significant performance enhancements. Additionally, compared to the improved YOLOv10l, the proposed method achieves increases of 1% and 2% in mAP@50 and mAP@75, respectively, showcasing its superior detection capability and robustness.

**Table 2.** Performance evaluation of different models on weed recognition tasks.

| Model | Precision | Recall | Accuracy | mAP@50 | mAP@75 | FPS |
|---|---|---|---|---|---|---|
| leafdetection [40] | 0.83 | 0.80 | 0.81 | 0.82 | 0.80 | 23.1 |
| DETR | 0.84 | 0.82 | 0.83 | 0.84 | 0.82 | 18.9 |
| TinySegformer [41] | 0.86 | 0.84 | 0.85 | 0.86 | 0.84 | 34.7 |
| YOLO v9l | 0.88 | 0.86 | 0.87 | 0.88 | 0.86 | 45.8 |
| YOLO v10l | 0.91 | 0.89 | 0.90 | 0.91 | 0.89 | 43.6 |
| FasterRCNN (VGG16) | 0.81 | 0.79 | 0.80 | 0.81 | 0.79 | 18.9 |
| FasterRCNN (Xception) | 0.84 | 0.82 | 0.82 | 0.84 | 0.80 | 23.5 |
| Proposed Method | 0.94 | 0.90 | 0.92 | 0.92 | 0.91 | 33.5 |

From a theoretical perspective, the performance differences among these models are primarily attributed to their mathematical properties and architectural designs. The leafdetection model utilizes traditional image segmentation and detection techniques, emphasizing lightweight architecture but lacking the ability to deeply capture contextual information, resulting in lower precision and recall. DETR incorporates a transformer architecture and self-attention mechanisms, allowing it to capture global feature relationships, leading to moderate improvements in precision and mAP. However, its computational complexity imposes performance limitations. TinySegformer combines lightweight design with hierarchical transformer structures, further enhancing global information extraction while reducing computational costs, thus surpassing DETR and leafdetection across multiple metrics. YOLOv9l and YOLOv10l, as efficient object detection models, balance precision and speed by optimizing receptive fields and feature extraction modules. Notably, YOLOv10l achieves improved performance through better multi-scale feature fusion, approaching the results of the proposed method. However, the proposed model introduces semi-supervised diffusion generation networks and generative attention mechanisms, which exhibit significant advantages in feature enhancement and multi-task optimization. Mathematically, the diffusion model generates high-quality pseudo-labels through sampling, effectively addressing the issue of insufficient labeled data. Simultaneously, the generative attention mechanism dynamically allocates feature weights, further improving feature extraction accuracy and efficiency. Consequently, the proposed method's performance advantages in complex scenarios can be attributed to the rationality of its mathematical design and the synergistic effects of its modules, providing an efficient and robust solution for smart agriculture.

#### 4.1.5. Results Analysis for Different Weed Types Using the Proposed Method

This experiment was conducted to evaluate the applicability and performance stability of the proposed model in detecting different types of weeds. The data in Table 3 reveal that the model's detection performance varies across weed types, but overall, it exhibits high detection precision and stability. The results indicate that the model performs best on "Setaria viridis," achieving a precision of 0.97 and mAP@50 and mAP@75 values of 0.95 and 0.94, respectively, demonstrating exceptional effectiveness in identifying weed types with prominent features. However, for "Amaranthus rudis," precision and recall drop to 0.90 and 0.87, respectively, and mAP metrics are relatively lower. This discrepancy highlights the model's limitations in handling weeds with less distinct visual features or those that closely resemble background elements. Such differences suggest that the morphological characteristics of different weed types and their complexity in relation to the background significantly impact detection performance.

**Table 3.** Detection performance for different weed types.

| Weed Type | Precision | Recall | Accuracy | mAP@50 | mAP@75 |
|---|---|---|---|---|---|
| Setaria viridis | 0.97 | 0.93 | 0.95 | 0.95 | 0.94 |
| Xanthium spinosum | 0.95 | 0.91 | 0.93 | 0.93 | 0.92 |
| Cyclachaena xanthiifolia | 0.93 | 0.89 | 0.91 | 0.91 | 0.90 |
| Xanthium italicum | 0.92 | 0.88 | 0.90 | 0.90 | 0.89 |
| Amaranthus rudis | 0.90 | 0.87 | 0.88 | 0.87 | 0.86 |

Theoretically, these results are closely related to the distribution of weed features and the mathematical properties of the model. The superior performance on "Setaria viridis" can be attributed to its highly distinguishable morphological characteristics, such as clear edges and textured details. During the feature extraction stage, the generative attention mechanism effectively captures and enhances these prominent features. Specifically, the mechanism dynamically adjusts the weights of generated features and real features, enhancing information representation for target regions. The attention weight matrix is computed as $W = \text{softmax}(F_g' \cdot F_r^\top)$, ensuring priority is given to distinguishing features. In contrast, for "Amaranthus rudis," with complex backgrounds or indistinct features, the model may face instability in generating pseudo-labels, which could hinder the optimization of pseudo-supervised loss $L_u$. Since $L_u$ depends on high-confidence pseudo-labels, samples failing to meet the confidence threshold $\tau$ may not be adequately utilized. Furthermore, the diffusion-generated features may be susceptible to noise interference in complex scenes, leading to misalignment between generated and real data. These mathematical constraints further suggest that while the proposed model significantly improves detection performance, additional optimizations may be required to enhance its robustness and generalization in challenging scenarios.

#### 4.1.6. Test on Other Aerial-Based Detection Tasks

To further evaluate the robustness of the proposed method, we borrowed the research approach from [43] and extended it by testing our model on aerial view datasets. Specifically, we selected the open Kaggle dataset [44] for testing to assess the applicability of the proposed method in a broader set of data sources. This experiment was conducted to verify how well the model performs on aerial imagery and to confirm its robustness in different types of datasets, which may have different feature characteristics compared to ground-level data. The results are shown in the Table 4:

**Table 4.** Performance test on other dataset.

| Model | Precision | Recall | Accuracy | mAP@50 | mAP@75 | FPS |
|---|---|---|---|---|---|---|
| DETR | 0.70 | 0.71 | 0.70 | 0.68 | 0.65 | 18.9 |
| YOLO v9l | 0.72 | 0.71 | 0.72 | 0.70 | 0.68 | 45.8 |
| YOLO v10l | 0.71 | 0.69 | 0.70 | 0.65 | 0.62 | 43.6 |
| FasterRCNN (VGG16) | 0.67 | 0.61 | 0.63 | 0.65 | 0.61 | 18.9 |
| FasterRCNN (Xception) | 0.67 | 0.62 | 0.64 | 0.65 | 0.62 | 23.5 |
| Proposed Method | 0.75 | 0.72 | 0.72 | 0.72 | 0.69 | 33.5 |

The results demonstrate that the proposed model performs well on both the Kaggle dataset and the dataset used in this paper. Although there is a slight drop in performance when tested on the Kaggle dataset, the overall results remain impressive, showing that the model maintains strong accuracy and generalization even when applied to aerial view data. The minor difference in performance between the two datasets may be attributed to the variations in image quality, acquisition conditions, and feature characteristics between ground-level and aerial imagery. However, the results suggest that the proposed model is capable of adapting to different data sources, further validating its robustness in various agricultural detection tasks. In conclusion, the experiment confirms that the proposed method, originally trained on ground-level data, is also highly applicable to aerial imagery. This strengthens the applicability of the model for large-scale weed detection tasks, especially in remote sensing and UAV-based applications. Future work may involve further optimizing the model to bridge the performance gap across different types of agricultural data sources and exploring additional datasets with more diverse environmental factors.

### *4.2. Discussion*

#### 4.2.1. Discussion on Different Data Augmentation Methods

Data augmentation plays a crucial role in enhancing the performance of deep learning models, particularly when labeled data are limited. In the context of agricultural image analysis, including weed detection, data augmentation techniques help increase the diversity and quantity of training samples, making the model more robust and improving its generalization ability. This section discusses the effects of different data augmentation methods, including CutOut, MixUp, and GridMask, on the performance of the proposed model. These methods were applied individually and in combination to assess their impact on key evaluation metrics, such as precision, recall, accuracy, and mAP@50/mAP@75. An ablation experiment was conducted to compare the performance of the model under different augmentation conditions. The results are shown in Table 5.

**Table 5.** Ablation experiment results on different data augmentation methods.

| Augmentation Method | Precision | Recall | Accuracy | mAP@50 | mAP@75 |
|---|---|---|---|---|---|
| No Augmentation | 0.85 | 0.80 | 0.82 | 0.85 | 0.83 |
| CutOut | 0.88 | 0.84 | 0.86 | 0.87 | 0.85 |
| MixUp | 0.90 | 0.86 | 0.88 | 0.89 | 0.87 |
| GridMask | 0.89 | 0.85 | 0.87 | 0.88 | 0.86 |
| CutOut + MixUp + GridMask | 0.94 | 0.90 | 0.92 | 0.92 | 0.91 |

The results indicate that data augmentation significantly improves the performance of the model across all metrics. The model's performance is enhanced by each individual augmentation method, with MixUp showing the most notable improvement in precision, recall, and mAP metrics. Specifically, MixUp contributes to a better balance between the diversity of data and the reduction of overfitting, which helps the model to generalize better. On the other hand, GridMask and CutOut also improve performance by forcing the model to focus on different parts of the image, promoting the learning of more robust features. When combined, CutOut, MixUp, and GridMask together yield the best results, with the highest values for precision, recall, accuracy, and mAP metrics, indicating the complementary nature of these augmentation methods. In summary, the results highlight the significant impact of data augmentation on the model's performance. Combining different augmentation techniques maximizes the benefits, as each method enhances different aspects of feature learning and robustness. These findings reinforce the importance of employing data augmentation strategies, particularly in agricultural image analysis, where data are often limited and diverse. Future work can explore additional augmentation techniques and investigate their effects on the model's generalization capability in more diverse agricultural environments.

#### 4.2.2. Ablation Study on Different Learning Methods

The objective of this experiment is to analyze the impact of different learning methods on the performance of the weed detection model and to verify the superiority of the proposed semi-supervised attention mechanism. The experimental results, as shown in Table 6, indicate that both GAN and VAE, as semi-supervised learning methods, exhibit improved detection accuracy compared to standard attention mechanisms.

**Table 6.** Ablation study on different learning methods.

| Model | Precision | Recall | Accuracy | mAP@50 | mAP@75 | FPS |
|---|---|---|---|---|---|---|
| GAN (Semi-Supervised) | 0.72 | 0.69 | 0.71 | 0.71 | 0.70 | 28.5 |
| VAE (Semi-Supervised) | 0.87 | 0.83 | 0.85 | 0.84 | 0.83 | 41.9 |
| Standard Self-Attention | 0.89 | 0.85 | 0.86 | 0.84 | 0.81 | 18.3 |
| CBAM | 0.85 | 0.81 | 0.83 | 0.83 | 0.82 | 31.6 |
| Semi-Supervised Attention | 0.94 | 0.90 | 0.92 | 0.92 | 0.91 | 33.5 |

### 4.3. Exploratory Data Analysis

In the exploratory data analysis (EDA) section of this study, the performance of different models on the weed detection task was compared, and the experimental results were visualized using violin plots and box plots, as shown in Figures 7 and 8. These analyses provide an intuitive understanding of the distribution of model performance and highlight the advantages of the proposed method.

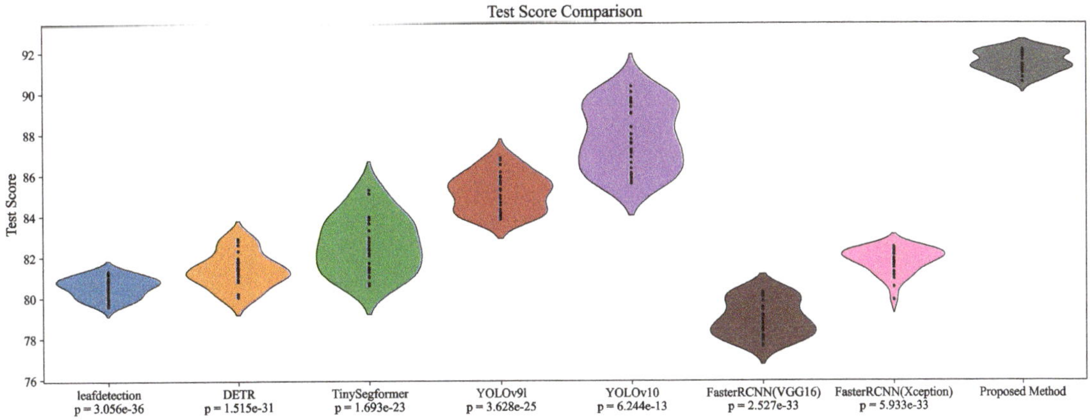


**Figure 7.** EDA violin plots.

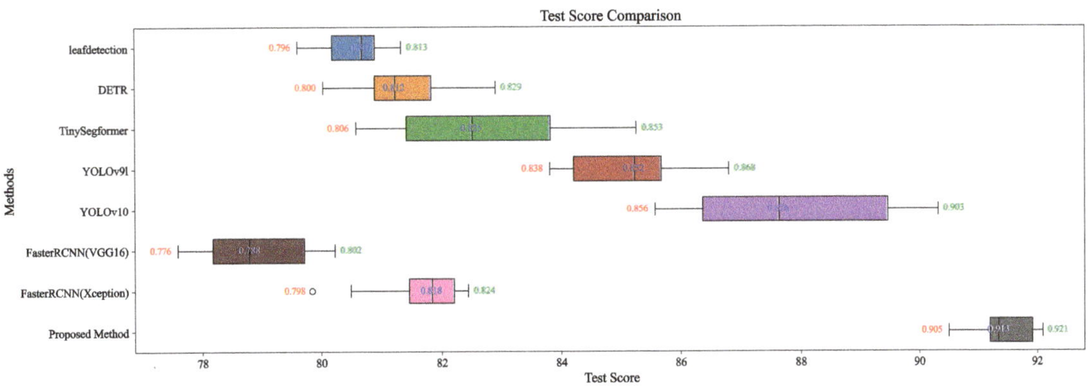


**Figure 8.** EDA box plots.

From the violin plot, significant differences in score distributions among the models can be observed. The traditional methods, leafdetection and DETR, exhibit relatively concentrated distributions but lower overall scores, mainly within the ranges of 80–82 and 81–83, respectively. This indicates their limitations in complex agricultural scenarios. TinySegformer, incorporating a lightweight transformer architecture, achieves a certain performance improvement over DETR, with its score range extending to 82–85. However, instability in detection results is still observed in some cases. YOLOv9 and YOLOv10l, as improved object detection models, enhance receptive fields and feature extraction capabilities, leading to significantly higher scores, with YOLOv10l achieving a score distribution between 85 and 90. Notably, the proposed method outperforms all other models, exhibiting the highest average score (close to 92) and the most compact distribution, indicating superior robustness across different test samples, lower variance, and higher prediction stability.

The box plot further illustrates the score ranges and outlier distribution of each model. It can be observed that leafdetection and Faster R-CNN (VGG16) have relatively large interquartile ranges (IQRs), suggesting that their performance varies considerably across different samples. In contrast, the proposed method has the smallest IQR, demonstrating the stability of its detection results. Furthermore, based on the p-value statistics, the proposed method exhibits a statistically significant advantage over other models, further validating its superior performance in the weed detection task.

Limitations and Future Work

Despite achieving significant performance improvements in weed detection tasks, several limitations remain that require further exploration in future research. First, the performance of the semi-supervised generative model heavily depends on the quality of the pseudo-labels, which are influenced by the complexity of unlabeled data and the generalization capacity of the generative model. When the background information is complex or the weed target features are ambiguous, the pseudo-labels may contain noise or bias, potentially impacting the training effectiveness and generalization performance of the model. Additionally, the data augmentation techniques employed in this study, such as CutOut, MixUp, and GridMask, while effectively enhancing the robustness and generalization capability of the model by increasing data diversity, may still face challenges under complex backgrounds or varying lighting conditions. These methods might not be the highest priority technique to rely on in such scenarios, and a more comprehensive and diversified augmentation pipeline could be devised to better handle the intricacies of agricultural image data.

Moreover, although the diffusion model demonstrates exceptional performance in generating high-quality and diverse data, its generation process involves multiple iterative computations, resulting in extended training times and high computational costs. This limitation may pose challenges in practical applications, particularly under resource and time constraints. Furthermore, in the proposed generative attention mechanism, the issue of generating features that may introduce noise has not been explicitly addressed. In situations where complex backgrounds or target features are not obvious, this noise may affect the performance of the model. Future work should explore ways to optimize the feature generation process and minimize the introduction of noise, ensuring more accurate feature generation and improving the model's robustness in complex scenarios. Finally, while the dataset established in this study covers various weed types and growth environments, certain weed species or environmental characteristics specific to other global regions may remain unaddressed. As a result, the model's generalization capability in these scenarios requires further validation.

## 5. Conclusions

Weed recognition and detection are critical research topics in smart agriculture, playing a significant role in improving agricultural productivity and reducing environmental pollution. To address challenges such as limited labeled data, diverse weed types, and high morphological similarity between weeds and crops in agricultural scenarios, a weed detection model based on a semi-supervised diffusion generative network is proposed. This model incorporates a generative attention mechanism and semi-diffusion loss, forming an efficient semi-supervised learning framework. Experimental results demonstrate that the proposed model significantly outperforms existing methods across multiple evaluation metrics, achieving a precision of 0.94, recall of 0.90, accuracy of 0.92, and mAP@50 and mAP@75 of 0.92 and 0.91, respectively. These results validate the robustness and detection capability of the model in complex agricultural scenarios. Compared to traditional methods such as DETR, the proposed method improves precision and recall by approximately 8%. Furthermore, compared to the enhanced YOLOv10, mAP metrics are increased by 1% to 2%, further demonstrating the model's superiority.The key innovations of this study are as follows: First, the introduction of the diffusion model for pseudo-label generation effectively utilizes unlabeled data, significantly improving performance under limited labeled data conditions. Second, the generative attention mechanism dynamically fuses generated features with real features, enhancing the representation of target regions. Third, the semi-diffusion loss combines supervised loss from labeled data with pseudo-supervised loss from unlabeled data, enabling collaborative multi-task optimization. Additionally, experimental analyses provide in-depth insights into the impact of different attention mechanisms and generative methods on weed detection performance, highlighting the crucial role of feature fusion between generated and real features in improving model performance. In conclusion, the proposed model and methodology offer an efficient and robust solution for weed detection in agricultural scenarios. Not only does the model surpass existing methods in performance, but it also provides new perspectives and practical foundations for further research on the application of semi-supervised learning in smart agriculture. Future studies can focus on improving the efficiency and generalization capability of the model and extending its application to broader agricultural scenarios, thereby advancing the development and practice of smart agriculture.

**Author Contributions:** Conceptualization, R.L., X.W. and Y.C.; Data curation, Y.Z. and X.T.; Formal analysis, Y.X. and C.J.; Funding acquisition, H.D. and S.Y.; Investigation, Y.X.; Methodology, R.L., X.W., Y.C. and C.J.; Project administration, H.D. and S.Y.; Resources, Y.Z., X.T. and Y.S.; Software, R.L., X.W. and Y.C.; Supervision, Y.S., H.D. and S.Y.; Validation, Y.X. and C.J.; Visualization, Y.Z., X.T. and Y.S.; Writing—original draft, R.L., X.W., Y.C., Y.X., Y.Z., X.T., C.J., Y.S., H.D. and S.Y. All authors have read and agreed to the published version of the manuscript.

**Funding:** This research was funded by The National Key Research and Development Program of China (2024YFC2607600).

**Data Availability Statement:** The data presented in this study are available on request from the corresponding author.

**Conflicts of Interest:** The authors declare no conflicts of interest.

## References

1. Kumar, A.; Sharma, P. Impact of climate variation on agricultural productivity and food security in rural India. *Soc. Sci. Res. Netw.* **2022**. [CrossRef]
2. Chouhan, G.K.; Verma, J.P.; Jaiswal, D.K.; Mukherjee, A.; Singh, S.; de Araujo Pereira, A.P.; Liu, H.; Abd_Allah, E.F.; Singh, B.K. Phytomicrobiome for promoting sustainable agriculture and food security: Opportunities, challenges, and solutions. *Microbiol. Res.* **2021**, *248*, 126763. [CrossRef] [PubMed]
3. Zhang, Y.; Yang, X.; Liu, Y.; Zhou, J.; Huang, Y.; Li, J.; Zhang, L.; Ma, Q. A time-series neural network for pig feeding behavior recognition and dangerous detection from videos. *Comput. Electron. Agric.* **2024**, *218*, 108710. [CrossRef]
4. Esposito, M.; Westbrook, A.S.; Maggio, A.; Cirillo, V.; DiTommaso, A. Neutral weed communities: The intersection between crop productivity, biodiversity, and weed ecosystem services. *Weed Sci.* **2023**, *71*, 301–311. [CrossRef]
5. Zhang, Y.; Wa, S.; Liu, Y.; Zhou, X.; Sun, P.; Ma, Q. High-accuracy detection of maize leaf diseases CNN based on multi-pathway activation function module. *Remote Sens.* **2021**, *13*, 4218. [CrossRef]
6. Monteiro, A.; Santos, S. Sustainable approach to weed management: The role of precision weed management. *Agronomy* **2022**, *12*, 118. [CrossRef]
7. Adeux, G.; Cordeau, S.; Antichi, D.; Carlesi, S.; Mazzoncini, M.; Munier-Jolain, N.; Bàrberi, P. Cover crops promote crop productivity but do not enhance weed management in tillage-based cropping systems. *Eur. J. Agron.* **2021**, *123*, 126221. [CrossRef]
8. Choudhary, A.K.; Yadav, D.; Sood, P.; Rahi, S.; Arya, K.; Thakur, S.; Lal, R.; Kumar, S.; Sharma, J.; Dass, A.; et al. Post-Emergence herbicides for effective weed management, enhanced wheat productivity, profitability and quality in North-Western Himalayas: A 'Participatory-Mode'Technology Development and Dissemination. *Sustainability* **2021**, *13*, 5425. [CrossRef]

9. Rai, N.; Zhang, Y.; Ram, B.G.; Schumacher, L.; Yellavajjala, R.K.; Bajwa, S.; Sun, X. Applications of deep learning in precision weed management: A review. *Comput. Electron. Agric.* **2023**, *206*, 107698. [CrossRef]
10. Vasileiou, M.; Kyrgiakos, L.S.; Kleisiari, C.; Kleftodimos, G.; Vlontzos, G.; Belhouchette, H.; Pardalos, P.M. Transforming weed management in sustainable agriculture with artificial intelligence: A systematic literature review towards weed identification and deep learning. *Crop Prot.* **2024**, *176*, 106522. [CrossRef]
11. Upadhyay, A.; Sunil, G.; Zhang, Y.; Koparan, C.; Sun, X. Development and evaluation of a machine vision and deep learning-based smart sprayer system for site-specific weed management in row crops: An edge computing approach. *J. Agric. Food Res.* **2024**, *18*, 101331. [CrossRef]
12. Saleh, A.; Olsen, A.; Wood, J.; Philippa, B.; Azghadi, M.R. Semi-Supervised Weed Detection for Rapid Deployment and Enhanced Efficiency. *arXiv* **2024**, arXiv:2405.07399.
13. Muvva, V.B.R.; Kumpati, R.; Skarka, W. Efficient Weed Detection Using CNN with an Autonomous Robot. In Proceedings of the 2024 2nd International Conference on Unmanned Vehicle Systems-Oman (UVS), Muscat, Oman, 12–14 February 2024; pp. 1–7.
14. Narayana, C.L.; Ramana, K.V. An efficient real-time weed detection technique using YOLOv7. *Int. J. Adv. Comput. Sci. Appl.* **2023**, *14*, 65. [CrossRef]
15. Li, J.; Chen, D.; Yin, X.; Li, Z. Performance evaluation of semi-supervised learning frameworks for multi-class weed detection. *Front. Plant Sci.* **2024**, *15*, 1396568. [CrossRef] [PubMed]
16. Zhang, X.; Cui, J.; Liu, H.; Han, Y.; Ai, H.; Dong, C.; Zhang, J.; Chu, Y. Weed identification in soybean seedling stage based on optimized Faster R-CNN algorithm. *Agriculture* **2023**, *13*, 175. [CrossRef]
17. Yan, J.; Wang, X. Unsupervised and semi-supervised learning: The next frontier in machine learning for plant systems biology. *Plant J.* **2022**, *111*, 1527–1538. [CrossRef] [PubMed]
18. Garibaldi-Márquez, F.; Flores, G.; Mercado-Ravell, D.A.; Ramírez-Pedraza, A.; Valentín-Coronado, L.M. Weed classification from natural corn field-multi-plant images based on shallow and deep learning. *Sensors* **2022**, *22*, 3021. [CrossRef] [PubMed]
19. Trabucco, B.; Doherty, K.; Gurinas, M.; Salakhutdinov, R. Effective data augmentation with diffusion models. *arXiv* **2023**, arXiv:2302.07944.
20. Li, Q.; Zhang, Y. Confidential Federated Learning for Heterogeneous Platforms against Client-Side Privacy Leakages. In Proceedings of the ACM Turing Award Celebration Conference 2024, Changsha, China, 5–7 July 2024; pp. 239–241.
21. Veeragandham, S.; Santhi, H. Optimization enabled Deep Quantum Neural Network for weed classification and density estimation. *Expert Syst. Appl.* **2024**, *243*, 122679. [CrossRef]
22. Belissent, N.; Peña, J.M.; Mesías-Ruiz, G.A.; Shawe-Taylor, J.; Pérez-Ortiz, M. Transfer and zero-shot learning for scalable weed detection and classification in UAV images. *Knowl.-Based Syst.* **2024**, *292*, 111586. [CrossRef]
23. Zhang, F.; Ren, F.; Li, J.; Zhang, X. Automatic stomata recognition and measurement based on improved YOLO deep learning model and entropy rate superpixel algorithm. *Ecol. Inform.* **2022**, *68*, 101521. [CrossRef]
24. Das, S.; Chatterjee, M.; Stephen, R.; Singh, A.K.; Siddique, A. Unveiling the Potential of YOLO v7 in the Herbal Medicine Industry: A Comparative Examination of YOLO Models for Medicinal Leaf Recognition. *Int. J. Eng. Res. Technol.* **2024**, *13*. [CrossRef]
25. Adhinata, F.D.; Wahyono; Sumiharto, R. A comprehensive survey on weed and crop classification using machine learning and deep learning. *Artif. Intell. Agric.* **2024**, *13*, 45–63. [CrossRef]
26. Sujatha, R.; Chatterjee, J.M.; Jhanjhi, N.; Brohi, S.N. Performance of deep learning vs machine learning in plant leaf disease detection. *Microprocess. Microsyst.* **2021**, *80*, 103615. [CrossRef]
27. Shorewala, S.; Ashfaque, A.; Sidharth, R.; Verma, U. Weed density and distribution estimation for precision agriculture using semi-supervised learning. *IEEE Access* **2021**, *9*, 27971–27986. [CrossRef]
28. Al-Badri, A.H.; Ismail, N.A.; Al-Dulaimi, K.; Salman, G.A.; Khan, A.; Al-Sabaawi, A.; Salam, M.S.H. Classification of weed using machine learning techniques: A review—challenges, current and future potential techniques. *J. Plant Dis. Prot.* **2022**, *129*, 745–768. [CrossRef]
29. Li, Y.; Chao, X. Semi-supervised few-shot learning approach for plant diseases recognition. *Plant Methods* **2021**, *17*, 1–10. [CrossRef] [PubMed]
30. Moreno, H.; Gómez, A.; Altares-López, S.; Ribeiro, A.; Andújar, D. Analysis of Stable Diffusion-derived fake weeds performance for training Convolutional Neural Networks. *Comput. Electron. Agric.* **2023**, *214*, 108324. [CrossRef]
31. Benchallal, F.; Hafiane, A.; Ragot, N.; Canals, R. ConvNeXt based semi-supervised approach with consistency regularization for weeds classification. *Expert Syst. Appl.* **2024**, *239*, 122222. [CrossRef]
32. Hu, R.; Su, W.H.; Li, J.L.; Peng, Y. Real-time lettuce-weed localization and weed severity classification based on lightweight YOLO convolutional neural networks for intelligent intra-row weed control. *Comput. Electron. Agric.* **2024**, *226*, 109404. [CrossRef]
33. Liu, T.; Jin, X.; Zhang, L.; Wang, J.; Chen, Y.; Hu, C.; Yu, J. Semi-supervised learning and attention mechanism for weed detection in wheat. *Crop Prot.* **2023**, *174*, 106389. [CrossRef]
34. Rizve, M.N.; Duarte, K.; Rawat, Y.S.; Shah, M. In defense of pseudo-labeling: An uncertainty-aware pseudo-label selection framework for semi-supervised learning. *arXiv* **2021**, arXiv:2101.06329.

35. Ajayi, O.G.; Ashi, J.; Guda, B. Performance evaluation of YOLO v5 model for automatic crop and weed classification on UAV images. *Smart Agric. Technol.* **2023**, *5*, 100231. [CrossRef]
36. Manikandakumar, M.; Karthikeyan, P. Weed classification using particle swarm optimization and deep learning models. *Comput. Syst. Sci. Eng.* **2023**, *44*, 913–927. [CrossRef]
37. Sunil, G.; Zhang, Y.; Koparan, C.; Ahmed, M.R.; Howatt, K.; Sun, X. Weed and crop species classification using computer vision and deep learning technologies in greenhouse conditions. *J. Agric. Food Res.* **2022**, *9*, 100325.
38. Wang, C.Y.; Yeh, I.H.; Liao, H.Y.M. Yolov9: Learning what you want to learn using programmable gradient information. *arXiv* **2024**, arXiv:2402.13616.
39. Wang, A.; Chen, H.; Liu, L.; Chen, K.; Lin, Z.; Han, J.; Ding, G. Yolov10: Real-time end-to-end object detection. *arXiv* **2024**, arXiv:2405.14458.
40. Zhang, Y.; Wa, S.; Zhang, L.; Lv, C. Automatic plant disease detection based on tranvolution detection network with GAN modules using leaf images. *Front. Plant Sci.* **2022**, *13*, 875693. [CrossRef] [PubMed]
41. Zhang, Y.; Lv, C. TinySegformer: A lightweight visual segmentation model for real-time agricultural pest detection. *Comput. Electron. Agric.* **2024**, *218*, 108740. [CrossRef]
42. Carion, N.; Massa, F.; Synnaeve, G.; Usunier, N.; Kirillov, A.; Zagoruyko, S. End-to-end object detection with transformers. In *European Conference on Computer Vision*; Springer: Berlin/Heidelberg, Germany, 2020; pp. 213–229.
43. Rehman, M.U.; Eesaar, H.; Abbas, Z.; Seneviratne, L.; Hussain, I.; Chong, K.T. Advanced drone-based weed detection using feature-enriched deep learning approach. *Knowl.-Based Syst.* **2024**, *305*, 112655. [CrossRef]
44. Zhang, Y.; Li, M.; Ma, X.; Wu, X.; Wang, Y. High-precision wheat head detection model based on one-stage network and GAN model. *Front. Plant Sci.* **2022**, *13*, 787852. [CrossRef] [PubMed]

agriculture

MDPI

*Article*

# Framework for Apple Phenotype Feature Extraction Using Instance Segmentation and Edge Attention Mechanism

**Zichong Wang [1], Weiyuan Cui [1,2], Chenjia Huang [1,3], Yuhao Zhou [1], Zihan Zhao [1,2], Yuchen Yue [1,2], Xinrui Dong [1,4] and Chunli Lv [1,*]**

1 China Agricultural University, Beijing 100083, China; wangzichong22@cau.edu.cn (Z.W.); 2400935081@stu.pku.edu.cn (W.C.); chenjiahuang@cau.edu.cn (C.H.); zhouyuhao23@cau.edu.cn (Y.Z.); zihanzhao23@cau.edu.cn (Z.Z.); yueyuchen@cau.edu.cn (Y.Y.); dongxinrui21@cau.edu.cn (X.D.)
2 National School of Development, Peking University, Beijing 100871, China
3 Faculty of Humanities, China University of Political Science and Law, Beijing 102249, China
4 School of Economics and Management, University of Science and Technology Beijing, Beijing 100083, China
* Correspondence: lvcl@cau.edu.cn

**Abstract:** A method for apple phenotypic feature extraction and growth anomaly identification based on deep learning and natural language processing technologies is proposed in this paper, aiming to enhance the accuracy of apple quality detection and anomaly prediction in agricultural production. This method integrates instance segmentation, edge perception mechanisms, attention mechanisms, and multimodal data fusion to accurately extract an apple's phenotypic features, such as its shape, color, and surface condition, while identifying potential anomalies which may arise during the growth process. Specifically, the edge transformer segmentation network is employed to combine deep convolutional networks (CNNs) with the Transformer architecture, enhancing feature extraction and modeling long-range dependencies across different regions of an image. The edge perception mechanism improves segmentation accuracy by focusing on the boundary regions of the apple, particularly in the case of complex shapes or surface damage. Additionally, the natural language processing (NLP) module analyzes agricultural domain knowledge, such as planting records and meteorological data, providing insights into potential causes of growth anomalies and enabling more accurate predictions. The experimental results demonstrate that the proposed method significantly outperformed traditional models across multiple metrics. Specifically, in the apple phenotypic feature extraction task, the model achieved exceptional performance, with accuracy of 0.95, recall of 0.91, precision of 0.93, and mean intersection over union (mIoU) of 0.92. Furthermore, in the growth anomaly identification task, the model also performed excellently, with a precision of 0.93, recall of 0.90, accuracy of 0.91, and mIoU of 0.89, further validating its efficiency and robustness in handling complex growth anomaly scenarios. The method's integration of image data with agricultural knowledge provides a comprehensive approach to both apple quality detection and growth anomaly prediction, offering reliable decision support for agricultural production. The proposed method, by integrating image data with agricultural domain knowledge, provides precise decision support for agricultural production, not only improving the efficiency and accuracy of apple quality detection but also offering reliable technical assurance for agricultural economic analysis.

**Keywords:** apple phenotype feature extraction; growth anomaly recognition; deep learning; agricultural economic analysis

 https://doi.org/10.3390/agriculture15030305

## 1. Introduction

Apple phenotype characteristics are an important aspect of agricultural research and production, including the apple shape index (the ratio of longitudinal to transverse diameters), apple size (based on the transverse diameter), color (surface coloration rate), and surface condition (such as freshness and damage) [1]. These characteristics are not only key indicators for evaluating apple quality but also crucial for guiding agricultural production, improving cultivation techniques, and enhancing the commercial value of apples [2]. In recent years, with the development of agricultural automation and intelligence, computer vision and deep learning-based object detection and semantic segmentation technologies have demonstrated great potential in extracting apple phenotype data [3]. Traditional methods for apple phenotype analysis typically rely on manual measurement and empirical judgment, which are not only inefficient but also subject to subjective biases [4]. In contrast, instance segmentation-based techniques can precisely locate the boundaries of apples and simultaneously perform automatic extraction of various phenotype features, enabling large-scale apple quality detection and analysis [5]. However, apple growth is often influenced by environmental factors (such as climate change and soil conditions), leading to growth anomalies such as deformed apple shapes and surface damage, which significantly impact an apple's quality and market value [6].

To achieve precise extraction of apple phenotype characteristics and efficient identification of growth anomalies, a comprehensive method combining instance segmentation and natural language processing (NLP) is proposed in this study [7,8]. On the one hand, the edge transformer segmentation network based on instance segmentation can precisely extract various phenotype features of apples [9]. On the other hand, the NLP module parses and analyzes agricultural text data (such as expert notes, planting records, and meteorological data) to reveal potential causes of and development trends in growth anomalies from multiple dimensions [10]. This integration of image analysis and text parsing not only enhances the comprehensiveness and accuracy of apple phenotype data extraction but also provides new insights for apple quality management and anomaly prediction [11]. In recent years, object detection technologies such as the YOLO series and Mask R-CNN have been widely applied in apple recognition and classification, demonstrating excellent performance in both real-time detection and accuracy [12]. Additionally, semantic segmentation techniques, such as UNet and DeeplabV3+, have been proven to be highly effective in apple surface feature extraction and disease detection tasks [13,14]. Complementing these technologies, the rapid development of NLP has provided new tools for automating the processing of agricultural text data. For instance, Anand et al. proposed a deep learning framework, AgriSegNet, for multi-scale, attention-based semantic segmentation using drone-acquired images to automatically detect agricultural field anomalies [15]. Zhang et al. proposed a segmentation method which outperforms traditional PSO clustering methods in terms of stability and accuracy. It can accurately and effectively segment agricultural product images in various complex environments, facilitating automated agricultural product picking robots [16]. Su et al. introduced a novel data augmentation framework based on random image cropping and patching (RICAP), which effectively improves segmentation accuracy. The proposed framework boosts the average accuracy of deep neural networks from 91.01% to 94.02% by enhancing the original RICAP approach [17]. Zhang et al. developed a pruning inference method which automatically deactivates part of the network structure based on different conditions, reducing network parameters and operations and significantly increasing the network speed. The proposed model achieved accuracy, recall, and mAP rates of 90.01%, 98.79%, and 97.43% in detecting apple flowers, respectively [18]. These advancements demonstrate the promising potential of integrating image analysis with NLP technologies in agricultural production. Therefore, the method proposed in this paper aims to address the limitations of single-image analysis methods and provides technical support for improving agricultural production efficiency and economic benefits through multimodal data fusion. The contributions of this paper are as follows:

- Integration of Instance Segmentation and Natural Language Processing: For the first time, instance segmentation technology is combined with natural language processing (NLP) to achieve multi-dimensional data fusion in apple phenotype feature extraction and anomaly identification. Instance segmentation ensures the precise extraction of critical phenotype features such as an apple's size, color, and surface condition by accurately delineating an apple's boundaries and surface features. Meanwhile, the NLP module analyzes agricultural text data (e.g., expert notes, planting records, and meteorological data) to reveal potential causes of growth anomalies, providing a comprehensive and accurate analysis which overcomes the limitations of traditional single-image analysis.
- Innovative Application of the Edge Transformer Segmentation Network: The edge transformer segmentation network introduced in this paper integrates Transformer mechanisms with edge-aware modules to better handle complex boundary information of apples. This innovation improves segmentation precision and robustness, especially when dealing with damaged or irregular apple shapes. The method shows excellent performance in extracting key phenotype features from apple surface characteristics and provides reliable support for apple quality assessment and growth anomaly monitoring.
- Multi-Modal Data Fusion for Anomaly Recognition: In contrast to traditional single-image analysis, this study proposes a method for multi-modal data fusion. By combining image data with agricultural text data (e.g., meteorological data and planting records), the NLP module conducts multi-dimensional analysis, leading to more accurate identification of growth anomalies (such as apple deformities and surface damage) and revealing potential causes from a broader context. This cross-modal data fusion offers new perspectives for anomaly prediction and apple quality management in agricultural production.

In the following sections, we provide a detailed overview of the proposed method and its components. Section 2 reviews the foundational methods in object detection, semantic segmentation, and related techniques, which serve as the building blocks for our approach. Section 3 introduces the materials and methods used in this study, including the dataset collection, preprocessing steps, and detailed architecture of the proposed model. Section 4 presents the experimental results and discusses the performance of our method in comparison to existing baseline models. Finally, Section 5 concludes this paper by summarizing the findings and discussing potential future research directions.

## 2. Related Work

Object detection, semantic segmentation, and natural language processing (NLP) are three critical technologies in modern computer science and artificial intelligence, with broad application prospects in agricultural fruit recognition and phenotype data analysis [19,20].

### 2.1. Object Detection

Object detection is a core task in computer vision, primarily aimed at locating and classifying target objects within an image [12,21]. The core of object detection lies in predicting the bounding box and category label for each target object [22]. The optimization objective of object detection generally consists of both classification loss and bounding box regression loss, with common loss functions including cross-entropy loss and intersection over union (IoU) loss [23]. The task of object detection can be divided into two main parts:

object localization and object classification. Object localization refers to accurately finding the location of the target in an image and determining its shape, usually by predicting the bounding box through a regression model. Object classification involves determining the category of the target based on its appearance and features. These two tasks are typically trained simultaneously by minimizing both the classification loss and localization loss. The difference between the predicted bounding box and the true values can be expressed as follows:

$$L_{\text{bbox}} = \sum_{i=1}^{N}\left(|x_i - \hat{x}_i| + |y_i - \hat{y}_i| + |w_i - \hat{w}_i| + \left|h_i - \hat{h}_i\right|\right), \tag{1}$$

where $x_i, y_i, w_i, h_i$ are the predicted values, $\hat{x}_i, \hat{y}_i, \hat{w}_i, \hat{h}_i$ are the ground truth values, and $N$ is the number of targets. Common methods for object detection include region proposal-based methods (such as the R-CNN series [24]) and regression-based methods (such as the YOLO series [25]). In agricultural fruit recognition tasks, object detection technology is widely applied to quickly locate the positions and categories of fruits [26]. These models learn the visual features of fruits in images, enabling efficient classification and localization of fruits amd allowing agricultural producers to monitor the growth status of crops in real time and optimize agricultural management. For instance, the Faster R-CNN model, by combining region proposal networks (RPNs) with convolutional neural networks (CNNs), enhances detection accuracy, particularly for smaller fruits like tomatoes and grapes [27].

### *2.2. Semantic Segmentation*

Semantic segmentation is another crucial task in computer vision, aimed at classifying each pixel in an image to achieve fine-grained segmentation of the target objects [28]. Unlike object detection, which focuses on bounding box prediction, semantic segmentation is concerned with pixel-level prediction, requiring more complex model architectures to capture the fine-grained features of an image [29]. The UNet model, through its encoder-decoder structure, progressively extracts deep features from an image and gradually recovers the spatial resolution to achieve fine pixel-level segmentation [30]. DeepLabV3+ integrates dilated convolutions (atrous convolutions) to expand the receptive field, effectively capturing contextual information in the image while enhancing segmentation accuracy through multi-scale features [31]. The loss function for semantic segmentation is typically defined based on the pixel-level cross-entropy:

$$L_{\text{seg}} = -\frac{1}{HW}\sum_{i=1}^{H}\sum_{j=1}^{W}\sum_{c=1}^{C} y_{ijc}\log\hat{y}_{ijc}, \tag{2}$$

where $H$ and $W$ represent the height and width of the image, $C$ is the number of categories, $y_{ijc}$ is the true label indicating that the pixel $(i,j)$ belongs to category $c$, and $\hat{y}_{ijc}$ is the predicted probability from the model. To address class imbalance, the weighted cross-entropy or Dice loss is often introduced to improve segmentation accuracy, particularly when handling smaller or less-common classes [32]. In the agricultural domain, semantic segmentation is widely applied in fruit phenotype analysis to extract the contours and surface features of fruits, such as measuring a fruit's size, shape, and surface damage [33–35].

### *2.3. Natural Language Processing*

NLP is a core technology for processing and understanding textual data, with significant applications in agricultural data analysis [36,37]. In agricultural text data analysis, text classification tasks can be used to classify fruit planting records by variety or growth stage, assisting agricultural experts in taking appropriate management measures according to different stages. NER can extract specific key information from large volumes of agricultural texts, such as climate conditions, soil types, and management practices, which significantly influence fruit growth and pest control [38–40]. Sentiment analysis can help agricultural managers understand the quality of crops, market demand, and other factors based on feedback from farmers or market evaluations, providing a foundation for subsequent agricultural planning and production. Text data embedding is the foundation of NLP, where word embeddings represent the semantic information of words [41]. Common text embedding technologies, such as Word2Vec and GloVe, capture the relationships between words through statistical methods. For example, word vectors **w** represent words, and contextual information is used to learn the semantic representation of each word in a specific context. The optimization objective can be expressed as follows:

$$L_{\text{NLP}} = -\sum_{i=1}^{N} \log p(w_i|\text{context}), \tag{3}$$

where $p(w_i|\text{context})$ represents the probability of predicting the word $w_i$ given the context and $N$ is the total number of words. As agricultural production increasingly relies on digital records, NLP technologies can also be used for automated agricultural log analysis, disease identification, and predictive analysis. For example, NLP-based models can automatically extract fruit growth patterns under different environmental conditions from historical planting records, providing data support for future agricultural decisions [10].

## 3. Materials and Methods

### 3.1. Image Construction

In this study, the collection of image datasets and image annotation are key steps in fruit phenotypic analysis and anomaly recognition tasks. To ensure the diversity and representativeness of the data, a large number of apple images were collected from multiple regions, covering different growth environments and climatic conditions. The image data were primarily collected from apple orchards in Changping District, Beijing, Qixia City, Yantai, Shandong Province from March 2023 to August 2024, with some images also sourced from the internet, as shown in Table 1, totaling 24,042 images. The image acquisition equipment and methods employed in this study are critical. The image acquisition method and samples are shown in Figure 1.

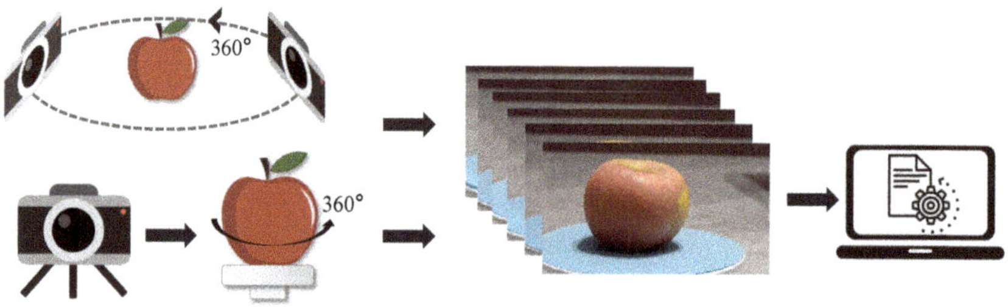


**Figure 1.** Image acquisition scheme and examples.

A Canon EOS 5D Mark IV camera, manufactured by Canon Inc., headquartered in Tokyo, Japan, was utilized due to its exceptional imaging quality and ability to capture fine details, meeting the requirements for precise acquisition of fruit details. This camera was paired with a Canon EF 100mm f/2.8L Macro IS USM macro lens, which is particularly suited for capturing high-precision images at close distances. To minimize the impact of shadows and reflections on the image quality, all images were captured under soft natural light conditions during early morning or late afternoon. However, in real-world agricultural settings, images may sometimes still contain shadows or overexposure due to fluctuating lighting conditions. To handle such cases, we employed image preprocessing techniques such as histogram equalization and contrast adjustment to reduce the effects of shadows and overexposure. Additionally, images with excessive overexposure or shadows which significantly obscured fruit features would be identified and excluded from the dataset during the quality control process. This ensured that only high-quality images suitable for phenotypic analysis were included in the final dataset. During image acquisition, particular attention was paid to capturing fruits from multiple angles to document their morphological characteristics and surface abnormalities. To ensure the dataset's representativeness, images were collected covering various fruit types, maturity stages, varieties, shapes, colors, and symptoms of different diseases. The images were obtained from apple orchards in Changping District, Beijing, and Qixia City, Yantai, Shandong Province. These two regions differ in terms of climate and soil conditions, resulting in diverse image backgrounds and fruit states. For image annotation, a semi-automated approach was adopted using the LabelMe tool. Annotators first manually outlined the fruit positions and drew bounding boxes around them. Subsequently, detailed annotations were added regarding each fruit's shape index, size, color grading, and surface conditions.

**Table 1.** Number of apple images for different data types.

| Index | Number |
|---|---|
| Fruit shape index | 6794 |
| Fruit size | 5703 |
| Color | 6003 |
| Surface condition | 5542 |

Building on the large-scale image data acquisition, additional data were collected by scraping open-source datasets and agriculture-related websites. These sources included expert notes, cultivation records, and meteorological data which, combined with the image data, contributed to the construction of a multimodal dataset. This dataset serves as a valuable resource for fruit phenotypic analysis and anomaly detection while also establishing a foundation for agricultural economic analysis [42,43]. By integrating expert notes, cultivation records, and meteorological data, it is possible to analyze the influence of environmental factors, management practices, and production decisions on fruit quality and yield during the growth process. This provides scientific guidance for agricultural production, optimizing management strategies, and enhancing economic benefits. Expert notes represent a significant component of the dataset, encompassing 20,432 entries documenting common issues and practical experiences throughout the apple cultivation process. These entries span various stages from planting to harvesting and include information on climate, soil, pest management, and irrigation techniques. Cultivation records provide detailed data on apple cultivation processes, including planting times, fertilization practices, irrigation frequencies, and soil treatments, amounting to 19,267 entries. Additionally, meteorological data are another vital source, consisting of 22,803 entries. The growth of apples is closely associated with climatic conditions, as factors such as temperature, humidity, and precipitation directly impact growth cycles, fruit quality, and pest outbreaks. In-depth analysis of these data sources offers valuable references for agricultural economic studies [44,45].

### 3.2. Data Preprocessing

Data preprocessing is the process of cleaning and adjusting raw images to eliminate noise, correct color biases, and crop regions of interest (ROIs) to improve data effectiveness. Common preprocessing operations in image processing include image cropping, flipping, rotation, resizing, denoising, and white balance correction, as shown in Figure 2.

**Figure 2.** Data preprocessing: (**A**) original image, (**B**) horizontal flip, (**C**) perspective transformation, (**D**) rotation, (**E**) translation, and (**F**) center crop.

Image cropping involves selecting ROIs to remove irrelevant background information, thereby reducing computational redundancy and emphasizing the target object. This method is particularly important in fruit phenotype analysis as cropping eliminates background interference, allowing the model to focus more on the features of the fruit itself. Denoising is achieved through techniques such as filtering to reduce random noise in the image, thereby improving image quality. Gaussian filtering is a commonly used denoising method. White balance correction is used to correct color distortion in images and restore the true color information of the fruit. This process adjusts the mean values of the red, green, and blue (RGB) channels of the image such that they align with the target values.

### 3.3. Data Augmentation

Data augmentation is the process of applying various transformations to the original images to generate a more diverse set of training samples, thereby enhancing the robustness and generalization ability of the model. Common data augmentation techniques include Cutout, Mixup, and CutMix. Cutout involves randomly masking a rectangular region on an image to simulate scenarios where the target is partially occluded. This augmentation technique effectively improves the model's prediction ability under occlusion. Let the size of the image $I$ be $(H, W)$ and a region of a size $h \times w$ be randomly occluded at position $(x_0, y_0)$. The augmented image $A$ can be expressed as follows:

$$A(x,y) = \begin{cases} 0, & \text{if } x_0 \leq x < x_0 + h,\ y_0 \leq y < y_0 + w, \\ I(x,y), & \text{otherwise.} \end{cases} \tag{4}$$

Mixup involves linearly mixing two images at a certain ratio, with the aim of improving the model's smoothness and prediction ability for unseen samples. The augmentation formula for Mixup is

$$A = \lambda I_1 + (1-\lambda) I_2, \quad y = \lambda y_1 + (1-\lambda) y_2, \tag{5}$$

where $\lambda \sim \text{Beta}(\alpha, \alpha)$ is the mixing coefficient sampled from a Beta distribution, $I_1$ and $I_2$ are the two original images, and $y_1$ and $y_2$ are the corresponding labels. CutMix combines the ideas of Cutout and Mixup by pasting a portion of one image onto another and adjusting the labels to reflect the proportion of the mixed region. Let $I_1$ and $I_2$ be the two images. The augmented image $A$ can be expressed as

$$A(x,y) = \begin{cases} I_1(x,y), & \text{if } (x,y) \in \text{Region1}, \\ I_2(x,y), & \text{otherwise.} \end{cases} \tag{6}$$

The label adjustment formula is

$$y = \lambda y_1 + (1-\lambda) y_2, \tag{7}$$

where $\lambda$ is the proportion of Region1.

### 3.4. Hyperparameters

Dataset partitioning is a key step in training and validating machine learning models. The goal is to appropriately allocate data for model training, validation, and testing to ensure the scientific and representative evaluation of the model's performance. In this study, the dataset was divided into training and validation sets with a ratio of 8:2. Additionally, to enhance the stability and robustness of the model, $K$-fold cross-validation was employed. Specifically, the dataset was divided into $K$ non-overlapping subsets, with $K-1$ subsets used for training and the remaining subset used for validation. After repeating this process $K$ times, the average performance was computed. The formula for calculating the average performance of $K$-fold cross-validation is

$$\text{Performance} = \frac{1}{K}\sum_{k=1}^{K} P_k, \tag{8}$$

where $P_k$ represents the performance metric (such as the accuracy or $mIoU$) for the $k$th validation. In fruit phenotype analysis, through proper data preprocessing, augmentation, and partitioning, a high-quality dataset was built to provide a reliable foundation for subsequent model training and performance evaluation. This method not only enhances the model's robustness but also provides important support for practical deployment.

### 3.5. Proposed Method

The method presented in this study, from its overall design to specific implementation, aims to extract fruit phenotype features and identify growth anomalies through a multimodal data fusion approach, combining instance segmentation and NLP techniques. The overall framework of the model is a continuous flow from input data to final output predictions, involving the collaborative work of multiple modules, as shown in Figure 3.

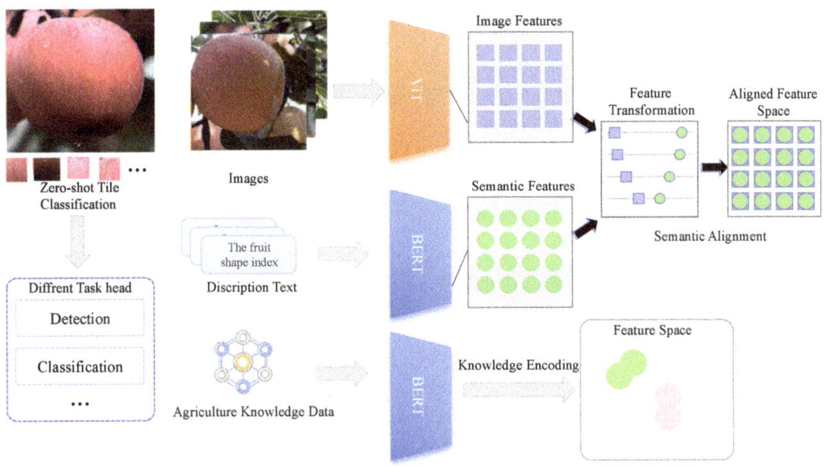


**Figure 3.** Flowchart of the whole process in the proposed method, where the Agriculture Knowledge Data block integrates domain-specific agricultural knowledge to enhance model performance and decision making.

First, after the input data were preprocessed, it entered the instance segmentation network for feature extraction from the images. The extracted image features were then processed by subsequent modules, such as the feature transformation and alignment stages, to refine and integrate the information. Concurrently, the Agricultural Knowledge Data block, which includes textual data such as planting records, climate data, and expert notes, was parsed by the NLP module. This NLP process extracted meaningful insights from the textual data to complement the visual features extracted from the images. The integration of both image and agricultural knowledge data allowed for more accurate fruit phenotype recognition and the identification of growth anomalies. Finally, the combined results from both the image-based and text-based analyses were used to make precise predictions regarding the fruit phenotype and detect potential growth anomalies.

### 3.5.1. Edge Transformer Segmentation Network

In this study, the proposed edge transformer segmentation network is a key component for fruit phenotype analysis and growth anomaly identification, combining agricultural images with agricultural knowledge (text data).

As shown in Figure 4, the network design incorporates a deep fusion of instance segmentation and NLP techniques, aiming to precisely extract fruit phenotype features through image segmentation and text analysis, while also integrating agricultural text data (such as planting records and meteorological data) for anomaly detection. The network input consists of two parts: (1) agricultural images, which provide visual information about the fruits, and (2) agricultural knowledge (text data), which supplies multi-dimensional information regarding the fruit's growth environment, climate change, and management practices. The fusion of this multimodal data helps enhance the prediction accuracy of fruit growth anomalies and provides robust support for fruit quality management and agricultural production decision making. The network implementation explicitly leverages the Transformer architecture to model cross-modal interactions rather than applying simple rules to a 2D matrix. Specifically, both image and text features are projected into a shared feature space, where multi-head self-attention (MHSA) is employed to dynamically learn the dependencies between different modalities. First, the network extracts basic features from the image using convolutional layers, generating an initial feature map. This feature map then passes through the edge-aware module, which is designed to enhance the network's focus on the fruit's edge areas, especially when the fruit has a complex shape or surface damage. The edge-aware module further strengthens the edge features in the image by combining traditional edge detection algorithms (such as Sobel or Canny) with convolutional operations. The module computes the edge feature map of an image and merges it with the feature map generated by the convolution layers, enhancing the sensitivity to fruit contours, cracks, and other detailed regions and thereby improving the segmentation accuracy. Next, the enhanced feature map enters the Transformer module for global information modeling. Unlike conventional convolution-based approaches, which rely on local receptive fields, the Transformer module effectively models long-range dependencies within an image through self-attention mechanisms. Given an input feature representation $F$, the self-attention operation is computed as follows:

$$Z = \text{softmax}\left(\frac{(W_Q F)(W_K F)^T}{\sqrt{d_k}}\right) W_V F, \tag{9}$$

where $W_Q$, $W_K$, and $W_V$ are the query, key, and value projection matrices, respectively, and $d_k$ is the dimensionality of the key vectors. By leveraging this global attention mechanism, the network learns contextual relationships across different regions of an image, particularly addressing cases where fruit morphology spans multiple spatial regions. In addition to image data, the network's input also includes textual data related to fruit growth. The NLP module parses and analyzes agricultural text data (such as climate records, planting records, and expert notes), providing additional support for predicting fruit growth anomalies. To ensure a seamless fusion of text and image data, textual information is encoded using a Transformer-based embedding model, such as BERT or a domain-specific language model. This process converts textual descriptions into dense feature vectors, which are then aligned with the visual embeddings via cross-attention layers. The NLP module employs a Transformer structure to process the text data, extracting key information related to fruit growth such as climate changes, fertilization management, and environmental factors. This information, combined with the image data, provides more comprehensive background knowledge for the network, helping the model better understand anomalies in the fruit's growth process while identifying its phenotype features. To further refine multimodal interactions, a cross-attention mechanism is incorporated, where text features serve as queries and image features serve as keys and values, thereby guiding the visual representation learning process based on domain knowledge. Ultimately, the edge transformer segmentation network, through the joint processing of image and text data, is capable of providing high-precision and robust segmentation results for fruit phenotype analysis and growth anomaly detection. Compared with traditional methods, the network design presented in this study fully accounts for the fruit's morphological features and surface conditions and the influence of external environments. This design, particularly when dealing with fruits with fuzzy edges, damage, or irregular shapes, demonstrates stronger accuracy and robustness. The multimodal data fusion approach not only enhances the accuracy of fruit quality assessment but also provides powerful technical support for intelligent decision making in agricultural production.

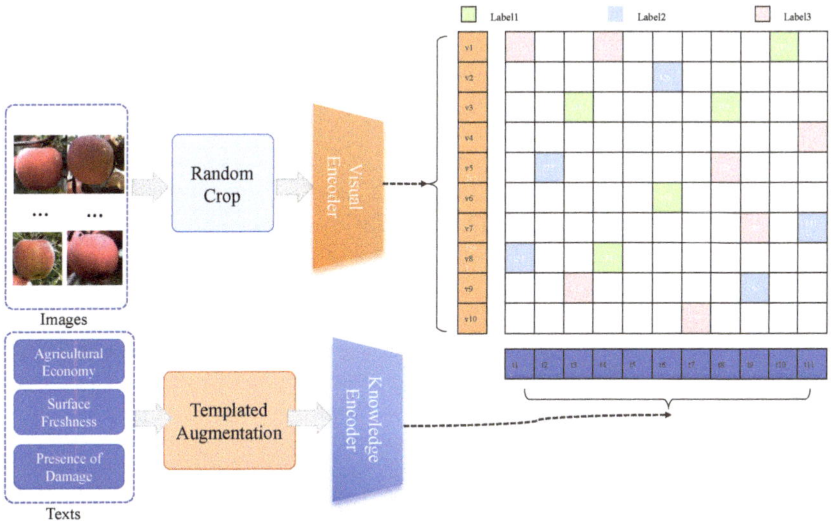


**Figure 4.** Architecture of edge transformer segmentation network.

### 3.5.2. Edge Attention Mechanism

The proposed edge attention mechanism module is an extension of the traditional self-attention mechanism, with a particular focus on edge information in the image to enhance the network's sensitivity to the boundary regions of an apple. The goal of this module is to introduce an edge-aware mechanism which increases the network's attention to the fruit's boundary regions, thereby optimizing segmentation results, particularly when fruit shapes are complex, the surface is damaged, or the boundaries are blurred. Compared with the traditional self-attention mechanism, the edge attention mechanism enables the network to prioritize the learning and enhancement of edge features during training, ensuring accurate segmentation in complex fruit shapes and damaged regions, as shown in Figure 5.

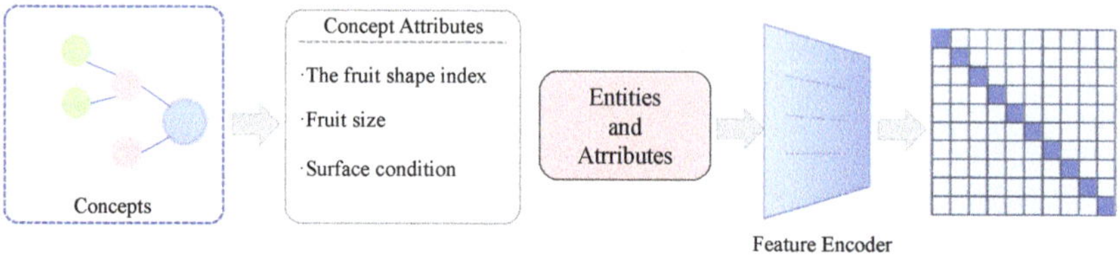


**Figure 5.** Flowchart of edge attention mechanism.

In the design of the edge attention mechanism, the core idea of the network architecture is to combine the advantages of the standard self-attention mechanism with the edge-aware module. Specifically, the network first uses traditional convolutional layers to extract feature maps and then employs edge detection algorithms to generate an edge map, followed by a weighted self-attention mechanism to increase the network's focus on the edge regions. The design of the number of layers in the edge attention mechanism includes multiple layers of self-attention mechanisms, with each layer containing a multi-head self-attention module and a feedforward neural network. The width and height of each layer remain consistent, ensuring spatial consistency of the feature maps across different layers. The specific design of the network is as follows. The input image size is $H \times W \times C$, where $H$ is the image height, $W$ is the image width, and $C$ is the number of input channels. In the self-attention layers, the output feature map remains $H \times W \times C$ and is further enhanced in the edge-aware module, improving the edge regions of the feature map. To enhance the self-attention mechanism with edge awareness, an edge-weighting function $E(i, j)$ is introduced, dynamically adjusting the attention computation such that pixels in the edge regions receive higher weights. The edge-enhanced attention is formulated as follows:

$$Z_{\text{edge}} = \text{softmax}\left(\frac{(W_Q F)(W_K F)^T}{\sqrt{d_k}}\right) W_V (F \cdot E). \tag{10}$$

where $E$ represents the edge feature map generated by the edge detection module, and its influence is adjusted through learnable weights. This formulation ensures that the network focuses more on fruit contours, even when dealing with irregular shapes or blurred edges, thereby improving segmentation precision. The parameters for each layer are as follows. The dimension of each self-attention head is $C/N$, where $N$ is the number of heads, typically set to eight. The output from each layer is processed through a feedforward neural network, with the width and height of the feature map maintained at $H \times W$ and the number of channels remaining being $C$. To incorporate edge information, a weighting mechanism is designed by combining the edge feature map $E$ with the feature map $F$ to enhance the attention weights on the edge regions. The specific weighted calculation is given by

$$\text{Edge Attention} = \frac{\exp(\text{Edge Feature})}{\sum_i \exp(\text{Edge Feature}_i)}, \tag{11}$$

where Edge Feature refers to the edge features generated by the edge detection module and $E$ represents the edge regions of the image. Through this mechanism, the edge attention mechanism prioritizes modeling the fruit's edges, ensuring segmentation precision, particularly in regions where the fruit shape is complex or damaged. The edge attention mechanism enhances the network's attention to the boundaries and surface damage of the fruit, while the edge transformer segmentation network handles extraction of the overall phenotype features from an image. These two modules work collaboratively in the segmentation process, improving segmentation precision. By enhancing the network's sensitivity to edge regions, the edge attention mechanism provides higher segmentation accuracy in detailed image parts, particularly in edges and damage areas. Meanwhile, the edge transformer segmentation network focuses on processing global information and fruit morphological features. This design ensures that when handling complex scenarios and fruit with detailed features, the model can perform segmentation with higher precision, especially when fruit surfaces exhibit cracks, spots, or rot, while still maintaining high recognition accuracy. Therefore, the edge attention mechanism not only improves the model's performance in segmentation tasks, but also, by combining edge-aware mechanisms with global context modeling, greatly enhances the model's adaptability and robustness in fruit phenotype analysis.

#### 3.5.3. Edge Loss Function

The proposed edge loss function in this study is a novel loss function designed to optimize edge accuracy in image segmentation tasks, especially for fruit phenotype analysis. Traditional loss functions, such as cross-entropy loss and Dice loss, are effective in optimizing models in most cases. However, they typically overlook the details of the edge regions, especially when dealing with complex shapes, blurred boundaries, or damaged areas, leading to unclear segmentation results at the boundaries. To address this issue, the edge loss function introduces an edge-aware mechanism which assigns more weight to the edge regions of an image, enabling finer segmentation. The core idea of the edge loss function is to incorporate specific optimization for the edge regions on top of traditional loss functions. Traditional loss functions are usually optimized at the pixel level across the entire image, neglecting the importance of edge regions in image segmentation. In fruit phenotype analysis, accurate segmentation of the fruit's contours, surface damage, and deformed areas is crucial for the quality of the results. The edge loss function, by weighting the loss of the edge regions, ensures that the model focuses more on the boundary areas during training, thereby improving segmentation precision. The mathematical formula for edge loss function can be expressed as follows:

$$L_{\text{Edge}} = \sum_{i=1}^{N} \left( |\hat{y}_i - y_i| \cdot \mathbb{K}_{\text{edge}}(i) \right), \tag{12}$$

where $\hat{y}_i$ is the predicted value, $y_i$ is the ground truth value, $N$ is the total number of pixels, and $\mathbb{K}_{\text{edge}}(i)$ is the indicator function. When pixel $i$ belongs to the edge region, $\mathbb{K}_{\text{edge}}(i) = 1$; otherwise, it is zero. This weighted loss allows the network to place more learning emphasis on the fruit's edge regions while reducing overemphasis on non-edge areas, resulting in more precise boundary segmentation. In traditional loss functions, such as cross-entropy loss, the formula is

$$L_{\text{CE}} = -\sum_{i=1}^{N} (y_i \log \hat{y}_i + (1 - y_i) \log(1 - \hat{y}_i)) \tag{13}$$

These loss functions typically optimize all pixels with equal weight, ignoring the uniqueness of edge regions in segmentation. The edge loss function, on the other hand, addresses this by adding specialized weighting for the edge areas, allowing the network to focus more on precise segmentation of the edges, especially when handling complex fruit shapes and surface damage. This design significantly improves the recognition accuracy of edge parts, particularly when fruit surfaces exhibit cracks, spots, or rot, preventing mis-segmentation due to blurred edges. The integration of the edge attention mechanism and edge loss function within the edge transformer segmentation network plays a critical role in fruit phenotype analysis and anomaly detection. The Transformer-based segmentation model captures both global context and fine-grained local details, while the edge-specific enhancements ensure that boundary information is accurately preserved and learned. Compared with traditional segmentation methods which rely solely on pixel-wise classification, this approach effectively models complex shape variations and enables precise contour extraction, making it particularly useful in agricultural applications where fruit shape and surface characteristics are crucial quality indicators. By incorporating both attention-based edge enhancement and an edge-sensitive loss function, the proposed approach not only improves the segmentation accuracy but also enhances robustness in real-world agricultural scenarios. The ability to distinguish subtle fruit defects and deformations makes this method valuable for automated quality assessment, early disease detection, and optimized resource management in smart agriculture.

### *3.6. Experimental Design*

#### 3.6.1. Hardware and Software Platforms

In this study, the choice and configuration of the hardware and software platform played a critical role in ensuring efficient execution of the experiments and the reliability of the results. On the hardware side, an NVIDIA A100 GPU was used, which is designed specifically for artificial intelligence and high-performance computing. Based on the Ampere architecture, the A100 supports multi-precision computations (including FP64, FP32, TF32, and FP16), with up to 6912 cores and 40 GB or 80 GB of memory, providing exceptional data processing capabilities. During large-scale data training and deep learning model execution, the A100 GPU significantly accelerates computation and supports parallel processing of multiple tasks. The experimental platform also includes a high-performance CPU, more than 256 GB of memory, and high-speed NVMe solid-state drives, ensuring efficient data loading, processing, and storage. For the software platform, the experiment was run on a Linux operating system, specifically Ubuntu 20.04 LTS, which is widely used for its stability and good support for deep learning frameworks. The deep learning models were developed and trained using the PyTorch framework version 1.12.0, with CUDA 11.6 and cuDNN 8.3 installed to fully leverage GPU acceleration. The Adam optimizer was chosen for its fast convergence and adaptability, making it one of the mainstream optimization algorithms in deep learning. The learning rate was set to 0.001, which was determined through several experimental adjustments to be the optimal value, ensuring quick convergence without oscillation. Additionally, the OpenCV and Albumentations libraries were used for efficient data preprocessing and augmentation, while model performance evaluation and visualization were performed using tools such as Matplotlib and Seaborn. The entire experimental environment was deployed through Docker containerization, which not only improved the reproducibility of the experiments but also facilitated cross-platform migration.

#### 3.6.2. Baseline Models

To comprehensively assess the performance of the proposed method, several classic deep learning models were chosen as baseline models, including Tiny-Segformer [23], Mask R-CNN [33], UNet [30], UNet++ [46], and DeepLabV3+ [31]. These models represent different technological directions and architectural characteristics in the field of image segmentation. Tiny-Segformer is a lightweight Transformer architecture which combines an efficient self-attention mechanism with convolution operations, maintaining computational efficiency while offering strong feature extraction capabilities, making it particularly suitable for resource-constrained scenarios. Mask R-CNN is a dual-task model based on object detection and segmentation capable of generating pixel-level segmentation masks in addition to bounding box detection. Its loss function includes the classification loss $L_{\text{cls}}$, bounding box regression loss $L_{\text{bbox}}$, and segmentation loss $L_{\text{mask}}$. Both UNet and its improved version, UNet++, use an encoder-decoder structure at their core, integrating multi-scale features through skip connections. These architectures are particularly suitable for fine-grained segmentation tasks in medical and agricultural imaging, with UNet++ further enhancing the network's expressive power by redesigning the skip connection modules. DeepLabV3+ uses dilated convolution (atrous convolution) and the Atrous Spatial Pyramid Pooling (ASPP) module to effectively capture multi-scale contextual information, with the loss function typically based on pixel-level cross-entropy.

To ensure a fair comparison, all baseline models were trained and tested on the same dataset, using the same image data for phenotype feature extraction and growth anomaly detection. However, since most of these models (such as UNet, UNet++, Mask R-CNN, and DeepLabV3+) are designed primarily for image-based segmentation tasks, they were not originally built to process multimodal information. Therefore, for these models, only the image input was utilized, without directly integrating meteorological data or other agricultural textual information. In contrast, the proposed method incorporates both image and textual data, leveraging a dedicated NLP module to process agricultural knowledge (such as planting records and meteorological data) and fusing it with visual features through an attention-based multimodal learning approach. To maintain a fair experimental set-up, Tiny-Segformer, which is a Transformer-based segmentation model, was extended with an NLP component similar to the one in the proposed system. However, due to its original lightweight design, its capacity for processing and integrating textual information remains more limited than the proposed method. These baseline models provide a reference standard for performance comparison in this study and help thoroughly verify the effectiveness of the proposed method.

#### 3.6.3. Evaluation Metrics

In this study, several evaluation metrics were used to comprehensively assess the model's performance, including the precision, recall, accuracy, and mean intersection over union (mIoU). These metrics measure the model's performance in fruit phenotype analysis and anomaly detection from different dimensions. Precision measures the proportion of true positive samples among all samples predicted to be positive, focusing on the correctness of the predictions, which is particularly important in high-precision scenarios. Recall measures the proportion of true positive samples which were correctly predicted to be positive, reflecting the model's ability to capture positive samples and serving as an important metric for evaluating false negatives. Accuracy represents the proportion of correctly classified samples among all predictions, suitable for evaluating the overall performance of the model. The mIoU, commonly used in semantic segmentation tasks, calculates the ratio of the intersection to the union between the predicted and ground truth regions, averaging this ratio across all categories to measure the global consistency of the model's segmentation results. The mathematical definitions of these evaluation metrics are as follows:

$$\text{Precision} = \frac{\text{TP}}{\text{TP} + \text{FP}}, \tag{14}$$

$$\text{Recall} = \frac{\text{TP}}{\text{TP} + \text{FN}}, \tag{15}$$

$$\text{IoU} = \frac{\text{Prediction} \cap \text{Ground Truth}}{\text{Prediction} \cup \text{Ground Truth}}, \tag{16}$$

$$\text{mIoU} = \frac{1}{C}\sum_{c=1}^{C} \text{IoU}_c, \tag{17}$$

where TP represents true positives, FP represents false positives, FN represents false negatives, TN represents true negatives, ∩ denotes the intersection, ∪ denotes the union, $C$ is the total number of classes, and $\text{IoU}_c$ is the intersection over union for class $c$.

## 4. Results and Discussion

### 4.1. Experimental Results of Phenotype Feature Extraction Models

The experimental design presented in this study aims to evaluate the performance of different deep learning models in the task of extracting fruit phenotype features. By comparing the precision, recall, accuracy, and mIoU metrics of various models, the study analyzed the advantages and shortcomings of each model in fine segmentation tasks, providing a theoretical foundation for subsequent model optimization. The models used in the experiments included UNet, Mask R-CNN, DeeplabV3+, UNet++, Tiny-Segformer, and the proposed method. Through these comparative experiments, the influence of the model architecture, loss functions, and optimization strategies on the accuracy of phenotype feature extraction could be deeply understood, offering theoretical support for intelligent fruit analysis in agricultural production.

As shown in Table 2, the experimental results demonstrate different levels of performance across all models in terms of precision, recall, accuracy, and the mIoU. The UNet model achieved a precision of 0.84, recall of 0.82, accuracy of 0.83, and mIoU of 0.80, indicating good performance in basic segmentation tasks but with room for improvement in handling finer details. Compared with UNet, Mask R-CNN showed improvements in its precision and recall, achieving scored of 0.86 and 0.83, respectively, with an accuracy and mIoU of 0.85 and 0.83, respectively, indicating better handling of object boundaries and details in the instance segmentation task. DeeplabV3+ introduced atrous convolution and spatial pyramid pooling modules, which enhanced the model's ability to capture multi-scale features, with precision, recall, and mIoU scores of 0.89, 0.86, and 0.86, respectively, demonstrating an advantage in processing multi-scale contextual information. UNet++ further improved the precision to 0.90, the recall to 0.88, and the mIoU to 0.87, showing that its enhanced skip connections improved detail recovery. Tiny-Segformer, a lightweight Transformer architecture, demonstrated strong feature extraction capabilities with efficient self-attention and convolution operations, achieving precision and recall scores of 0.92 and 0.89, respectively, and an mIoU of 0.89, indicating that the model could provide strong feature extraction while maintaining computational efficiency. The proposed method outperformed all other models, with a precision of 0.95, recall of 0.91, accuracy of 0.93, and mIoU of 0.92, demonstrating that the model, through combining edge perception mechanisms and global contextual modeling, provided higher precision and robustness in the complex task of fruit phenotype feature extraction. From a theoretical perspective, the architectures of the models significantly influenced the experimental results. Both UNet and UNet++ employ encoder-decoder structures and fuse multi-scale features through skip connections, but UNet++ further enhances network expressiveness through improved skip connections, leading to better performance in detail recovery. Mask R-CNN, a dual-task model for object detection and segmentation, not only provides bounding box detection but also generates pixel-level segmentation masks, which contribute to better segmentation results when handling object boundaries and complex structures. DeeplabV3+ effectively captures multi-scale contextual information through atrous convolution and spatial pyramid pooling modules, which is particularly advantageous when dealing with complex backgrounds and large objects. The lightweight Transformer architecture of Tiny-Segformer, which combines efficient self-attention mechanisms with convolution operations, enables the model to extract strong global features while maintaining computational efficiency, which is why it achieves high precision. The proposed method combines the advantages of these models by incorporating edge perception mechanisms and global contextual information modeling, enabling more precise capture of fruit contours, surface damage, and other detailed features, which is why it outperformed the other models across all evaluation metrics. These results highlight the significant impact of the model structure, loss functions, and optimization strategies on segmentation performance, especially in complex scenarios. The models which incorporated edge information and a global context had greater robustness and accuracy.

**Table 2.** Experimental results of phenotype feature extraction models.

| Model | Precision | Recall | Accuracy | mIoU |
|---|---|---|---|---|
| UNet | 0.84 | 0.82 | 0.83 | 0.80 |
| Mask R-CNN | 0.86 | 0.83 | 0.85 | 0.83 |
| DeeplabV3+ | 0.89 | 0.86 | 0.87 | 0.86 |
| UNet++ | 0.90 | 0.88 | 0.99 | 0.87 |
| Tiny-Segformer | 0.92 | 0.89 | 0.91 | 0.89 |
| Proposed Method | 0.95 | 0.91 | 0.93 | 0.92 |

### *4.2. Experimental Results of Growth Anomaly Recognition Models*

The design of this experiment aims to evaluate the performance of various deep learning models in the task of growth anomaly recognition, particularly in identifying abnormal conditions during fruit growth, such as pest damage and cracks. As shown in Table 3, the experiment compared the performance of different models based on various metrics to analyze their advantages and limitations in handling growth anomalies and to provide a theoretical basis for practical applications in fruit quality monitoring and pest warning systems.

**Table 3.** Experimental results of growth anomaly recognition models.

| Model | Precision | Recall | Accuracy | mIoU |
|---|---|---|---|---|
| UNet | 0.82 | 0.80 | 0.81 | 0.79 |
| Mask R-CNN | 0.84 | 0.82 | 0.83 | 0.81 |
| DeeplabV3+ | 0.87 | 0.84 | 0.85 | 0.82 |
| UNet++ | 0.89 | 0.87 | 0.88 | 0.84 |
| Tiny-Segformer | 0.91 | 0.88 | 0.90 | 0.86 |
| Proposed Method | 0.93 | 0.90 | 0.91 | 0.89 |

From the experimental results, it is evident that all models exhibited varying degrees of performance in the growth anomaly recognition task. The UNet model achieved a precision of 0.82, recall of 0.80, accuracy of 0.81, and mIoU of 0.79, demonstrating basic performance. Compared with UNet, Mask R-CNN showed improvements in precision, recall, and mIoU, with values of 0.84, 0.82, and 0.81, respectively. This indicates that its dual-task structure (object detection and instance segmentation) played an active role in recognizing fruit growth anomalies. DeeplabV3+, with the incorporation of atrous convolution and spatial pyramid pooling modules, showed an advantage in handling multi-scale contextual information, achieving a precision of 0.87, recall of 0.84, and mIoU of 0.82. This suggests that DeeplabV3+ performs better than the previous models in complex scenarios. UNet++, with its improved skip connection module, demonstrated excellent performance in terms of both precision (0.89) and recall (0.87), with an mIoU of 0.84, confirming its advantage in detail recovery and multi-scale information fusion. Tiny-Segformer, a lightweight Transformer-based architecture, combined self-attention mechanisms and convolution operations, achieving a precision of 0.91, recall of 0.88, and mIoU of 0.86, indicating its powerful feature extraction ability and global information modeling capabilities. The proposed method outperformed all other models in all metrics, with a precision of 0.93, recall of 0.90, accuracy of 0.91, and mIoU of 0.89, demonstrating that combining edge perception mechanisms and global information modeling significantly improves recognition accuracy and robustness in handling complex growth anomaly scenarios. From a theoretical analysis perspective, the architectural characteristics of the models directly influenced the experimental results. Both UNet and UNet++ use an encoder-decoder structure and fuse multi-scale features through skip connections. However, UNet++ further optimizes skip connections, improving its ability to recover details, which is why it performed better in growth anomaly recognition. Mask R-CNN, with its dual-task structure for object detection and instance segmentation, can simultaneously precisely segment a fruit's location and area, making it superior to UNet in handling anomalies with clear boundaries. DeeplabV3+ benefits from atrous convolution and spatial pyramid pooling modules, enabling it to capture richer contextual information in multi-scale contexts and thereby improving its ability to recognize complex backgrounds and irregular anomalies. Tiny-Segformer, through the combination of efficient self-attention mechanisms and convolution operations, excels in feature extraction and global information modeling, allowing it to better capture long-range dependencies, which contributed to its high precision and recall scores. The proposed method introduces an edge perception mechanism, enabling the network to focus more on fruit edges and anomaly regions. By combining this mechanism with global information modeling, the model effectively improves performance in complex growth anomaly scenarios. The edge perception mechanism strengthens the precise recognition of anomaly regions, while the Transformer architecture enhances the understanding of the global context, allowing the model to maintain high precision when dealing with complex fruit shapes and surface damage. These experimental results suggest that optimizing the model architecture, enhancing feature extraction capabilities, and introducing edge perception mechanisms are crucial for improving the accuracy and robustness of growth anomaly recognition.

### *4.3. Accuracy Results of Different Models for Various Phenotype Features*

As shown in Table 4, the design of this experiment aimed to evaluate the performance of various deep learning models in extracting and recognizing fruit phenotype features such as the fruit shape index, fruit size, color, and surface state. By comparing the accuracy of different models on these phenotype features, the goal was to analyze the performance differences among models when handling various fruit features and to investigate the impact of different model architectures on feature extraction and recognition accuracy. The analysis of the experimental results provides theoretical support for selecting the optimal model in fruit phenotype analysis tasks and reveals the role of model architecture and optimization strategies in enhancing fruit feature extraction accuracy.

**Table 4.** Accuracy results of different models for various phenotype features.

| Model | Fruit Shape Index | Fruit Size | Color | Surface State |
|---|---|---|---|---|
| UNet | 0.81 | 0.82 | 0.83 | 0.84 |
| Mask R-CNN | 0.83 | 0.84 | 0.85 | 0.87 |
| DeeplabV3+ | 0.85 | 0.87 | 0.87 | 0.89 |
| UNet++ | 0.86 | 0.89 | 0.89 | 0.91 |
| Tiny-Segformer | 0.90 | 0.91 | 0.92 | 0.92 |
| Proposed Method | 0.92 | 0.93 | 0.94 | 0.95 |

The experimental results show significant differences in the performance of all models for different phenotype features. The UNet model exhibited relatively lower accuracy scores across all phenotype features, with values of 0.81, 0.82, 0.83, and 0.84, indicating its basic ability in fruit feature extraction but struggles in capturing complex feature relationships due to its relatively simple architecture. Mask R-CNN showed improvement over UNet, especially in recognizing fruit surface states, with an accuracy of 0.87 compared with 0.84 for UNet. This suggests that the model's dual-task structure (object detection and instance segmentation) enhances boundary detail extraction in the instance segmentation task. DeeplabV3+ incorporates atrous convolution and spatial pyramid pooling modules, which better handle multi-scale features, leading to improved recognition accuracy for the fruit size and surface state, with values of 0.87 and 0.89, respectively. UNet++ improves upon the skip connection in the encoder-decoder structure, further enhancing the accuracy of all phenotype features, especially the fruit size and surface state, with accuracy values of 0.89 and 0.91, respectively, indicating its advantage in handling complex structural features. Tiny-Segformer, a lightweight Transformer-based model, significantly improved the accuracy across all features, especially fruit color and surface state, reaching 0.92 for both, demonstrating the advantage of the self-attention mechanism in global feature modeling. The proposed method outperformed all other models in feature extraction, with accuracy values of 0.92, 0.93, 0.94, and 0.95 for the various phenotype features, especially in surface state recognition. By combining edge perception mechanisms with global contextual information modeling, the proposed model significantly enhanced its detail capture, allowing it to more precisely identify the surface features of the fruit. From a theoretical analysis perspective, the differences in the experimental results were directly influenced by the model architectures. Both UNet and UNet++ adopt an encoder-decoder structure and fuse multi-scale features through skip connections, but UNet++ further optimizes skip connections, enhancing its ability to integrate multi-scale features, which led to better performance in fruit phenotype feature extraction. Mask R-CNN, as a multi-task learning model, leverages object detection mechanisms to effectively extract object boundaries and generate accurate segmentation masks. This ability enabled the model to achieve better results when handling surface state and complex fruit shape features. DeeplabV3+ uses atrous convolution and spatial pyramid pooling modules, which expand the receptive field and capture multi-scale contextual information, giving it an advantage in handling large objects and complex backgrounds. Tiny-Segformer combines self-attention mechanisms with convolution operations, allowing it to better capture long-range dependencies and efficiently utilize global information during feature extraction, resulting in improved performance in terms of fruit color and shape recognition. The proposed method, with the introduction of an edge perception mechanism, not only improved the segmentation accuracy of the surface state but also effectively captured subtle changes in fruit shape, demonstrating strong robustness and high precision when identifying complex fruit shapes and surface damage. These results indicate that model architecture innovation, enhanced feature extraction capabilities, and effective utilization of edge information are crucial factors in improving the accuracy of phenotype feature extraction.

### 4.4. Ablation Experiment with Different Attention for Phenotype Features

As shown in Table 5, the design of this experiment aimed to evaluate the impact of different attention mechanisms on model performance, particularly in the precise extraction of features and recognition of anomalies in fruit phenotype analysis tasks. Specifically, the experiment compared the performance of the standard self-attention mechanism, the channel and spatial attention mechanism (CBAM), and the proposed improved attention mechanism across various metrics. The objective of these comparisons was to clarify the effectiveness of different attention mechanisms in capturing image features, thereby verifying whether the proposed method could effectively enhance model performance, particularly in the tasks of fruit phenotype feature extraction and growth anomaly recognition.

**Table 5.** Ablation experiment with Different attention for phenotype feature.

| Model | Precision | Recall | Accuracy | mIoU |
|---|---|---|---|---|
| Standard Self-Attention | 0.76 | 0.72 | 0.74 | 0.71 |
| CBAM | 0.85 | 0.81 | 0.83 | 0.80 |
| Proposed Method | 0.95 | 0.91 | 0.93 | 0.92 |

From the experimental results, it can be observed that the standard self-attention mechanism exhibited relatively worse performance, with a precision of 0.76, recall of 0.72, accuracy of 0.74, and mIoU of 0.71. This suggests that while the standard self-attention mechanism can capture global information, its ability to focus on local features and details is limited. In comparison, the CBAM showed significant improvements across all metrics, with a precision of 0.85, recall of 0.81, accuracy of 0.83, and mIoU of 0.80. This indicates that the combination of channel and spatial attention mechanisms effectively enhanced the model's focus on different regions and features of an image, especially improving sensitivity to fruit features and anomaly areas. The proposed method, which combines edge perception mechanisms and global contextual information modeling, achieved the best results in terms of precision (0.95), recall (0.91), accuracy (0.93), and mIoU (0.92), demonstrating that the method effectively enhances feature capture in edge and complex regions, improving the precision of fruit phenotype feature extraction and growth anomaly recognition. From a theoretical perspective, the standard self-attention mechanism models global information by calculating the similarity between image features, but it lacks specialized attention to key local regions. As a result, it may overlook edge information and details when processing complex image features. The CBAM improves upon the standard self-attention mechanism by introducing attention mechanisms in both the channel and spatial dimensions, enhancing the model's attention to different channels and spatial regions. This improvement effectively boosts the model's ability to extract features, particularly in handling the details of fruit shapes and surface states. The proposed method further refines this approach by introducing an edge perception mechanism, which enables the network to focus on the edge regions of a fruit during training. This is especially crucial when dealing with surface damage or deformities, as it allows for precise identification of these complex regions. Mathematically, the standard self-attention mechanism typically models global information by calculating the relationships between each pixel in the input feature map, but its expression of local details, particularly the surface details of a fruit, is weak. The CBAM, by introducing attention mechanisms in both the channel and spatial dimensions, not only allows the model to focus on global features but also enables it to weight the important regions of an image, enhancing its ability to recognize details. The proposed method improves this further by incorporating an edge perception mechanism, which focuses more attention on edge regions, maintaining high segmentation accuracy even in cases where a fruit shape is complex, the surface is damaged, or boundaries are unclear. This design illustrates the advantage of the self-attention mechanism in combining both local and global features, significantly enhancing the model's robustness and precision, especially when dealing with complex image tasks.

### 4.5. Application in Agricultural Economics

The model proposed in this study, by integrating agricultural knowledge with deep learning techniques, has made significant contributions to the agricultural economy. Through the precise extraction of fruit phenotypic features and the effective identification of growth anomalies, the model not only enhanced the ability to monitor fruit quality but also provided data support for quantifiable economic assessments in agricultural production. In the context of agricultural economics, the quality of fruit directly influences the market value and production efficiency. Therefore, accurate fruit quality assessment is crucial for improving agricultural productivity and economic benefits. By comprehensively analyzing the shape, surface condition, size, and color of a fruit, the model provides agricultural producers with detailed fruit quality information, enabling farmers to monitor crop growth in real time and take timely measures to optimize production processes and reduce losses. This model further contributes to economic optimization by reducing post-harvest losses through precise defect detection and classification, allowing for better sorting and market positioning of agricultural products.

Additionally, the incorporation of agricultural knowledge for anomaly detection allows the model to address the impact of environmental factors such as climate change and soil conditions on fruit growth, offering targeted warnings and recommendations. This not only helps improve the sustainability of agricultural production but also reduces the occurrence of pests and diseases, thus minimizing pesticide usage and promoting the development of green agriculture. By leveraging real-time phenotypic analysis, producers can adopt data-driven strategies to adjust cultivation practices dynamically, aligning resource investment with the predicted yield and market demand. Practical application of the model in agricultural economics has facilitated the optimization of resource allocation. By accurately identifying anomalies in fruit growth and pest occurrences, producers can precisely deploy fertilization, irrigation, and pest control measures, avoiding excessive resource use and further reducing production costs. The model's ability to integrate phenotypic analysis with environmental and economic parameters provides a foundation for cost-benefit assessments, allowing for more informed decision making regarding investment in precision farming technologies. In large-scale agricultural production, this intelligent fruit monitoring technology significantly increases labor productivity, reduces labor costs, and enhances the digitalization and automation levels of the agricultural supply chain, promoting the modernization of agriculture.

### 4.6. Future Work in Smart Agriculture

In future work, the focus will be placed on the practical application of the proposed model in real-world agricultural production. To achieve this, efforts will be directed toward optimizing the model for deployment on edge computing devices such as the Jetson Nano, enabling real-time processing with reduced computational complexity, as shown in Table 6.

Table 6. Performance on different hardware platforms.

| Hardware Platform | Precision | Recall | Accuracy | mIoU | FPS |
|---|---|---|---|---|---|
| GPU platform (baseline) | 0.95 | 0.91 | 0.93 | 0.92 | 47.1 |
| Jetson Nano (lightweight version) | 0.91 | 0.89 | 0.90 | 0.89 | 30.5 |
| Huawei P40 pro (lightweight version) | 0.88 | 0.85 | 0.85 | 0.83 | 18.3 |

By designing a lightweight version of the proposed edge transformer segmentation network, the feasibility of implementing the model in resource-constrained environments, such as automated agricultural machinery and intelligent monitoring systems, will be explored. This will allow for real-time fruit phenotype analysis and anomaly detection in the field, providing immediate feedback for decision making in agricultural management. Additionally, the integration of the model into unmanned aerial vehicles (UAVs) and robotic harvesting systems will be investigated to enhance precision agriculture practices. These developments will further bridge the gap between theoretical advancements and practical applications, ensuring that the proposed methodology contributes to improving agricultural efficiency, reducing resource waste, and supporting intelligent agricultural decision making in real-world scenarios.

## 5. Conclusions

With the rapid development of intelligent agriculture, efficiently and accurately assessing fruit growth status and quality in apple production has become a critical factor in enhancing agricultural productivity and economic benefits. This study aimed to propose a deep learning-based approach for apple phenotype analysis and growth anomaly recognition by integrating instance segmentation, NLP, and innovative attention mechanisms. The method addresses the limitations of traditional fruit quality detection and anomaly recognition techniques, providing robust technical support for intelligent agriculture and agricultural economic analysis.

The proposed approach introduces several innovations. First, a comprehensive method combining instance segmentation and NLP is presented. By integrating image analysis and text parsing, the method accurately extracts fruit phenotype features, such as the fruit shape, color, and surface condition, while simultaneously utilizing agricultural textual data, including meteorological information and cultivation records, to identify growth anomalies. This multimodal data fusion overcomes the limitations of traditional image-based methods, enabling a holistic improvement in the accuracy of fruit quality detection and anomaly prediction from multiple perspectives. Moreover, this study introduced innovative edge attention modules and edge loss mechanisms, which enhance the model's focus on fruit edge regions and refine its handling of abnormal areas. These advancements significantly improve the model's performance in scenarios involving complex fruit morphology, surface damage, and growth anomalies. Through these innovative designs, the proposed method not only enhances the precision of fruit phenotype feature extraction but also provides more accurate and reliable data support for agricultural production decision making. The experimental results demonstrate that the proposed approach achieved significant improvements in accuracy and offers new perspectives and technological support for agricultural economic analysis.

**Author Contributions:** Conceptualization, Z.W., W.C., C.H. and C.L.; data curation, Y.Z. and Z.Z.; formal analysis, Y.Z., Y.Y. and X.D.; funding acquisition, C.L.; investigation, Y.Y.; methodology, Z.W., W.C. and C.H.; project administration, C.L.; resources, Y.Z. and Z.Z.; software, Z.W., W.C., C.H. and X.D.; supervision, C.L.; validation, Z.Z.; visualization, Y.Y. and X.D.; writing—original draft, Z.W.,

W.C., C.H., Y.Z., Z.Z., Y.Y., X.D. and C.L. Z.W., W.C. and C.H. contributed evenly to this work. All authors have read and agreed to the published version of the manuscript.

**Funding:** This research was funded by The National Key Research and Development Program of China (2024YFC2607600).

**Data Availability Statement:** The data presented in this study are available on Github at https://github.com/user837498178/apple-agriculture (accessed on 26 January 2025).

**Conflicts of Interest:** The authors declare no conflicts of interest.

## References

1. Zhang, Y.; Cao, Y.F.; Huo, H.L.; Xu, J.Y.; Tian, L.M.; Dong, X.G.; Dan, Q.; Chao, L. An assessment of the genetic diversity of pear (*Pyrus* L.) Germplasm Resour. Based Fruit Phenotypic Trait. *J. Integr. Agric.* **2022**, *21*, 2275–2290. [CrossRef]
2. Medda, S.; Mulas, M. Fruit quality characters of myrtle (*Myrtus communis* L.) selections: Review of a domestication process. *Sustainability* **2021**, *13*, 8785. [CrossRef]
3. Zhang, Y.; Wa, S.; Sun, P.; Wang, Y. Pear defect detection method based on resnet and dcgan. *Information* **2021**, *12*, 397. [CrossRef]
4. Zhang, Y.; Lv, C. TinySegformer: A lightweight visual segmentation model for real-time agricultural pest detection. *Comput. Electron. Agric.* **2024**, *218*, 108740. [CrossRef]
5. Zhang, Y.; Liu, X.; Wa, S.; Chen, S.; Ma, Q. GANsformer: A detection network for aerial images with high performance combining convolutional network and transformer. *Remote Sens.* **2022**, *14*, 923. [CrossRef]
6. Cui, M.; Pham, M.D.; Hwang, H.; Chun, C. Flower development and fruit malformation in strawberries after short-term exposure to high or low temperature. *Sci. Hortic.* **2021**, *288*, 110308. [CrossRef]
7. Gu, W.; Bai, S.; Kong, L. A review on 2D instance segmentation based on deep neural networks. *Image Vis. Comput.* **2022**, *120*, 104401. [CrossRef]
8. Kazakova, M.A.; Sultanova, A.P. Analysis of natural language processing technology: Modern problems and approaches. *Adv. Eng. Res.* **2022**, *22*, 169–176. [CrossRef]
9. Bergen, L.; O'Donnell, T.; Bahdanau, D. Systematic generalization with edge transformers. *Adv. Neural Inf. Process. Syst.* **2021**, *34*, 1390–1402.
10. Zhao, P.; Wang, W.; Liu, H.; Han, M. Recognition of the agricultural named entities with multifeature fusion based on albert. *IEEE Access* **2022**, *10*, 98936–98943. [CrossRef]
11. Andriyanov, N. Combining Text and Image Analysis Methods for Solving Multimodal Classification Problems. *Pattern Recognit. Image Anal.* **2022**, *32*, 489–494. [CrossRef]
12. Zhang, Y.; Wa, S.; Liu, Y.; Zhou, X.; Sun, P.; Ma, Q. High-accuracy detection of maize leaf diseases CNN based on multi-pathway activation function module. *Remote Sens.* **2021**, *13*, 4218. [CrossRef]
13. Zhang, Y.; Wa, S.; Zhang, L.; Lv, C. Automatic plant disease detection based on tranvolution detection network with GAN modules using leaf images. *Front. Plant Sci.* **2022**, *13*, 875693. [CrossRef] [PubMed]
14. Luo, Z.; Yang, W.; Yuan, Y.; Gou, R.; Li, X. Semantic segmentation of agricultural images: A survey. *Inf. Process. Agric.* **2024**, *11*, 172–186. [CrossRef]
15. Anand, T.; Sinha, S.; Mandal, M.; Chamola, V.; Yu, F.R. AgriSegNet: Deep aerial semantic segmentation framework for IoT-assisted precision agriculture. *IEEE Sensors J.* **2021**, *21*, 17581–17590. [CrossRef]
16. Zhang, H.; Peng, Q. PSO and K-means-based semantic segmentation toward agricultural products. *Future Gener. Comput. Syst.* **2022**, *126*, 82–87. [CrossRef]
17. Su, D.; Kong, H.; Qiao, Y.; Sukkarieh, S. Data augmentation for deep learning based semantic segmentation and crop-weed classification in agricultural robotics. *Comput. Electron. Agric.* **2021**, *190*, 106418. [CrossRef]
18. Zhang, Y.; He, S.; Wa, S.; Zong, Z.; Liu, Y. Using generative module and pruning inference for the fast and accurate detection of apple flower in natural environments. *Information* **2021**, *12*, 495. [CrossRef]
19. Lin, X.; Wa, S.; Zhang, Y.; Ma, Q. A dilated segmentation network with the morphological correction method in farming area image Series. *Remote Sens.* **2022**, *14*, 1771. [CrossRef]
20. Zhou, X.; Chen, S.; Ren, Y.; Zhang, Y.; Fu, J.; Fan, D.; Lin, J.; Wang, Q. Atrous Pyramid GAN Segmentation Network for Fish Images with High Performance. *Electronics* **2022**, *11*, 911. [CrossRef]
21. Kaur, R.; Singh, S. A comprehensive review of object detection with deep learning. *Digit. Signal Process.* **2023**, *132*, 103812. [CrossRef]
22. Diwan, T.; Anirudh, G.; Tembhurne, J.V. Object detection using YOLO: Challenges, architectural successors, datasets and applications. *Multimed. Tools Appl.* **2023**, *82*, 9243–9275. [CrossRef]
23. Amit, Y.; Felzenszwalb, P.; Girshick, R. Object Detection. In *Computer Vision: A Reference Guide*; Springer: Berlin/Heidelberg, Germany, 2021; pp. 875–883.

24. Bharati, P.; Pramanik, A. Deep Learning Techniques—R-CNN to Mask R-CNN: A Survey. In *Computational Intelligence in Pattern Recognition: Proceedings of CIPR 2019*; Springer: Berlin/Heidelberg, Germany, 2020; pp. 657–668.
25. Jiang, P.; Ergu, D.; Liu, F.; Cai, Y.; Ma, B. A Review of Yolo algorithm developments. *Procedia Comput. Sci.* **2022**, *199*, 1066–1073. [CrossRef]
26. Khan, S.; AlSuwaidan, L. Agricultural monitoring system in video surveillance object detection using feature extraction and classification by deep learning techniques. *Comput. Electr. Eng.* **2022**, *102*, 108201. [CrossRef]
27. Ren, S.; He, K.; Girshick, R.; Sun, J. Faster R-CNN: Towards real-time object detection with region proposal networks. *IEEE Trans. Pattern Anal. Mach. Intell.* **2016**, *39*, 1137–1149. [CrossRef]
28. Mo, Y.; Wu, Y.; Yang, X.; Liu, F.; Liao, Y. Review the state-of-the-art technologies of semantic segmentation based on deep learning. *Neurocomputing* **2022**, *493*, 626–646. [CrossRef]
29. Strudel, R.; Garcia, R.; Laptev, I.; Schmid, C. Segmenter: Transformer for Semantic Segmentation. In Proceedings of the IEEE/CVF International Conference on Computer Vision, Virtual, 11–17 October 2021; pp. 7262–7272.
30. Ronneberger, O.; Fischer, P.; Brox, T. U-net: Convolutional Networks for Biomedical Image Segmentation. In Proceedings of the Medical Image Computing and Computer-Assisted Intervention–MICCAI 2015: 18th international Conference, Proceedings, Part III 18, Munich, Germany, 5–9 October 2015; Springer: Berlin/Heidelberg, Germany, 2015; pp. 234–241.
31. Chen, L.C.; Zhu, Y.; Papandreou, G.; Schroff, F.; Adam, H. Encoder-Decoder with Atrous Separable Convolution for Semantic Image Segmentation. In Proceedings of the European Conference on Computer Vision (ECCV), Munich, Germany, 8–14 September 2018; pp. 801–818.
32. De Boer, P.T.; Kroese, D.P.; Mannor, S.; Rubinstein, R.Y. A tutorial on the cross-entropy method. *Ann. Oper. Res.* **2005**, *134*, 19–67. [CrossRef]
33. He, K.; Gkioxari, G.; Dollár, P.; Girshick, R. Mask r-cnn. In Proceedings of the IEEE International Conference on Computer Vision, Venice, Italy, 22–29 October 2017; pp. 2961–2969.
34. Xu, S.; Shen, J.; Wei, Y.; Li, Y.; He, Y.; Hu, H.; Feng, X. Automatic plant phenotyping analysis of Melon (*Cucumis melo* L.) Germplasm Resour. Using Deep Learn. Methods Comput. Vision. *Plant Methods* **2024**, *20*, 166. [CrossRef] [PubMed]
35. Wang, C.; Du, P.; Wu, H.; Li, J.; Zhao, C.; Zhu, H. A cucumber leaf disease severity classification method based on the fusion of DeepLabV3+ and U-Net. *Comput. Electron. Agric.* **2021**, *189*, 106373. [CrossRef]
36. Min, B.; Ross, H.; Sulem, E.; Veyseh, A.P.B.; Nguyen, T.H.; Sainz, O.; Agirre, E.; Heintz, I.; Roth, D. Recent advances in natural language processing via large pre-trained language models: A survey. *ACM Comput. Surv.* **2023**, *56*, 1–40. [CrossRef]
37. Rezayi, S.; Liu, Z.; Wu, Z.; Dhakal, C.; Ge, B.; Dai, H.; Mai, G.; Liu, N.; Zhen, C.; Liu, T.; et al. Exploring new frontiers in agricultural nlp: Investigating the potential of large language models for food applications. *IEEE Trans. Big Data* **2024**. . [CrossRef]
38. Espinoza, S.; Aguilera, C.; Rojas, L.; Campos, P.G. Analysis of Fruit Images With Deep Learning: A Systematic Literature Review and Future Directions. *IEEE Access* **2023**, *12*, 3837–3859. [CrossRef]
39. Duan, J.L.; Lai, L.Q.; Yang, Z.; Luo, Z.J.; Yuan, H.T. Multi-feature language-image model for fruit quality image classification. *Comput. Electron. Agric.* **2024**, *227*, 109462. [CrossRef]
40. Apostolopoulos, I.D.; Tzani, M.; Aznaouridis, S.I. A general machine learning model for assessing fruit quality using deep image features. *AI* **2023**, *4*, 812–830. [CrossRef]
41. Ibtissam, B. Automatic Date Fruit Sorting System Based on Machine Learning and Visual Features. Ph.D. Thesis, University of Biskra, Biskra, Algeria, 2024.
42. Ahmed, A.A.; Reddy, G.H. A mobile-based system for detecting plant leaf diseases using deep learning. *AgriEngineering* **2021**, *3*, 478–493. [CrossRef]
43. Vishnoi, V.K.; Kumar, K.; Kumar, B.; Mohan, S.; Khan, A.A. Detection of apple plant diseases using leaf images through convolutional neural network. *IEEE Access* **2022**, *11*, 6594–6609. [CrossRef]
44. Khan, A.I.; Quadri, S.; Banday, S.; Shah, J.L. Deep diagnosis: A real-time apple leaf disease detection system based on deep learning. *Comput. Electron. Agric.* **2022**, *198*, 107093. [CrossRef]
45. Rahman, S.U.; Alam, F.; Ahmad, N.; Arshad, S. Image processing based system for the detection, identification and treatment of tomato leaf diseases. *Multimed. Tools Appl.* **2023**, *82*, 9431–9445. [CrossRef]
46. Zhou, Z.; Rahman Siddiquee, M.M.; Tajbakhsh, N.; Liang, J. Unet++: A Nested U-Net Architecture for Medical Image Segmentation. In Proceedings of the Deep Learning in Medical Image Analysis and Multimodal Learning for Clinical Decision Support: 4th International Workshop, DLMIA 2018, and 8th International Workshop, ML-CDS 2018, Held in Conjunction with MICCAI 2018, Proceedings 4, Granada, Spain, 20 September 2018; Springer: Berlin/Heidelberg, Germany, 2018; pp. 3–11.

agriculture

MDPI

*Article*

# Quality Grading of *Oudemansiella raphanipes* Using Three-Teacher Knowledge Distillation with Cascaded Structure for LightWeight Neural Networks

**Haoxuan Chen [1,2], Huamao Huang [2], Yangyang Peng [1], Hui Zhou [1], Haiying Hu [3,*] and Ming Liu [1,*]**

1 Guangdong Key Laboratory for New Technology Research of Vegetables, Vegetable Research Institute, Guangdong Academy of Agricultural Sciences, Guangzhou 510640, China; hohinchan@foxmail.com (H.C.); pengyangyang@gdaas.cn (Y.P.); zhouhui@gdaas.cn (H.Z.)

2 School of Physics and Optoelectronics, South China University of Technology, Guangzhou 510640, China; schhm@scut.edu.cn

3 School of Civil Engineering and Transportation, South China University of Technology, Guangzhou 510640, China

* Correspondence: cthyhu@scut.edu.cn (H.H.); liuming@gdaas.cn (M.L.)

**Abstract:** *Oudemansiella raphanipes* is valued for its rich nutritional content and medicinal properties, but traditional manual grading methods are time-consuming and labor-intensive. To address this, deep learning techniques are employed to automate the grading process, and knowledge distillation (KD) is used to enhance the accuracy of a small-parameter model while maintaining a low resource occupation and fast response speed in resource-limited devices. This study employs a three-teacher KD framework and investigates three cascaded structures: the parallel model, the standard series model, and the series model with residual connections (residual-series model). The student model used is a lightweight ShuffleNet V2 0.5x, while the teacher models are VGG16, ResNet50, and Xception. Our experiments show that the cascaded structures result in improved performance indices, compared with the traditional ensemble model with equal weights; in particular, the residual-series model outperforms the other models, achieving a grading accuracy of 99.7% on the testing dataset with an average inference time of 5.51 ms. The findings of this study have the potential for broader application of KD in resource-limited environments for automated quality grading.

**Keywords:** *Oudemansiella raphanipes*; quality grading; knowledge distillation; multi-teacher model

## 1. Introduction

China was the first country in the world to recognize and cultivate the edible mushroom, and it is now the world's largest producer, consumer, and exporter of this type of crop. In recent years, China's edible mushroom sector has consistently maintained a growth trend, and edible mushrooms have emerged as China's "fifth-largest crop" after grain, oil, vegetables, and fruits in terms of agricultural production [1]. *Oudemansiella raphanipes* is a well-known edible mushroom renowned for its rich nutritional content, which comprises proteins, amino acids, vitamins, and trace elements. It can also serve as a medicine, containing polyphenols, polysaccharides, flavonoids, and various other bioactive compounds that are beneficial for enhancing immunity, regulating body functions, repairing damaged organs, and providing analgesia and anti-inflammatory effects [2,3]. The increasing awareness of consumers concerning quality leads to significant price variations among different grades. However, traditional methods of artificial quality grading are time-consuming and labor-intensive, significantly limiting the economic benefits. Therefore, there is an urgent need for an algorithm capable of automatically assessing and grading the quality of *Oudemansiella raphanipes*.

With advancements in deep learning, convolutional neural networks (CNNs) have emerged as powerful tools for extracting high-dimensional features from objects for image classification [4]. In the context of mushroom quality grading, several studies have demonstrated their effectiveness. Tongkai Li et al. discussed a quality grading algorithm for *Oudemansiella raphanipes* using transfer learning and MobileNetV2, exhibiting a test accuracy of 98.75% [5]. Yanqiang Wu et al. proposed a size-grading method of antler mushrooms using YOLOv5 and PSPNet, achieving a detection accuracy of 94% [6]. Yinhua Zuo et al. introduced a quality grading model for *Pleurotus cristatus* based on an improved EfficientNet with a self-attention module, resulting in an average recognition accuracy of 91.5% [7]. Li Wang et al. proposed a grade recognition method for dried shiitake mushrooms using an improved VGG network (D-VGG), achieving a high classification accuracy of 96.21% [8]. Lei Shi et al. presented a lightweight grading detection method for Oyster mushrooms using OMC-YOLO, improved based on YOLOv8n, with an mAP50 value of 94.95% [9]. Ziyuan Wei et al. employed a convolutional autoencoder–support vector machine model for the quality classification of *Pleurotus eryngii*, achieving an accuracy of 91.58% [10].

However, it is essential to develop a model that balances classification accuracy, resource occupation, and response speed, especially for resource-limited devices. Although large-parameter models like VGGNet [11] and ResNet [12] offer high accuracy, they fall short in terms of resource efficiency and response speed. Conversely, small-parameter models such as Xception [13] and ShuffleNet [14,15] have smaller resource occupation and faster calculation speeds but may lack satisfactory classification accuracy. To address this challenge, knowledge distillation (KD) algorithms can be employed to enhance the accuracy of small-parameter models while maintaining their speed advantages [16,17]. KD is a technique in deep learning, aimed at transferring knowledge from large-parameter models to small-parameter models. The large-parameter model, typically characterized by high accuracy but significant resource occupation and slower response speed, is used as the teacher model, whereas the small-parameter model, typically featuring low resource occupation and fast response speed but lower accuracy, is used as the student model. Without KD, the student model learns knowledge solely from the ground truth (referred to as hard labels). With KD, in addition to the ground truth, the student model also learns knowledge from the teacher model, whose softmax output (referred to as soft labels) is utilized to train the student model. Consequently, the small-parameter student model can achieve similar accuracy to the large-parameter teacher model while maintaining low resource occupation and fast response speed. This approach was first proposed to utilize a single-teacher model [16,17]. Since multiple teachers can facilitate the transfer of a broader range of knowledge and mitigate the impact of a single low-quality teacher, multi-teacher KD has been proposed [18–26]. In this context, several teacher models are combined together with certain weights to form an ensemble model. As a result, the weights assigned to each teacher model directly influence the performance of the distilled student model. A simple scheme is to adopt equal weights for each teacher model [18]. However, equal weights fail to distinguish between high-quality and low-quality teacher models, potentially resulting in unsatisfactory performance. A variety of methods have been developed to adjust the weights, such as manually tuning weights [19] and adaptive weights [20–24]. To eliminate the necessity of selecting weights, there are other methods to select teacher models, including a voting strategy to choose the majority side [18], randomly selecting a teacher model [25], and applying reinforcement learning to filter out inappropriate teacher models [26]. However, some teacher models were discarded and consequently went to waste. The challenge lies in fully utilizing all three teacher models while mitigating the adverse effects posed by low-quality teacher models, which remains an unresolved issue.

In this paper, a three-teacher KD algorithm utilizing cascaded teacher models is proposed for quality grading of *Oudemansiella raphanipes*, with the aim of improving the grading accuracy while maintaining a low resource occupation and fast response speed. The main contributions of this paper are as follows:

(1) Three cascaded structures, including the parallel model, the standard series model, and the series model modified with residual connections (hereafter denoted as the residual-series model), were investigated.

(2) Compared with the student model distilled with a single teacher or an ensemble three-teacher model with equal weights (hereafter denoted as the equal-weight three-teacher model), these cascaded structures exhibit improved performance indices. In particular, the residual-series model outperforms the other models, achieving the highest grading accuracy.

(3) The superiority of the residual-series model is also demonstrated by comparison with recently published model compression techniques, through its ability to transfer to edge devices with limited computing resources, and by its generalization ability on the public mushroom dataset.

## 2. Materials and Methods

### 2.1. Overall Process

The overall process is sketched in Figure 1. The first step is to create datasets by taking photos of the manually graded *Oudemansiella raphanipes*. Subsequently, the training dataset is sent to a high-performance computing device for training using the KD algorithm, where a small-parameter student model learns knowledge from large-parameter teacher models. After that, the performance of the well-trained student model is evaluated on the testing dataset using the same training device or on a personal computer. Finally, the verified student model can be transferred to an edge device with limited computing resources for practical application. Ultimately, the edge device can use its camera to take photos of an *Oudemansiella raphanipes* and accurately grade its quality using the lightweight student model.

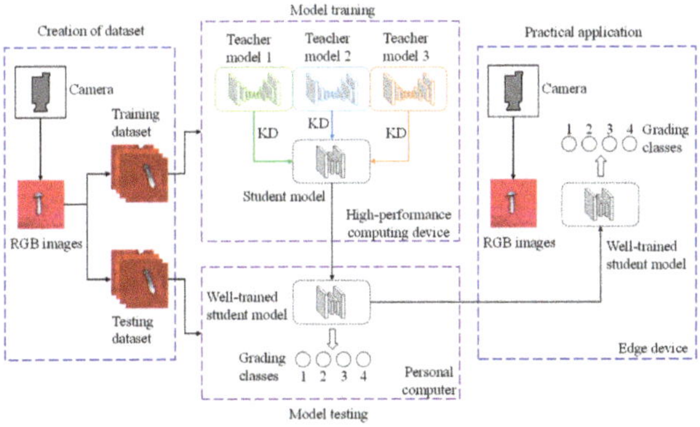


**Figure 1.** Sketch of the overall process.

### 2.2. Datasets

The *Oudemansiella raphanipes* were purchased from the farmer's market in April 2024, in Guangzhou, Guangdong Province, China. The datasets were collected in the same month at the Vegetable Research Institute, Guangdong Academy of Agricultural Sciences. According to the quality, including size, shape, color, and texture, the *Oudemansiella raphanipes* were divided into four classes: Class 1, Class 2, Class 3, and Class 4, based on the T/GVEAIA 009.3-2019 standard [27]. Three experts inspected each class and ensured that the classification was correct. Figure 2a presents the experimental setup for image acquisition. Each *Oudemansiella raphanipes* was put into a shielding box, in which a light source (four 45-watt fluorescent lamps with a color temperature of 5000 K) was illuminated from the top and a color camera (Canon 760D (Canon Inc., Tokyo, Japan) with EOS Utility Version 3.12.0 software) took the images with the vertical shooting height of 45 cm. The application scenario involves industrial automation selection equipment in a factory. A pile of *Oudemansiella raphanipes* is distributed onto an array of containers through a sieve, with each container holding one *Oudemansiella raphanipes*. The containers are transported one by one to the image acquisition module for quality grating, and then *Oudemansiella raphanipes* are collected into different boxes based on their quality grades. The resolution of each image is 3984 × 2656. All these images were cropped to a small square of 600 × 600 pixels, as shown in Figure 2b–e.

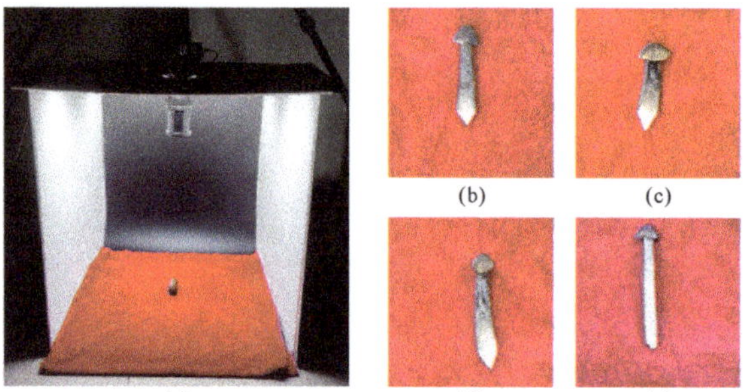

(a) (b) (c) (d) (e)

**Figure 2.** (**a**) Experimental setup for image acquisition. Images of *Oudemansiella raphanipes*. (**b**) Class 1, (**c**) Class 2, (**d**) Class 3, and (**e**) Class 4.

A total of 3000 square images of *Oudemansiella raphanipes* were evenly collected from the four classes. The images from each class were randomly divided into a training dataset and a testing dataset at a ratio of 8:2. Subsequently, the training datasets, originally consisting of 2400 images (600 images per class), was expanded to 38,400 images (9600 images per class) using an open-source tool, imgaug [28]. To enhance the diversity of the training dataset, the following augmentation techniques were employed, with their application order being random:

(1) Mirroring 50% of the images;

(2) Flipping 20% of the images left to right;

(3) Cropping a random subset of images with a cropping range of 90% to 100% of their original size;

(4) Applying affine transformations to 30% of the images, including the following: scaling the images to between 80% and 120% of their original size, translating them by ±20% of their dimensions, rotating them by ±45°, and applying shear transformations by ±16°;

(5) Randomly applying up to five of the following techniques to the images, including representing part of the image using superpixels, enhancing with Gaussian blur, mean blur, or median blur, adding an embossing effect, sharpening, performing edge detection and overlaying the binarized edges on the original image, adding Gaussian noise, setting 1% to 10% of the pixels to black or covering 3% to 15% of the image area with black squares that are 2% to 5% of the original image size, inverting the intensity of 1% to 10% of the pixels, randomly adding or subtracting an intensity value between −10 and 10 to each

pixel, multiplying the intensity value of each pixel by a random number between 0.5 and 1.5, halving or doubling the overall contrast of the image, converting the RGB image to grayscale and then adding the resultant values (multiplied by an alpha parameter) to the original image, and warping local regions of the image.

This approach can enhance the robustness of the deep learning model, enabling it to better handle real-world data. Figure 3 presents examples of augmented images. We also conducted a thorough examination to ensure the absence of redundancy in the augmented dataset. The testing dataset, comprising 600 images (150 images per class), remained in its original form without augmentation.

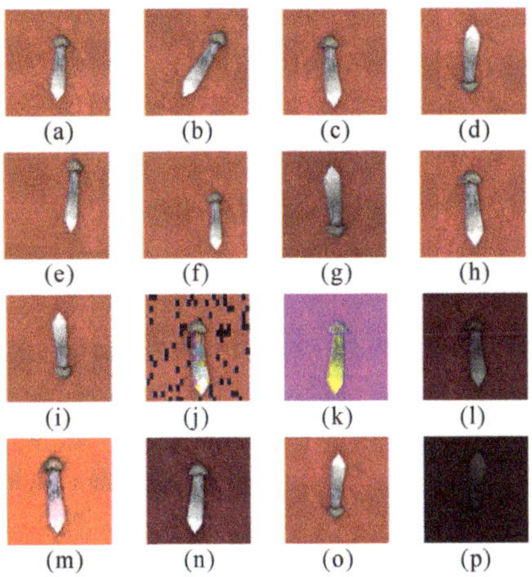


**Figure 3.** Examples of (**a**) original image and (**b**–**p**) augmented images. (**b**) Affine transformation, (**c**) left–right flipping, (**d**) up–down flipping, (**e**) adding average blur after affine transformation, (**f**) adding median blur after affine transformation, (**g**) sharpening after up–down flipping, (**h**) adding an embossing effect after left–right flipping, (**i**) adding Gaussian noise after up–down flipping, (**j**) setting pixels to black after left–right flipping, (**k**) inverting pixel intensities, (**l**) multiplying pixels by random numbers between 0.5 and 1.5 after rotation, (**m**) changing the contrast after left–right flipping, (**n**) transforming RGB into a gray image and then multiplying alpha to add to the original image, (**o**) adding Gaussian blur after up–down flipping, and (**p**) adding edge detection after up–down flipping.

### 2.3. *Three-Teacher Knowledge Distillation*

#### 2.3.1. Basic Structure

The basic structure of the KD module is illustrated in Figure 4. During the training process, the training dataset is sent to both the pre-trained teacher model and the student model in the KD module. The teacher model produces soft labels by converting its logit, $z_i$, to a probability, $q_i$, for the $i$-th class using the following softmax function [16]:

$$q_i = \frac{\exp(z_i/T)}{\sum_j \exp(z_j/T)} \quad (1)$$

where $T$ is the distillation temperature. A larger $T$ leads to a smoother probability distribution, causing the student model to learn more knowledge from negative labels. Similarly, the student model produces a probability, $p_i$, from its logit, $v_i$, for the $i$-th class using Equation (1). The Kullback–Leibler (KL) divergence from the probability distribution of the student model, $(p_1, p_2, p_3, p_4)$, to the soft labels of the teacher model, $(q_1, q_2, q_3, q_4)$, is adopted to quantify the match of the two models' predictions. Therefore, the soft loss (also named the distillation loss), $L_{soft}$, is computed by the following KL divergence equation:

$$L_{soft} = \sum_i q_i \log\left(\frac{q_i}{p_i}\right) \tag{2}$$

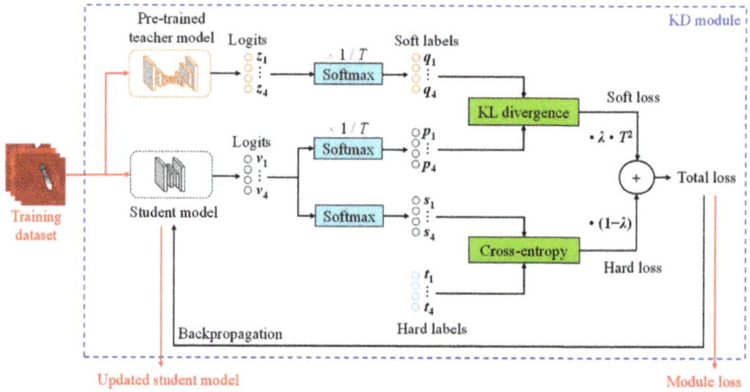


**Figure 4.** Basic structure of a KD module.

Equation (2) is a wrapped function in TensorFlow. In addition to learning knowledge from the teacher model, the student model also learns from the ground truth. The student model produces a probability, $s_i$, from its logit, $v_i$, for the $i$-th class using the standard softmax function [16]:

$$s_i = \frac{\exp(v_i)}{\sum_j \exp(v_j)} \tag{3}$$

Equation (3) is the degenerate form of Equation (1) at $T = 1$. The cross-entropy between the probability distribution of the student model, $(s_1, s_2, s_3, s_4)$, and the hard labels of the ground truth, $(t_1, t_2, t_3, t_4)$, is employed to represent the hard loss (also named the student loss), $L_{hard}$, using the following equation:

$$L_{hard} = -\sum_i t_i \log(s_i) \tag{4}$$

Equation (4) is a wrapped function in TensorFlow. As a result, after passing through a KD module, the total loss of the module, $L_{module}$, can be described as follows:

$$L_{module} = \lambda \bullet T^2 \bullet L_{soft} + (1-\lambda) \bullet L_{hard} \tag{5}$$

where $\lambda$ is the factor to balance the soft loss and hard loss. The student model updates its network weight parameters through backpropagation (BP) to minimize the total loss. By multiplying the coefficient of $T^2$ with $L_{soft}$, the impact of soft loss on the gradient during the BP process will increase, thereby ensuring the consistency of the amplitudes of both the soft loss and the hard loss throughout the training process.

### 2.3.2. Traditional Ensemble Model

For a traditional three-teacher KD model, during the training process, the training dataset is sent to all three pre-trained teacher models and the student model. Each teacher model produces soft labels using Equation (1), and then combines those soft labels with the probability distribution generated by the student model to calculate the soft loss, $L_{soft\_n}$ ($n$ = 1, 2, 3), using Equation (2). Then, the three soft losses are weighted summed as $\sum L_{soft\_n} \bullet t_n$, where $t_n$ is the weight of the soft loss for $n$-th teacher model and $\sum t_n = 1$. On the other hand, the student model produces a probability distribution using Equation (3), and then combines those probabilities with the hard labels generated by the ground truth to calculate the hard loss using Equation (4). As a result, after passing through a traditional three-teacher KD model, the total loss of the ensemble model, $L_{ensemble}$, can be described as follows:

$$L_{ensemble} = \lambda \bullet T^2 \bullet \sum_n (L_{soft\_n} \bullet t_n) + (1 - \lambda) \bullet L_{hard} \tag{6}$$

There are numerous possible combinations of weights for the three teacher models, making the selection of weights a difficult task. A simple scheme is to adopt equal weights for each teacher model, that is, $t_1 = t_2 = t_3 = 1/3$.

### 2.3.3. Parallel Model

A scheme that eliminates the necessity of selecting weights is to parallelize all models. As depicted in Figure 5, during the training process, the training dataset is sent to all three KD modules. In each KD module, following Section 2.3.1 and Figure 4, the pre-trained teacher model and the student model produce a module loss using Equation (5); based on the module loss and the BP process, the student model updates its network weight parameters. Subsequently, by comparing the three module losses, only the KD module possessing the minimum loss is selected and reserved, while the other two KD modules with larger losses are discarded. As a result, in this parallel three-teacher KD model, the student model learns knowledge from only one selected "best" teacher model within each epoch. Consequently, this parallel model is expected to outperform the equal-weight model.

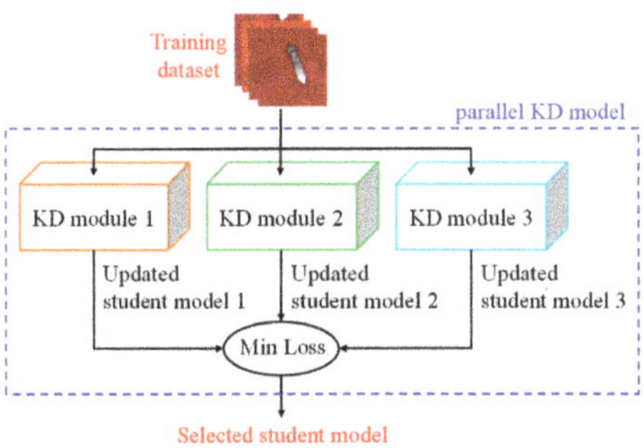


**Figure 5.** Schematic diagram of the parallel three-teacher KD model.

### 2.3.4. Series Model and Its Residual Connections

To fully utilize all three teacher models, the series architecture can be adopted. As presented in Figure 6a, during the training process, the training dataset is sent to the first KD module. Following Section 2.3.1 and Figure 4, the first pre-trained teacher model and the student model produce a module loss using Equation (5); based on this module loss and the BP process, the student model is updated to $M_{module\_1}$. Next, the training dataset is sent into the second KD module. After a similar processing sequence, the second pre-trained teacher model and the updated student model, $M_{module\_1}$, generate another module loss, leading to the update of the student model to $M_{module\_2}$. Subsequently, the training dataset is passed to the third KD module. Upon completion of the same processing steps, the third pre-trained teacher model and the updated student model, $M_{module\_2}$, yield a module loss, resulting in the final update of the student model to $M_{module\_3}$. As a result, in this standard series three-teacher KD model, the student model sequentially acquires knowledge from each teacher model within each epoch. Consequently, this standard series model is expected to outperform the parallel model, if all teacher models can transfer high-quality knowledge effectively. However, the inclusion of a low-quality teacher model can degrade performance.

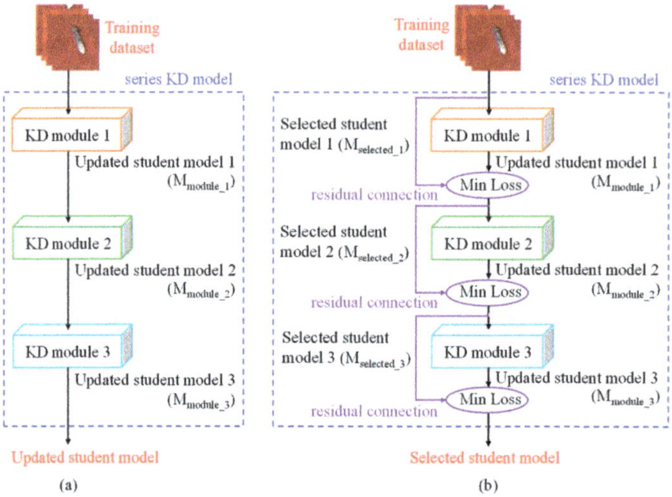


**Figure 6.** Schematic diagram of the series three-teacher KD model (**a**) without and (**b**) with residual connections.

To eliminate the influence of the low-quality teacher model, the residual connections can be incorporated into the series KD model. As illustrated in Figure 6b, the selected student model from the previous epoch, $M_{selected\_1}$, serves dual purposes in the first KD module: it is used to update the student model and, through a residual connection, to compare with the newly updated student model, $M_{module\_1}$. The model with the minimum module loss, either $M_{selected\_1}$ or $M_{module\_1}$, is retained as the selected student model, $M_{selected\_2}$. This process is repeated in the second KD module. $M_{selected\_2}$ is utilized to update the student model and compared with $M_{module\_2}$. The model possessing the minimum loss, either $M_{selected\_2}$ or $M_{module\_2}$, is reserved as the selected student model, $M_{selected\_3}$. Similarly, in the third KD module, $M_{selected\_3}$ is employed for updating and compared with $M_{module\_3}$. The model with the minimum loss, either $M_{selected\_3}$ or $M_{module\_3}$, is chosen as the final selected student model, $M_{selected}$. The $M_{selected}$ is then used for the subsequent epoch. As a result, in this residual-series model, the student model progressively acquires knowledge from each teacher model within each epoch while discarding the low-quality knowledge, ensuring that each subsequent KD module performs at least as well as its predecessor. Consequently, this residual-series model is expected to outperform both the parallel model and the standard series model.

### 2.4. Performance Indices

The grading accuracy and average inference time were used for comparing the models on the training dataset.

Additionally, to more accurately measure the models' performance, the mean accuracy, mean precision, mean recall, mean F1 score, and kappa coefficient were calculated on the testing dataset, using the following equations:

$$\text{Accuracy} = \frac{TP + TN}{TP + TN + FP + FN} \quad (7)$$

$$\text{Precision} = \frac{TP}{TP + FP} \tag{8}$$

$$\text{Recall} = \frac{TP}{TP + FN} \tag{9}$$

$$\text{F1 Score} = 2 \bullet \frac{\text{Precision} \bullet \text{Recall}}{\text{Precision} + \text{Recall}} \tag{10}$$

$$\text{Kappa coefficient} = \frac{P_o - P_e}{1 - P_e} \tag{11}$$

where *TP* (true positives) represents the number of samples correctly predicted by the model as positive cases, *TN* (true negatives) indicates the number of samples correctly predicted by the model as negative cases, *FP* (false positives) indicates the number of samples incorrectly predicted by the model as positive cases, *FN* (false negatives) indicates the number of samples incorrectly predicted by the model as negative cases, $P_o$ is the testing accuracy, and $P_e$ represents the sum of the actual number and predicted number of respective categories divided by the square of the total number of samples.

## 3. Experiments

### 3.1. Computing Resources and Parameters

Since the purpose of this study is to evaluate the performance of KD in classification tasks, a typical published ShuffleNet V2 0.5x lightweight model [15] was selected as the student model, while three typical published large-parameter models, VGG16 [11], ResNet50 [12], and Xception [13], served as teacher models. The resources occupied by each model are shown in Table 1. Prior to training, each image was resized to meet the input specifications of the model.

**Table 1.** Resource occupation.

| Model | Number of Parameters | Size of the Model |
| --- | --- | --- |
| VGG16 | 134.26 M | 512 M |
| ResNet50 | 23.58 M | 94.4 M |
| Xception | 20.90 M | 83.8 M |
| ShuffleNet v2 0.5x | 0.35 M | 1.43 M |

The hardware platform used in this experiment was a personal computer, with an Intel Core i7-12700KF (3.61 GHz), 64 GB of memory, and an NVIDIA GeForce RTX 3080Ti GPU (NVIDIA Corporation, Santa Clara, CA, USA). The software environment consisted of the Windows 10 (64-bit) operating system, Python version 3.8, and TensorFlow version 2.4.0.

All models were trained for 30 epochs. The widely used Adam optimizer [29] and a step learning rate reduction strategy were utilized. The initial learning rate was set to 0.004, and the learning rate was adjusted to 60% of its current value every five epochs. Different learning rate descent strategies, such as cosine decay, adaptive descent, and the step reduction strategy, were tried. Both the cosine decay and adaptive descent strategies exhibited varying degrees of overfitting, while the step reduction strategy did not overfit during training and demonstrated superior performance. In the KD module, the temperature coefficient, $T$, was set to 5. When the student model has a relatively small number of parameters, a relatively lower value of $T$ suffices, because a model with fewer parameters may not be able to learn all the knowledge from the teacher model, and it is appropriate to ignore some information related to negative labels. Additionally, the balance factor, $\lambda$, was set to 0.8 to enhance the effects of the soft loss.

### 3.2. Results

#### 3.2.1. Single-Teacher KD

Figure 7 presents the loss curves for the original student model and single-teacher KD models on both the training and validation datasets. During the training process, there is no evidence of underfitting or overfitting. For the best epoch in the training process, the grading accuracy and inference time of the single-teacher KD model are presented in Table 2. For the sake of comparison, the results of each pre-trained teacher model and the original student model are also given. The pre-trained teacher models reach a high grading accuracy (the lowest value was 98.3%) but have a large inference time (the smallest value was 65 ms), whereas the original student model only achieves a low grading accuracy (86.5%) but has a small inference time (5.03 ms). Through single-teacher KD, the distilled student model obtains an improved grading accuracy (the highest value is 92.3%) while maintaining the small inference time (about 5 ms).

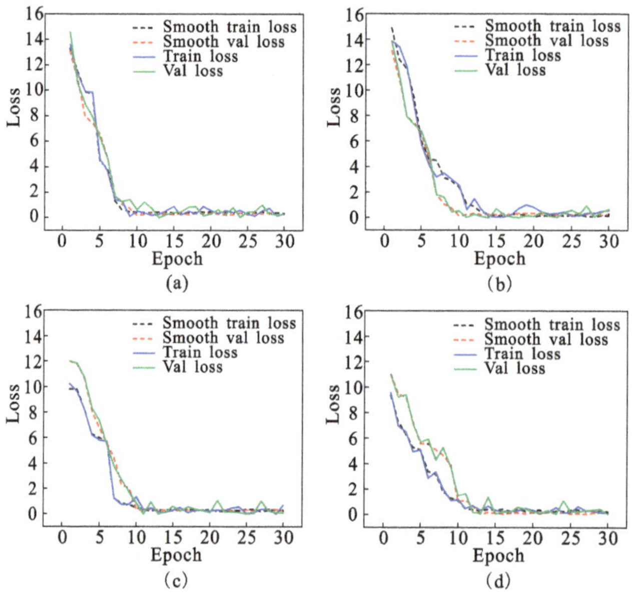


**Figure 7.** Loss curves for (**a**) the original student model and (**b–d**) single-teacher KD models on the training and validation datasets. (**b**) VGG16, (**c**) ResNet50, and (**d**) Xception.

**Table 2.** Performance indices for pre-trained teacher models, the original student model, and single-teacher KD models on the training dataset.

| Model | Grading Accuracy | Average Inference Time |
|---|---|---|
| VGG16 (Pre-trained teacher model) | 99.1% | 71 ms |
| ResNet50 (Pre-trained teacher model) | 98.3% | 65 ms |
| Xception (Pre-trained teacher model) | 99.6% | 135 ms |
| Shufflenet v2 (Original student model) | 86.5% | 5.03 ms |
| Shufflenet v2 distilled by VGG16 | 92.3% | 5.14 ms |
| Shufflenet v2 distilled by ResNet50 | 92.0% | 4.99 ms |
| Shufflenet v2 distilled by Xception | 91.5% | 5.72 ms |

The performance of the single-teacher KD models was evaluated on the testing dataset. The confusion matrices resulting from these evaluations are illustrated in Figure 8. For comparison, the results of the original student model are also provided. It seems that the main issue with the grading is the inability to distinguish samples between Class 1 and

Class 2, and between Class 3 and Class 4. Compared to the original undistilled student model, the single-distillation model with VGG16 significantly reduces incorrect grading between these classes. Additionally, the single-distillation model with ResNet50 further reduces incorrect grading between Class 1 and Class 2. These results indicate that VGG16 and ResNet50 are high-performing teachers. However, the single-distillation model with Xception, intended to enhance the grading accuracy, results in incorrect grading between Class 2 and Class 4. This suggests that Xception is a low-performing teacher.

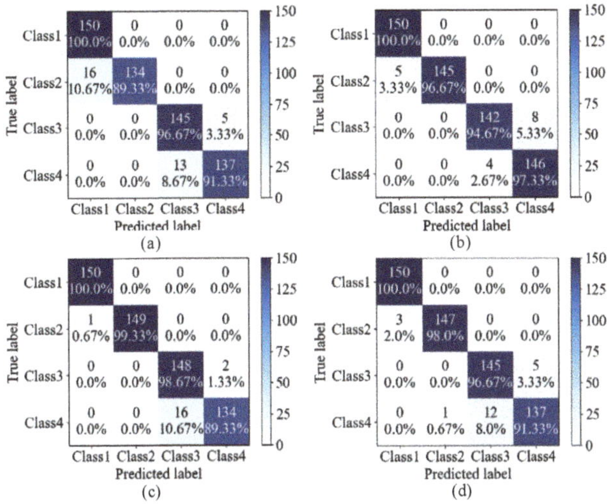


**Figure 8.** Confusion matrices of prediction results for (**a**) the original student model and (**b–d**) single-teacher KD models on the testing dataset. (**b**) VGG16, (**c**) ResNet50, and (**d**) Xception.

The various performance indices evaluated from the confusion matrices are presented in Table 3. Compared to the original student model (for example, with a mean F1 score of 94.5% and a Kappa coefficient of 0.9244), the distilled student model obtained through single-teacher KD achieves higher values for the performance indices (with the best mean F1 score of 97.1% and Kappa coefficient of 0.9622).

**Table 3.** Performance indices for the original student model and single-teacher KD models on the testing dataset.

| Model | Mean Accuracy | Mean Precision | Mean Recall | Mean F1 Score | Kappa Coefficient |
|---|---|---|---|---|---|
| Shufflenet v2 (original student model) | 94.3% | 94.6% | 94.3% | 94.5% | 0.9244 |
| Shufflenet v2 distilled by VGG16 | 97.1% | 97.2% | 97.1% | 97.1% | 0.9622 |
| Shufflenet v2 distilled by ResNet50 | 96.8% | 97.0% | 96.8% | 96.9% | 0.9577 |
| Shufflenet v2 distilled by Xception | 96.5% | 96.5% | 96.5% | 96.5% | 0.9533 |

3.2.2. Three-Teacher KD with Equal-Weight Ensemble Model

It is supposed that the KD model with multiple teachers can facilitate the transfer of a broader range of knowledge. The traditional equal-weight three-teacher KD was adopted to enhance grading accuracy. Figure 9 presents the loss curves for the equal-weight three-teacher KD model on both the training and validation datasets, showing no evidence of underfitting or overfitting during the training process. For the best epoch in the training process, the grading accuracy and inference time are presented in Table 4. It is shown that the equal-weight model only attains a grading accuracy of 92.3% and an inference time of about 5.5 ms, not superior to the results of the student model distilled by a single teacher. The reason can be explained by the fact that the equal weights may lead to the underestimation of the knowledge from the high-performing teachers, while overemphasizing the knowledge from the low-performing teachers. This will affect the student model's ability to effectively learn from multiple teachers.

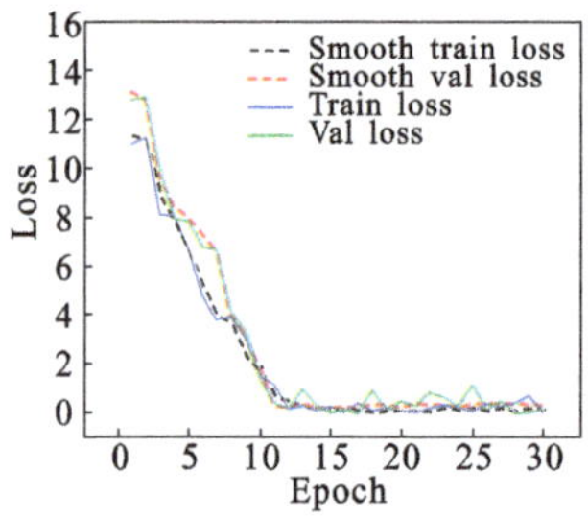


**Figure 9.** Loss curves for the equal-weight three-teacher KD model on the training and validation datasets.

**Table 4.** Performance indices for the equal-weight three-teacher KD model on the training dataset.

| Model | Grading Accuracy | Average Inference Time |
|---|---|---|
| Equal-weight model | 92.3% | 5.62 ms |

The performance of the equal-weight model was also evaluated on the testing dataset. The confusion matrix and various indices are illustrated in Figure 10 and Table 5, respectively. Compared to the best single-distillation model in Section 3.2.1, the implementation of the equal-weight three-teacher KD leads to an improvement in grading accuracy, especially in distinguishing samples between Class 3 and Class 4. Consequently, the mean F1 score and Kappa coefficient are increased to 98.2% and 0.9777, respectively.

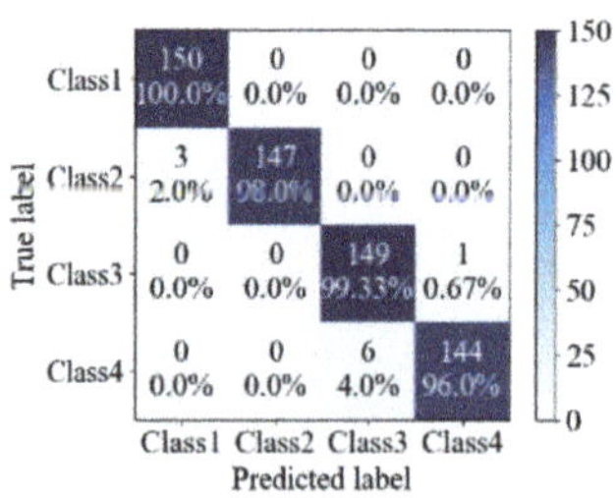


**Figure 10.** Confusion matrix of prediction results for the equal-weight three-teacher KD model on the testing dataset.

**Table 5.** Performance indices for the equal-weight three-teacher KD model on the testing dataset.

| Model | Mean Accuracy | Mean Precision | Mean Recall | Mean F1 Score | Kappa Coefficient |
|---|---|---|---|---|---|
| Equal-weight model | 98.3% | 98.3% | 98.3% | 98.2% | 0.9777 |

3.2.3. Three-Teacher KD with a Cascaded Structure

The three-teacher KD models with a cascaded structure, including the parallel model, the standard series model, and the residual-series model, were investigated. In the standard series three-teacher KD model, the student sequentially acquires knowledge from each teacher within an epoch. After the student model learns from a teacher model and its weights are updated, the student model proceeds to learn from the next teacher model. This implies that the knowledge acquired from the previous teacher model will influence the student model's learning from the subsequent teacher model. As shown in Table 6, the grading accuracy is influenced by the model sequence. Most standard series model outperform the traditional equal-weight model, with the best grading accuracy of 94.1% achieved on the Xception→VGG16→ResNet50 sequence. All residual-series models demonstrate higher grading accuracy than their standard series counterparts, reaching a best value of 94.5% on the same sequence. In the following of this paper, the best standard series and residual-series models are used for comparison with other cascaded structures.

**Table 6.** Grading accuracy for series models with different sequences on the training dataset.

| | Standard Series Model | Residual-Series Model |
|---|---|---|
| VGG16→ResNet50→Xception | 93.2% | 93.2% |
| VGG16→Xception→ResNet50 | 92.8% | 94.0% |
| ResNet50→VGG16→Xception | 92.5% | 94.1% |
| ResNet50→Xception→VGG16 | 93.1% | 93.2% |
| Xception→VGG16→ResNet50 | 94.1% | 94.5% |
| Xception→ResNet50→VGG16 | 91.6% | 94.3% |

Figure 11 presents the loss curves for the parallel model, the best standard series model, and the best residual-series model on both the training and validation datasets, showing no evidence of underfitting or overfitting during the training process. For the best epoch in the training process, the grading accuracy and inference time of the three-teacher KD models with a cascaded structure are presented in Table 7. For the parallel three-teacher KD model, the student model learns knowledge from the best one of the three pre-trained teacher models within each epoch, thereby improving the grading accuracy to 94.4%. For the best standard series three-teacher KD model, the student model sequentially acquires knowledge from a specific order of teacher models within each epoch, resulting in a grading accuracy of 94.1%. For the residual-series three-teacher KD model, the student model progressively acquires "good" knowledge while discarding the "bad" knowledge within each epoch, thereby ensuring that each subsequent KD module performs at least as well as its predecessor. Thus, compared to the standard series model, the grading accuracy for the best residual-series model is improved to 94.5%. For all the distilled student models, the inference time is about 5.5 ms.

The performance of these three-teacher KD models was also evaluated on the testing dataset. The results indicate that all performance indices are improved. The confusion matrices resulting from these evaluations are illustrated in Figure 12, and the various indices are presented in Table 8. When compared to the equal-weight model, the parallel model and the best standard series model have fewer incorrect grading predictions, both in distinguishing samples between Class 1 and Class 2, and between Class 3 and Class 4. These two models obtain the same improved mean F1 score of 99.1% and Kappa coefficient of 0.9888, respectively. Furthermore, the residual-series model exhibits the highest performance indices, achieving a mean F1 score of 99.9% and a Kappa coefficient of 0.9955. The high performance achieved on the testing dataset, which surpasses that of the training datasets, is largely attributed to the augmentation technique. Data augmentation in the training datasets can enhance the robustness of the model, thereby enabling it to better handle the testing dataset in which augmentation was not applied. These results also demonstrate the effectiveness of the cascaded structures, including the parallel model, the standard series model, and the demonstrated superiority of the residual-series model. Figure 12d,e present two misclassified samples using the residual-series model. The potential reason for their misclassification from Class 4 to Class 3 is that their stalks are full and have a good ratio to their caps. To improve the accuracy, more samples should be captured for training in the future to enhance the feature extraction.

**Table 7.** Performance indices for various three-teacher KD models with a cascaded structure on the training dataset.

| Model | Grading Accuracy | Average Inference Time |
|---|---|---|
| Parallel model | 94.4% | 5.52 ms |
| Best standard series model | 94.1% | 5.71 ms |
| Best residual-series model | 94.5% | 5.51 ms |

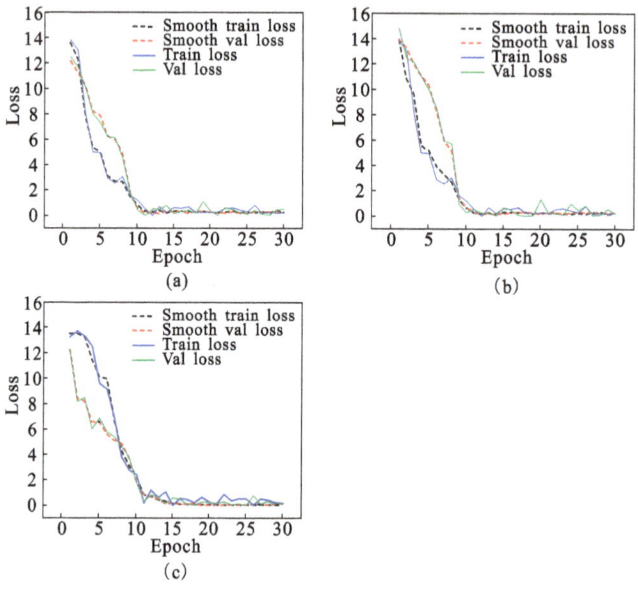


**Figure 11.** Loss curves for three-teacher KD models with a cascaded structure on the training and validation datasets. (**a**) The parallel model, (**b**) the best standard series model, and (**c**) the best residual-series model.

**Table 8.** Performance indices for various three-teacher KD models with a cascaded structure on the testing dataset.

| Model | Mean Accuracy | Mean Precision | Mean Recall | Mean F1 Score | Kappa Coefficient |
|---|---|---|---|---|---|
| Parallel model | 99.1% | 99.1% | 99.1% | 99.1% | 0.9888 |
| Best standard series model | 99.1% | 99.2% | 99.1% | 99.1% | 0.9888 |
| Best residual-series model | 99.7% | 99.7% | 99.9% | 99.9% | 0.9955 |

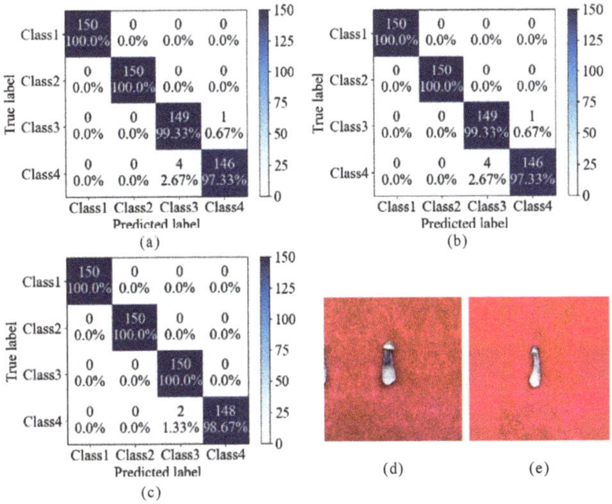


**Figure 12.** Confusion matrices of prediction results for three-teacher KD models with a cascaded structure on the testing dataset: (**a**) the parallel model, (**b**) the best standard series model, and (**c**) the best residual-series model. The (**d**) first and (**e**) second misclassified samples in (**c**).

## 4. Discussions

### 4.1. Comparison with Existing Technologies

To show the effectiveness of our proposed model, two existing technologies, including a filter pruning method [30] and a logit standardization KD [31], were compared with the best residual-series model. Using the same datasets described in Section 2.2, as well as the computing resources and parameters described in Section 3.1, the three models were trained and evaluated. The number of parameters, grading accuracy, and inference time evaluated on the testing dataset are tabulated in Table 9. The results show that the best residual-series model in this study can achieve the highest grading accuracy while maintaining the smallest parameters and fastest response speed.

**Table 9.** Comparison of the performance indices of filter pruning, logit standardization KD, and our proposed residual-series model on the testing dataset.

| Model | Number of Parameters | Grading Accuracy | Average Inference Time |
|---|---|---|---|
| VGG16 | 134.26 M | 99.1% | 71 ms |
| VGG16 compressed using the filter pruning described in [30] | 14.14 M | 98.3% | 13 ms |
| Shufflenet v2 distilled by the logit standardization KD described in [31] | 0.35 M | 99.0% | 5.50 ms |
| Shufflenet v2 distilled by the best residual-series model in this study | 0.35 M | 99.7% | 5.51 ms |

### 4.2. Model's Performance Evaluated on an Edge Device

A lightweight model is generally more favorable for deployment on edge devices, as it requires less memory and computational resources. This can be particularly important in real-world applications where energy consumption and hardware limitations are critical factors. Our proposed residual-series model is suitable for deployment on edge devices such as the Raspberry Pi, enabling real-time inference even with limited hardware resources.

The proposed residual-series model was transferred to a Raspberry Pi 4 Model B, equipped with a Broadcom BCM2711 Quad core 64-bit SoC (ARM Cortex-A72 architecture, 1.5 GHz), LPDDR4-2400 SDRAM (4 GB) (Raspberry Pi Ltd, Cambridge, UK). The software environment consisted of the Raspbian operating system based on Linux, Python version 3.7, and TensorFlow version 2.2.0. It was then evaluated on the testing dataset. Due to limited hardware resources, the inference time for the Raspberry Pi is significantly longer compared to that of the personal computer. As indicated in Table 10, the personal computer can detect images from a high-speed camera at a frame rate of 180 FPS in real time, whereas the Raspberry Pi can only detect images from a low-speed camera at a frame rate of 10 FPS in real time. However, both systems still represent a significant improvement over manual selection (with a typical grading time of approximately 1 s, resulting in an efficiency that is one-tenth that of the Raspberry Pi) and lead to reduced human resources and labor costs. Apart from inference time, all other performance metrics presented in Table 10 remain the same as those evaluated using the personal computer.

**Table 10.** Comparison of the performance indices for the best residual-series model on the testing dataset using a computer and an edge device.

| Model | Mean Accuracy | Mean Precision | Mean Recall | Mean F1 Score | Kappa Coefficient | Average Inference Time |
|---|---|---|---|---|---|---|
| A personal computer [1] | 99.7% | 99.7% | 99.9% | 99.9% | 0.9955 | 5.51 ms |
| An edge device [2] | 99.7% | 99.7% | 99.9% | 99.9% | 0.9955 | 87.49 ms |

[1] A personal computer described in Section 3.1; [2] an edge device described in Section 4.2.

### 4.3. Model's Performance Evaluated on a Dataset from a Public Database

To verify the generalization ability of our proposed model, four types of mushrooms, including Amanita, Apioperdon, Boletus, and Ceriporus, were randomly selected from a mushroom dataset in Kaggle [32]. For each type of mushroom, 1200, 150, and 150 images were randomly selected as the training, validation, and testing datasets, respectively. Using the same computing resources and parameters described in Section 3.1, the three-teacher KD models, including the equal-weight model, the parallel model, the standard series model, and the residual-series model, were trained and evaluated. The performance of these models for the best epoch in the training process and that of the well-trained models evaluated on the testing dataset are presented in Tables 11 and 12, respectively. Similar to Section 3.2.3, the parallel and standard series models can improve grading accuracy, and the residual-series model can achieve the highest performance, while their inference times are almost the same.

**Table 11.** Performance indices for the equal-weight model and the three-teacher KD models with a cascaded structure on the training dataset.

| Model | Grading Accuracy | Average Inference Time |
|---|---|---|
| Equal-weight model | 87.67% | 5.52 ms |
| Parallel model | 89.52% | 5.52 ms |
| Best standard series model | 88.92% | 5.61 ms |
| Best residual-series model | 91.83% | 5.49 ms |

**Table 12.** Performance indices for the equal-weight model and the three-teacher KD models with a cascaded structure on the testing dataset.

| Model | Mean Accuracy | Mean Precision | Mean Recall | Mean F1 Score | Kappa Coefficient |
|---|---|---|---|---|---|
| Equal-weight model | 87.7% | 87.7% | 87.7% | 87.7% | 0.9015 |
| Parallel model | 93.1% | 93.1% | 93.1% | 93.1% | 0.9628 |
| Best standard series model | 93.3% | 93.2% | 93.3% | 93.3% | 0.9603 |
| Best residual-series model | 95.0% | 95.0% | 95.0% | 95.0% | 0.9714 |

## 5. Conclusions

A three-teacher KD algorithm with a cascaded structure was proposed to improve the quality grading of *Oudemansiella raphanipes* using lightweight models in resource-limited devices. The cascaded three-teacher models, including the parallel model, the standard series model, and the residual-series model, outperformed single-teacher model and traditional equal-weight three-teacher model, demonstrating improved grading accuracy while maintaining fast inference times. The residual-series model, in particular, demonstrated a remarkable accuracy of 99.7%, making it a promising candidate for automated quality grading of *Oudemansiella raphanipes*. It was also successfully deployed to the Raspberry Pi, despite its limited computing resources. Compared with manual grading, the Raspberry Pi exhibits a significant improvement in efficiency by a factor of ten, while maintaining high accuracy. This will result in significant economic benefits for the industry owing to increased efficiency but reduced human resources and labor costs. Future work could focus on extending this approach to other types of agricultural products and exploring more advanced cascaded structures to further enhance performance.

**Author Contributions:** Conceptualization, H.H. (Huamao Huang) and M.L.; methodology, H.C. and H.H. (Huamao Huang); investigation, H.C.; data curation, Y.P. and H.Z.; data analysis, H.C. and H.H. (Haiying Hu); figures, H.H. (Haiying Hu); writing—original draft preparation, H.C. and H.H. (Huamao Huang); writing—review and editing, H.H. (Haiying Hu), M.L., Y.P. and H.Z.; supervision, H.H. (Huamao Huang); project administration, H.H. (Haiying Hu) and M.L.; funding acquisition, H.H. (Huamao Huang) and M.L. All authors have read and agreed to the published version of the manuscript.

**Funding:** This research was funded by the Guangzhou Key Research and Development Program (2024B03J1359).

**Institutional Review Board Statement:** Not applicable.

**Data Availability Statement:** The original contributions presented in this study are included in the article. Further inquiries can be directed to the corresponding authors.

**Conflicts of Interest:** The authors declare no conflicts of interest.

## References

1. Yuan, X.-H.; Fu, Y.-P.; Xiao, S.-J.; Li, C.-T.; Wang, D.; Li, Y. Research progress on mushroom phenotyping. *Mycosystema* **2021**, *40*, 721–742.
2. Fang, Q.; Lai, Y.; Zhang, D.; Lei, H.; Wang, F.; Guo, X.; Song, C. Gut microbiota regulation and prebiotic properties of polysaccharides from Oudemansiella raphanipes mushroom. *World J. Microbiol. Biotechnol.* **2023**, *39*, 167. [CrossRef] [PubMed]
3. Wei, Q.; Zhong, X.; Haruna, M.H.; Liu, S.; Zhou, F.; Chen, M. Evaluation of different agricultural wastes for the production of polysaccharides from Oudemansiella raphanipes and its anti-oxidant properties. *Food Sci. Nutr.* **2022**, *10*, 3444–3452. [CrossRef] [PubMed]
4. Zhao, X.; Wang, L.; Zhang, Y.; Han, X.; Deveci, M.; Parmar, M. A review of convolutional neural networks in computer vision. *Artif. Intell. Rev.* **2024**, *7*, 99. [CrossRef]
5. Li, T.; Huang, H.; Peng, Y.; Zhou, H.; Hu, H.; Liu, M. Quality grading algorithm of Oudemansiella raphanipes based on transfer learning and MobileNetV2. *Horticulturae* **2022**, *8*, 1119. [CrossRef]
6. Wu, Y.; Sun, Y.; Zhang, S.; Liu, X.; Zhou, K.; Hou, J. A size-grading method of Antler Mushrooms using YOLOv5 and PSPNet. *Agronomy* **2022**, *12*, 2601. [CrossRef]
7. Zuo, Y.; Zhao, M. SA-EfficientNet: Quality grading model of Stropharia rugoso-annulate. In Proceedings of the 2022 International Conference on Computer Engineering and Artificial Intelligence (ICCEAI), Shijiazhuang, China, 22–24 July 2022; pp. 358–362.
8. Wang, L.; Dong, P.; Wang, Q.; Jia, K.; Niu, Q. Dried shiitake mushroom grade recognition using D-VGG network and machine vision. *Front. Nutr.* **2023**, *10*, 1247075. [CrossRef] [PubMed]
9. Shi, L.; Wei, Z.; You, H.; Wang, J.; Bai, Z.; Yu, H.; Ji, R.; Bi, C. OMC-YOLO: A lightweight grading detection method for Oyster Mushrooms. *Horticulturae* **2024**, *10*, 742. [CrossRef]

10. Wei, Z.; Liu, H.; Xu, J.; Li, Y.; Hu, J.; Tian, S. Quality grading method for Pleurotus eryngii during postharvest storage based on hyperspectral imaging and multiple quality indicators. *Food Control* **2024**, *166*, 110763. [CrossRef]
11. Karen, S.; Zisserman, A. Very deep convolutional networks for large-scale image recognition. *arXiv* **2014**, arXiv:1409.1556.
12. He, K.; Zhang, X.; Ren, S.; He, J.S. Deep residual learning for image recognition. In Proceedings of the 2016 IEEE Conference on Computer Vision and Pattern Recognition (CVPR), Las Vegas, NV, USA, 27–30 June 2016; pp. 770–778.
13. Chollet, F. Xception: Deep learning with depthwise separable convolutions. In Proceedings of the 2017 IEEE Conference on Computer Vision and Pattern Recognition (CVPR), Honolulu, HI, USA, 21–26 July 2017; pp. 1251–1258.
14. Zhang, X.; Zhou, X.; Lin, M.; Sun, J. ShuffleNet: An extremely efficient convolutional neural network for mobile devices. In Proceedings of the 2018 IEEE/CVF Conference on Computer Vision and Pattern Recognition (CVPR), Salt Lake City, UT, USA, 18–23 June 2018; pp. 6848–6856.
15. Ma, N.; Zhang, X.; Zheng, H.-T.; Sun, J. ShuffleNet V2: Practical guidelines for efficient CNN architecture design. In Proceedings of the European Conference on Computer Vision (ECCV), Munich, Germany, 8–14 September 2018; pp. 116–131.
16. Hinton, G.; Vinyals, O.; Dean, J. Distilling the knowledge in a neural network. *arXiv* **2015**, arXiv:1503.02531.
17. Gou, J.; Yu, B.; Maybank, S.J.; Tao, D. Knowledge distillation: A survey. *Int. J. Comput. Vis.* **2021**, *129*, 1789–1819. [CrossRef]
18. You, S.; Xu, C.; Xu, C.; Tao, D. Learning from multiple teacher networks. In Proceedings of the 23rd ACM SIGKDD International Conference on Knowledge Discovery and Data Mining (KDD), Halifax, NS, Canada, 13–17 August 2017; pp. 1285–1294.
19. Shao, H.; Zhong, D. Multi-target cross-dataset palmprint recognition via distilling from multi-teacher. *IEEE Trans. Instrum. Meas.* **2023**, *72*, 5017814. [CrossRef]
20. Liu, Y.; Zhang, W.; Wang, J. Adaptive multi-teacher multi-level knowledge distillation. *Neurocomputing* **2020**, *415*, 106–113. [CrossRef]
21. Zhang, H.; Chen, D.; Wang, C. Confidence-aware multi-teacher knowledge distillation. In Proceedings of the 2022 IEEE International Conference on Acoustics, Speech and Signal Processing (ICASSP), Singapore, 7–13 May 2022; pp. 4498–4502.
22. Shang, R.; Li, W.; Zhu, S.; Jiao, L.; Li, Y. Multi-teacher knowledge distillation based on joint guidance of probe and adaptive corrector. *Neural Netw.* **2023**, *164*, 345–356. [CrossRef] [PubMed]
23. Chen, Z.; Deng, L.; Gou, J.; Wang, C.; Li, J.; Li, D. Building and road detection from remote sensing images based on weights adaptive multi-teacher collaborative distillation using a fused knowledge. *Int. J. Appl. Earth Obs. Geoinf.* **2023**, *124*, 103522. [CrossRef]
24. Lin, Y.-E.; Yin, S.; Ding, Y.; Liang, X. ATMKD: Adaptive temperature guided multi-teacher knowledge distillation. *Multimed. Syst.* **2024**, *30*, 292. [CrossRef]
25. Fukuda, T.; Suzuki, M.; Kurata, G.; Thomas, S.; Cui, J.; Ramabhadran, B. Efficient knowledge distillation from an ensemble of teachers. In Proceedings of the 18th Annual Conference of the International-Speech-Communication-Association (INTERSPEECH 2017), Stockholm, Sweden, 20–24 August 2017; pp. 3697–3701.
26. Yuan, F.; Shou, L.; Pei, J.; Lin, W.; Gong, M.; Fu, Y.; Jiang, D. Reinforced multi-teacher selection for knowledge distillation. In Proceedings of the 35th AAAI Conference on Artificial Intelligence/33rd Conference on Innovative Applications of Artificial Intelligence/11th Symposium on Educational Advances in Artificial Intelligence, Virtual, 2–9 February 2021; pp. 14284–14291.
27. *T/GVEAIA 009.3-2019*; Standards for Edible and Medicinal Mushrooms—Oudemansiella raphanipes—Part 3: Grading Standards for Oudemansiella raphanipes. Available online: https://max.book118.com/html/2019/0726/8051014034002037 (accessed on 1 August 2024).
28. Jung, A.B.; Wada, K.; Crall, J.; Tanaka, S.; Graving, J.; Reinders, C.; Yadav, S.; Banerjee, J.; Vecsei, G.; Kraft, A.; et al. Imgaug. 2020. Available online: https://github.com/aleju/imgaug (accessed on 1 August 2024).
29. Kingma, D.P.; Ba, J.L. Adam: A Method for Stochastic Optimization. In Proceedings of the International Conference on Learning Representations 2015, San Diego, CA, USA, 7–9 May 2015.
30. Lin, M.; Ji, R.; Wang, Y.; Zhang, Y.; Zhang, B.; Tian, Y.; Shao, L. Hrank: Filter pruning using high-rank feature map. In Proceedings of the IEEE/CVF Conference on Computer Vision and Pattern Recognition 2020, Seattle, WA, USA, 13–19 June 2020.
31. Sun, S.; Ren, W.; Li, J.; Wang, R.; Cao, X. Logit standardization in knowledge distillation. In Proceedings of the IEEE/CVF Conference on Computer Vision and Pattern Recognition 2024, Seattle, WA, USA, 16–22 June 2024.
32. Mushrooms Classification-Common Genus's Images. Available online: https://www.kaggle.com/datasets/maysee/mushrooms-classification-common-genuss-images (accessed on 5 January 2025).

agriculture

MDPI

Article

# Identifying Tomato Growth Stages in Protected Agriculture with StyleGAN3–Synthetic Images and Vision Transformer

**Yao Huo †, Yongbo Liu †, Peng He *, Liang Hu, Wenbo Gao and Le Gu**

Institute of Agricultural Information and Rural Economy, Sichuan Academy of Agricultural Sciences, Chengdu 610011, China

* Correspondence: hepeng78@scsaas.cn

† These authors contributed equally to this work.

**Abstract:** In protected agriculture, accurately identifying the key growth stages of tomatoes plays a significant role in achieving efficient management and high-precision production. However, traditional approaches often face challenges like non-standardized data collection, unbalanced datasets, low recognition efficiency, and limited accuracy. This paper proposes an innovative solution combining generative adversarial networks (GANs) and deep learning techniques to address these challenges. Specifically, the StyleGAN3 model is employed to generate high-quality images of tomato growth stages, effectively augmenting the original dataset with a broader range of images. This augmented dataset is then processed using a Vision Transformer (ViT) model for intelligent recognition of tomato growth stages within a protected agricultural environment. The proposed method was tested on 2723 images, demonstrating that the generated images are nearly indistinguishable from real images. The combined training approach incorporating both generated and original images produced superior recognition results compared to training with only the original images. The validation set achieved an accuracy of 99.6%, while the test set achieved 98.39%, marking improvements of 22.85%, 3.57%, and 3.21% over AlexNet, DenseNet50, and VGG16, respectively. The average detection speed was 9.5 ms. This method provides a highly effective means of identifying tomato growth stages in protected environments and offers valuable insights for improving the efficiency and quality of protected crop production.

**Keywords:** StyleGAN3; ViT; deep learning; tomato

## 1. Introduction

As one of the most important economic crops globally, the tomato's growth stages require careful monitoring and identification to safeguard both yield and quality [1]. The accurate recognition of critical developmental periods—such as the sapling, flowering, fruiting, and maturity stages—offers essential insights for precise management and decision-making regarding crop growth. This process is vital for improving agricultural efficiency and maximizing economic returns [2].

At present, the identification and monitoring of tomato growth stages primarily depend on manual observations based on the experience of practitioners. This method is not only inefficient and costly, but it is also highly vulnerable to the subjectivity of the observer and the influence of activities within the facility, which can compromise both the quality and yield of tomatoes. While technologies like inspection robots and smart imaging systems have emerged, their high costs and the difficulty of maintaining such equipment have hindered their widespread application, particularly in facilities with limited financial resources. Consequently, there is an increasing need for research into digital technology-based methods for identifying tomato growth stages, which would offer real-time, convenient, and efficient solutions for monitoring.

Deep learning, an essential branch of machine learning, was introduced by Hinton et al. [3,4] in 2006, sparking a significant wave of research within the neural network domain. Due to its exceptional capabilities in feature learning and classification, deep learning has become a foundational tool in several fields, including object detection, natural language processing [5], text information matching [6], medical image segmentation and classification [7,8], and video semantic segmentation [9,10]. In recent years, deep learning has also seen growing applications in the agricultural sector, particularly in the identification and classification of crop diseases and pests [11,12]. Convolutional neural networks (CNNs) [13], which are among the most widely adopted and effective models in this context, were successfully applied to crop recognition and classification tasks, with well-known networks like AlexNet [14] and GoogLeNet [15] leading the way.

The introduction of the Vision Transformer (ViT) model [16] in 2020 marked a breakthrough in the field of crop monitoring and classification. Compared to traditional CNN-based approaches, ViT offers enhanced accuracy in tasks like crop disease recognition and classification. Bai Yupeng et al. [17] developed a wheat disease recognition algorithm based on ViT, which achieved an impressive average recognition accuracy of over 95% for three different types of wheat diseases. This performance surpassed that of the AlexNet and VGG16 [18] models by 6.68% and 4.94%, respectively. In another study, Wang Yang et al. [19] addressed the poor robustness of deep convolutional neural networks (DCNNs) to noise by incorporating additional modules, such as augmented block serialization and masked multi-head attention, into the standard ViT model for tomato disease classification. Their modified ViT model achieved a classification accuracy of 99.63% on a tomato dataset, improving over 6% compared to classic models like ResNet50 [20]. However, despite these advancements, the use of ViT models in crop growth stage recognition remains limited. Most researchers still rely on more established deep learning models for such tasks. Rasti et al. [21] explored three machine learning approaches for identifying the growth stages of 11 wheat and 10 barley species: a five-layer Conver Net self-training model, a transfer learning-based VGG19 model, and support vector machines. Their findings indicated that the transfer learning-based model was the most effective, with an accuracy rate exceeding 99% and a substantial reduction in training time. Meanwhile, Tan et al. [22] conducted a comparative experiment using traditional machine learning methods and deep learning approaches to recognize the growth stages of rice. Their results showed that deep learning methods outperformed traditional methods, with the EfficientNetB4 model achieving the highest recognition accuracy, surpassing 99% in both accuracy and average precision.

Due to practical constraints, the amount of manually collected image data is limited. The tomato dataset suffers from issues such as class imbalance. However, the Vision Transformer (ViT) model requires a sufficient amount of data to achieve optimal recognition performance [23]. Traditional data augmentation techniques, such as rotation, translation, and cropping, are inefficient and only generate images with low signal-to-noise ratios and limited diversity. To address overfitting and other challenges, generative adversarial networks (GANs) [24] can produce clearer and more realistic samples, but traditional GANs suffer from issues such as training instability and model collapse [25]. StyleGAN3 [26], an improved version of GAN introduced by NVIDIA in 2021 based on StyleGAN2 [27], resolves problems like feature entanglement and is capable of generating higher-resolution, higher-quality images. However, StyleGAN3 also requires substantial computational resources to achieve satisfactory training results, typically necessitating more than 5000 iterations. This process imposes high demands on hardware performance and requires significant time and computational power to allow the model to adequately learn and capture the complex distributional characteristics of the data thereby generating high-quality images. Considering these factors, this work initially employs transfer learning by using a pre-trained model as the foundation for our custom StyleGAN model, which significantly reduces the training time. By further training the model on the existing dataset to generate more realistic synthetic images, we alleviate the data scarcity issue commonly faced by the ViT (Vision Transformer) model. This approach not only enhances the quality of the synthetic images but also expands the diversity of the training dataset, providing the model with a larger variety of learning samples. Ultimately, this contributes to improving the accuracy of tomato growth stage recognition. Through this optimization strategy, we aim to enhance the model's generalization ability and robustness, particularly in practical scenarios where data are limited.

The paper is organized into the following four sections: Section 1 provides an introduction, outlining the current state of tomato growth stage recognition in protected agriculture, along with a review of recent advancements in deep learning models for crop recognition and classification. Section 2 describes the dataset used in this work, as well as the StyleGAN3 and ViT models. Section 3 presents the experimental results and a comparative analysis with existing methods. Sections 4 and 5 are dedicated to the discussion and conclusion, respectively.

## 2. Materials and Methods

This section provides an overview of the dataset used in this work, along with a description of the models employed. The main contributions of this paper are divided into two parts: first, high-quality synthetic images are generated using the StyleGAN3 generative adversarial network; second, the ViT model is applied to tomato growth stage recognition and classification through transfer learning.

### 2.1. Dataset

Tomato images were collected from the Modern Agricultural Science and Technology Innovation Demonstration Park at the Sichuan Academy of Agricultural Sciences. This facility is located in the Xindu District of Chengdu, Sichuan Province, China, in the central depression of the Chengdu Plain, with an altitude of 510 m and a subtropical humid climate. The growth of tomatoes involves four stages, sapling, flowering, fruiting, and maturity, which occur throughout the year.

In this work, a total of 1200 images were collected across 4 growth stages, sapling, flowering, fruiting, and maturity, with 300 images for each stage. The collection time of the tomato images is in 2024. Based on the originally collected data, 2723 synthetic images were generated using GAN. All images were resized to a final resolution of 224 × 224. Example images for the four categories, along with specific details, are presented in Figure 1 and Table 1.

**Table 1.** Class-wise image distribution of growing period tomato dataset.

| Category | Number of Images (Original) | Number of Images (Synthetic) |
|---|---|---|
| Sapling | 300 | 450 |
| Flower | 300 | 396 |
| Fructification | 300 | 327 |
| Mature | 300 | 350 |

**Figure 1.** Four categories of tomato growth period.

## 2.2. StyleGAN3

Generative adversarial networks (GANs) operate by introducing two adversarial neural networks: the generator and the discriminator. These networks are optimized through adversarial training, where the generator creates realistic data samples, and the discriminator distinguishes between real and synthetic samples. Due to their robust generative capabilities and wide-ranging application potential, GANs have become a pivotal research area in fields such as computer vision, image processing, and machine learning. Over time, several GAN variants have been proposed, including Conditional GAN (CGAN) [28], Deep Convolutional GAN (DCGAN) [29], and BigGAN [30], each with its advantages and drawbacks. For instance, although CGAN and DCGAN are faster to train, they generate lower-resolution images with visible blurring at the edges, rendering them unsuitable for high-resolution tasks like generating tomato images. While BigGAN can produce high-resolution images, its considerable hardware requirements and long training times make it impractical for the present application. In contrast, StyleGAN3 offers a significant advantage: it can generate high-resolution images at 1024 × 1024 pixels and can be trained efficiently on a single GPU, thanks to its optimized CUDA cores, which enhance training speed and memory utilization.

As previously noted, StyleGAN3 effectively resolves the feature entanglement problem in StyleGAN2, where the fine details of generated images fail to move coherently with the object's shape, leading to inconsistencies in rotation and translation. The root cause of this issue lies in the generator architecture of existing models, which employs convolutions, nonlinear activations, and upsampling layers, failing to achieve adequate equivariance. In contrast, StyleGAN3 takes a signal processing approach, tracing the problem back to aliasing within the generator network, and proposes a solution. By treating all signals as continuous, StyleGAN3 introduces minor yet effective architectural modifications that prevent unintended information leakage during the hierarchical synthesis process, enhancing the model's overall performance.

StyleGAN3 achieves rotational invariance through two key modifications. First, all 3 × 3 convolution layers across the network are replaced with 1 × 1 convolutions, ensuring that only downsampling and upsampling operations propagate information between pixels thus preventing unnecessary information leakage into the hierarchical synthesis process. Second, the sinc downsampling filters are replaced with radial symmetric jinc filters, except for two critical layers. These changes significantly improve rotational invariance without compromising FID. Additionally, during the early stages of training, a Gaussian filter is applied to all images seen by the discriminator to prevent early stage training collapse. The generator may occasionally introduce high-frequency signals with small delays, degrading the discriminator's performance. A schematic of the generator is shown in Figure 2.

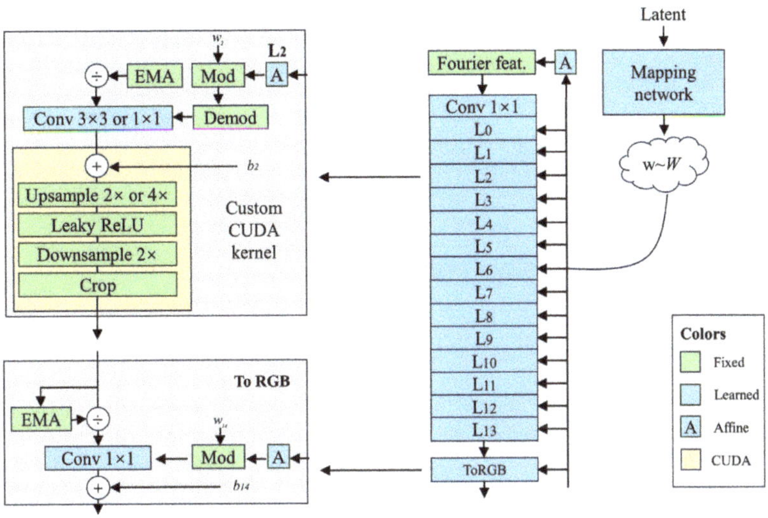


**Figure 2.** StyleGAN3 generator structure.

In this work, we employ StyleGAN3 to generate images representing different growth stages of tomatoes to address the limitations of the dataset. To generate these images, we first train a model on a tomato dataset. The training parameters are as follows: the configuration is set to stylegan-t (translation equiv); the total batch size is 32; gamma is set to 32 (R1 regularization weight); and the model is trained for 1000 kimg (total number of training iterations). Furthermore, to improve the monitoring and tracking of the training process, the snapshot frequency (snap) is set to 10.

### 2.3. Vision Transformer (ViT)

In this work, we employ the Vision Transformer (ViT) model to recognize the growth stages of tomatoes through transfer learning. The ViT architecture introduces the concept of image patches. Initially, the image is divided into non-overlapping patches of equal size, which are then encoded into sequence data using a positional encoding function. These sequence data are subsequently fed into a Transformer encoder. The output from the Transformer is processed through a fully connected layer followed by a softmax layer to produce the final classification result. The complete framework of the Vision Transformer is presented in Figure 3.

For tomato classification, we utilize the ViT-Base model, which segments the input image into 16 × 16 pixel patches. These patches are then processed through 12 encoder layers, each consisting of encoder blocks, stacked a total of 12 times. After passing through the embedding layer, each resulting vector has a dimension of 768. The fully connected layer includes 3072 units, and the multi-head self-attention mechanism incorporates 12 attention heads. The model contains 86 M parameters, and its architecture is detailed in Table 2.

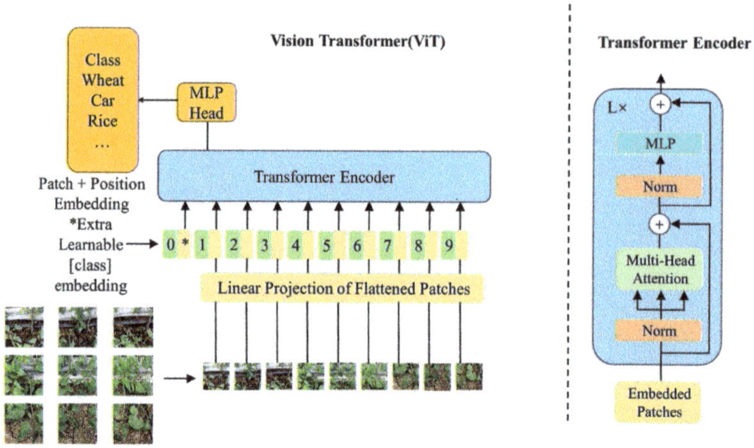


**Figure 3.** ViT model structure.

**Table 2.** Different versions of ViT model.

| Model | Patch Size | Layers | Hidden Size D | MLP Size | Heads | Params |
|---|---|---|---|---|---|---|
| ViT-Base | 16 × 16 | 12 | 768 | 3072 | 12 | 86 M |
| ViT-Large | 16 × 16 | 24 | 1024 | 4096 | 16 | 307 M |
| ViT-Huge | 14 × 14 | 32 | 1280 | 5120 | 16 | 632 M |

### *2.4. Experimental Platform*

The experiment utilized the PyCharm 2024.1.1 development environment alongside the PyTorch deep learning framework, with Windows 11 as the operating system. The hardware configuration included an Intel i5-12500 processor and an NVIDIA A5000 GPU with 24 GB of onboard memory. All comparative experiments were conducted using identical system specifications.

### *2.5. Experimental Setup and Evaluation Metrics*

This work involved three experimental stages. In the first stage, we employed the StyleGAN3 model to train on each category of the tomato dataset for 1000 epochs, ultimately generating models corresponding to four tomato growth stages. Following training, we generated 1523 synthetic images representing the tomato growth stages and conducted a quality assessment of these images. The second stage involved training and validating the ViT-Base model on both the original tomato dataset and an augmented dataset that included the synthetic images, allowing us to compare the recognition performance of the two datasets. In the third stage, we selected a set of classic classification models and compared their recognition accuracy for tomato growth stages against the ViT-Base model.

To evaluate the performance of the models, we utilized a range of metrics, including accuracy, precision, recall, $F1$ score, and confusion matrix. The formulas for these evaluation metrics are as follows:

$$Accuracy = \frac{TP + TN}{TP + TN + FP + FN} \quad (1)$$

$$Precision = \frac{TP}{TP + FP} \quad (2)$$

$$Recall = \frac{TP}{TP + FN} \quad (3)$$

$$F1 = \frac{2TP}{2TP + FP + FN} \quad (4)$$

In the context of classification tasks, true positive ($TP$), false positive ($FP$), and false negative ($FN$) are essential metrics for evaluating the performance of predictive models. More precisely, $TP$ represents the number of instances that are genuinely positive and are correctly classified as such by the model; $FP$ denotes the number of instances that are actually negative but are mistakenly classified as positive; and $FN$ indicates the number of actual positive instances that are erroneously classified as negative. These metrics are of paramount importance in performance analysis as they provide valuable insight into the model's classification accuracy, its ability to discriminate between classes, and the nature of its errors.

In this work, we use the Peak Signal-to-Noise Ratio ($PSNR$) to quantitatively evaluate the quality of the generated images. $PSNR$ is a well established metric in the fields of image and video processing, primarily used to measure the level of noise in an image and to assess the quality of the image in comparison to a reference. This metric is also widely employed to evaluate the performance of image processing algorithms by providing an objective measure of the image's fidelity. $PSNR$ is calculated based on the Mean Squared Error ($MSE$), which represents the average squared difference between the original and generated images. The following formulas define the calculations of $PSNR$ and $MSE$:

$$PSNR = 20 \cdot \log_{10}\left(\frac{MAX_1}{\sqrt{MSE}}\right) \quad (5)$$

$$MSE = \frac{1}{mn}\sum_{i=0}^{m-1}\sum_{j=0}^{n-1}[I(i,j) - K(i,j)]^2. \quad (6)$$

In Equation (5), $MSE$ represents the Mean Squared Error, a measure of the average squared differences between the original and generated images. $MAX$ refers to the maximum value of the pixel intensities in the image, which provides an upper bound for the image's dynamic range. In Equation (6), $I$ and $K$ are used to denote the original and generated images, respectively, while $m$ and $n$ correspond to the number of rows and columns in the images. These parameters are essential for computing the image quality metrics, which evaluate the fidelity of the generated image relative to the original.

## 3. Implementation and Result

This section begins by outlining the hardware, software, and evaluation metrics used in the experiments. It then evaluates the effects of using original and synthetic tomato images, analyzing the performance of tomato growth stage recognition with and without synthetic images. The section concludes by comparing the proposed model against several classical models.

### 3.1. Performance of StyleGAN3

The images generated using StyleGAN3, alongside the original tomato images, are presented in Figure 4. As illustrated, the generated images of the four tomato categories closely resemble the original images, demonstrating the model's ability to produce highly realistic results. However, it is important to highlight that StyleGAN3 requires significant computational resources and energy consumption, and the complexity of the images from the fruiting and maturity stages is considerably greater than that of the earlier stages. Achieving optimal training results for these stages necessitates over 20,000 epochs. Furthermore, the constraints of having only one high-performance GPU in our experimental setup will make the training time longer; the average training time per iteration is 4 min35 s, so to reduce the total training time, we sought to minimize the impact of background noise on the model and enhance the quality of the generated images. To achieve this, we employed the Rembg tool, based on U2-Net architecture, to remove the background from the images thereby isolating the foreground object for improved output.

**Figure 4.** The original tomato images and the StyleGAN3-generated images: The first to fourth columns represent the four categories of tomato growth stages. The first row shows the original images of the four tomato growth stages, while the second to fourth rows display the generated images corresponding to each category.

Table 3 provides a detailed statistical analysis of the PSNR between the original and generated tomato images. The findings clearly demonstrate that the quality of the images generated using StyleGAN3 closely approximates the quality of the real, original images. StyleGAN3 produces images of varying quality depending on the category. For the sapling and flower categories, the PSNR values are approximately 28 dB, suggesting that the generated images are somewhat distorted compared to the originals. These distortions are likely subtle and may not be easily detectable by the human eye. In contrast, the fructification and mature categories show a notable enhancement in image quality, with PSNR values increasing to around 39 dB. This significant rise in PSNR indicates that the generated images preserve more fine details, with minimal distortion, and closely mirror the original images in terms of visual fidelity.

**Table 3.** PSNR values (in dB) between original images and GAN-generated images for four categories.

| Category | PSNR Original Images | PSNR Original and StyleGAN3 Images |
|---|---|---|
| Sapling | 27.903 | 28.018 |
| Flower | 27.925 | 27.932 |
| Fructification | 38.78 | 39.422 |
| Mature | 37.389 | 38.784 |

From Table 4, it is evident that StyleGAN3 exhibits varying levels of image quality in terms of SSIM (Structural Similarity Index) across different categories. SSIM is a metric that evaluates the structural, luminance, and contrast similarities between two images. A higher SSIM value, closer to 1, indicates greater structural similarity between the original and generated images. In the sapling and flower categories, the SSIM values are notably low, at 0.066 and 0.454, respectively. However, these low values do not necessarily indicate poor image quality, as SSIM primarily evaluates structural aspects of the images. The sapling and flower categories contain more complex elements, such as leaves and stems, which may contribute to lower SSIM values. Therefore, the SSIM values of the sapling and flower categories should be interpreted in conjunction with the PSNR values and human visual judgment. On the other hand, in the fructification and mature categories, there is a marked increase in SSIM values. In the fructification category, the SSIM rises from 0.833 to 0.85, while in the mature category, it slightly decreases from 0.85 to 0.84. These values indicate that the generated images in both categories retain a high degree of structural similarity to the original images. Despite the slight decrease in SSIM in the mature category, the image quality remains high, with minimal distortion.

**Table 4.** SSIM values between original images and GAN-generated images for four categories.

| Category | SSIM Original Images | SSIM Original and StyleGAN3 Images |
|---|---|---|
| Sapling | 0.066 | 0.068 |
| Flower | 0.454 | 0.384 |
| Fructification | 0.833 | 0.85 |
| Mature | 0.85 | 0.84 |

### 3.2. *Performance of ViT Model*

As previously mentioned, to evaluate the classification performance of the tomato classification model, we selected accuracy (Acc), precision, recall, and F1 score as the evaluation metrics for the model's quality. Table 5 presents the classification performance of the ViT-Base model on both the original images and the combined dataset of original and generated images. The evaluation metrics for the ViT-Base model with the inclusion of generated images are 98.39%, 98.47%, 98.39%, and 98.39% for accuracy, precision, recall, and F1 score, respectively. These results represent improvements of 3.81%, 3.48%, 3.81%, and 3.98% over the performance achieved using only the original images. This demonstrates that incorporating generated images into the ViT model significantly enhances the model's classification ability and generalization, especially under data imbalance, while also mitigating the risk of overfitting.

**Table 5.** Impact of synthetic images on Vit-Base.

| Model | Accuracy | Precision | Recall | F1 Score |
|---|---|---|---|---|
| ViT-Base | 94.58% | 94.99% | 94.58% | 94.41% |
| ViT-Base + Synthetic images | 98.39% | 98.47% | 98.39% | 98.39% |

Figure 5 presents the confusion matrix results for the classification task, which was performed using both original and StyleGAN-generated images. The matrix indicates that classification performance—particularly in distinguishing true positives from false positives between the flowering and sapling stages—was notably reduced when only the original images were used. In contrast, a significant improvement in the model's ability to accurately classify these stages was observed with the inclusion of StyleGAN-generated images, resulting in a marked enhancement in overall classification performance. Specifically, the confusion matrix for the combined dataset, comprising both original and generated images, demonstrates a substantial increase in overall accuracy, as well as a more balanced distribution between true positives and false positives. These results emphasize the considerable advantages of augmenting the dataset with generated images, highlighting their critical role in enhancing the robustness and effectiveness of the classification model.

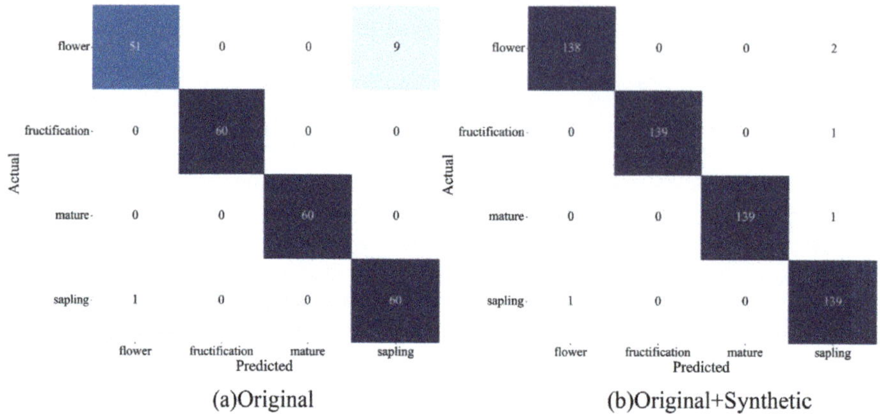


**Figure 5.** Confusion matrix of tomato growth classification using proposed method.

Figure 6 illustrates the AUC curves during training for ViT-Base, comparing the original dataset (Figure 6a) with the dataset augmented by generated images (Figure 6b). AUC, which assesses the model's performance at various classification thresholds, provides a more comprehensive evaluation by minimizing the effect of a single threshold. The results show that adding generated images significantly improves the performance of ViT-Base in the four-class classification task, with fewer fluctuations compared to using only the original dataset, and the VIT model can better perform its performance. This suggests that images generated by StyleGAN3 can effectively enhance the model's classification accuracy.

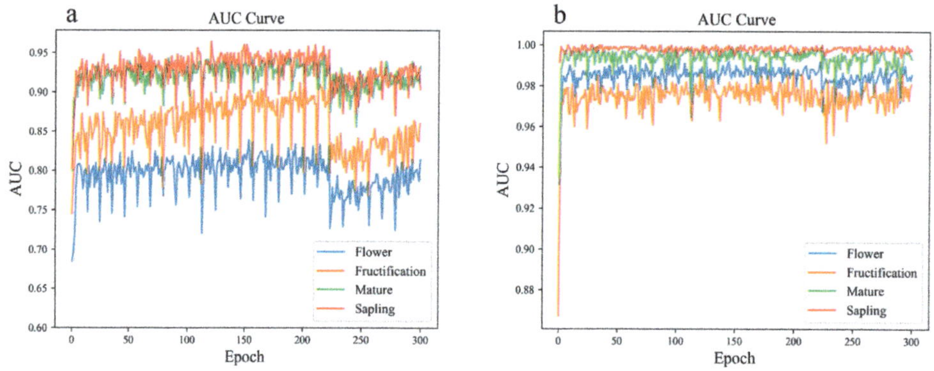


**Figure 6.** AUC comparison of ViT-Base performance with original dataset (**a**) and dataset augmented with generated image (**b**) during training.

### *3.3. Performance Comparison*

In this subsection, we first perform a detailed curve analysis of four models, examining their training and validation performance regarding accuracy and loss. Specifically, we investigate the ViT-Base model, augmented with generated images, and compare its four-class classification performance against that of three widely used models—AlexNet, DenseNet50 [31], and VGG16—enhanced with generated images. As presented in Figure 7, the curve analysis reveals that the ViT-Base model achieves the most consistent and favorable performance. It demonstrates smooth convergence, the fastest reduction in loss, and an impressive validation accuracy of 99.6%. In contrast, the other models show varying levels of instability, with AlexNet exhibiting notable overfitting issues. These findings suggest that the ViT-Base model not only outperforms the others in terms of accuracy but also maintains greater stability throughout the training process.

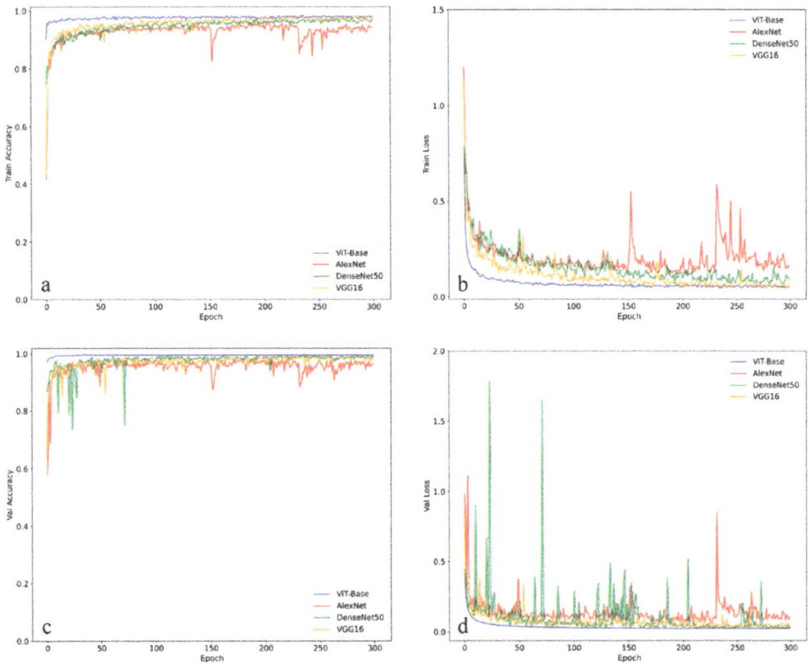


**Figure 7.** Performance of training parameters for four models. (**a**) Training accuracy, (**b**) training loss, (**c**) validation accuracy, (**d**) validation loss.

Table 6 demonstrates that the ViT-Base model achieves the highest performance, with accuracy improvements of 22.85%, 3.57%, and 3.21% over AlexNet, DenseNet50, and VGG16, respectively. This superior performance can be largely attributed to the ViT-Base model requirement for a large dataset to fully realize its classification potential. Notably, the extensive set of images generated by the StyleGAN3 model plays a crucial role in compensating for some of the inherent limitations of the ViT-Base model. By augmenting the training data with these synthetic images, we were able to address the model data requirements thus enabling it to achieve better results. Furthermore, a comparison with Table 4 reveals that, when augmented with generated images, the DenseNet50 and VGG16 models perform comparably to, or even outperform, the ViT-Base model trained solely on the original images. These results underscore the significant impact of data augmentation in the strategic use of StyleGAN3-generated images and in enhancing the performance of deep learning models.

**Table 6.** Performance comparison of different models.

| Model | Accuracy | Precision | Recall | F1 Score |
| --- | --- | --- | --- | --- |
| ViT-Base | 98.39% | 98.47% | 98.39% | 98.39% |
| AlexNet | 75.54% | 84.88% | 75.54% | 68.10% |
| DenseNet50 | 94.82% | 95.61% | 94.82% | 94.77% |
| VGG16 | 95.18% | 95.59% | 95.18% | 95.13% |

## 4. Discussion

This work presents a method for recognizing the growth stages of tomatoes, leveraging the StyleGAN3 model to generate synthetic images in conjunction with the ViT-Base model. The experimental results in Section 3 demonstrate that the generated images closely resemble real images, providing a solid foundation for their use in the recognition task. By incorporating these generated images into the ViT-Base model, superior performance is achieved in identifying the various growth stages of tomatoes. Through comparative experiments, it can be seen that the performance of the ViT-Base model combined with generated images for the identification of the tomato growth stage is significantly better than that of AlexNet, DenseNet50, VGG16, and other classical models combined with generated images. ViT-Base was shown to outperform traditional CNN models in some computer vision tasks such as image classification, object detection, and segmentation, especially when large-scale datasets are available. ViT's self-attention mechanism is able to capture global dependencies, which is difficult to achieve with convolutional operations in some tasks. On the other hand, DenseNet50 and VGG16 benefit more from data augmentation primarily due to their deeper architectures and greater model complexities. Their ability to learn more detailed and generalized features, coupled with a larger number of parameters and layers, enables them to fully leverage the diversity introduced by augmentation. In contrast, AlexNet, being a shallower network with fewer parameters, is less capable of fully exploiting the increased data diversity, resulting in comparatively smaller performance gains.

When selecting the ViT model for tomato classification, there was initial uncertainty regarding whether to opt for the more recent and improved Swin model, which has demonstrated superior performance in certain tasks. However, following in-depth discussions among the authors, we determined that ViT was more suitable for our work's primary focus on classification. Unlike tasks such as segmentation and detection, which are prediction-intensive and in which the Swin model excels, our work did not involve such tasks. Although ViT requires longer training times compared to Swin, it maintains high accuracy in classification tasks, particularly when enhanced by data. Moreover, the ViT model's relatively simple architecture, coupled with the availability of extensive pre-trained models, made it a more practical choice for training purposes thereby enabling more efficient model development for classification.

Although the StyleGAN3-generated images in this work are generally close to real images, their quality in noisy scenes is still insufficient, requiring preprocessing as is typical in many studies. For instance, the tomato fruiting and maturity stage datasets in this research were processed to remove background noise. Additionally, the long training times of StyleGAN3 pose challenges, especially with limited computational resources, leading to significant time and energy consumption. Transfer learning or cloud-based resources can be employed to mitigate these issues. Therefore, a critical area for future research will be developing methods to directly generate high-quality images quickly, without the need for preprocessing.

For an extended period, our research team has been leveraging computer vision technology to address practical problems in agricultural production. In the course of our work, we have also sought to extend our research to more complex production environments, such as open fields and orchards. As part of this effort, we utilized images collected from cameras in apple and citrus orchards to create a foundational dataset. By applying the StyleGAN3 model developed in this study for data augmentation, we successfully built growth stage recognition models tailored to specific apple and citrus varieties. While the controlled conditions in facility agriculture create a relatively stable environment, the complexity of orchards presents additional challenges. Although our approach showed some promising results in orchard environments and demonstrated its potential applicability in other scenarios, the actual test outcomes were not as effective as those seen in the controlled tomato production environment. This gap indicates that further refinements are needed, and improving this aspect will be a primary focus of our future research efforts.

## 5. Conclusions

This work begins by leveraging the StyleGAN3 generative adversarial network to generate images, addressing issues such as limited training data and data imbalances. Following this, the ViT-Base model is employed to recognize the growth stages of greenhouse tomatoes, producing promising results in terms of classification performance. The primary conclusions are as follows:

(1) The quality of images generated by StyleGAN3 is nearly identical to that of real images, with an average generation time of 153 milliseconds per image. This method proves to be an effective data augmentation solution, especially for cases with limited training data, such as small sample datasets. However, when computational resources are limited, transfer learning or background denoising techniques are necessary to reduce training time.
(2) By generating images through a generative adversarial network (GAN) and then applying the ViT-Base model for tomato growth stage recognition, this method achieves superior performance compared to direct recognition using original images. The combination of ViT with generated images reached an accuracy of 98.39% on the test set and an average detection speed of 9.5 milliseconds. When compared to AlexNet, DenseNet50, and VGG16, this method showed improvements in accuracy by 22.85%, 3.57%, and 3.21%, respectively, demonstrating its enhanced effectiveness in classification tasks.
(3) In areas with limited access to intelligent devices, the use of images generated by a generative adversarial network (GAN) significantly reduces the labor demands and inconsistencies of manual image collection. Furthermore, applying the ViT-Base model for tomato growth stage recognition can provide crucial data for informed decision-making and the precise management of tomato growth conditions. This approach offers considerable economic value by improving the efficiency of tomato production. Additionally, it can be extended to other similar crop categories, such as apples and citrus, in addition to tomatoes.

**Author Contributions:** Conceptualization, Y.L., P.H. and Y.H.; methodology, P.H.; software, Y.H., Y.L., L.H. and L.G.; validation, Y.H. and W.G.; data curation, Y.H.; writing—original draft, Y.H. and Y.L.; writing—review and editing, Y.L. and Y.H. All authors have read and agreed to the published version of the manuscript.

**Funding:** This work was supported by the Modern Agricultural Industrial Technology System Sichuan Grains Innovation Team special fund (SCCXTD-2024-20); the Special Fund for Independent Innovation of Sichuan Province—Research on the Application of Spatiotemporal Big Data Analysis in Agricultural Production Services (2022ZZCX034); the Key R&D Project of Sichuan Province Science and Technology Program; the 'Innovative Knowledge Service Platform for Crop and Livestock Breeding Research' (2021YFYZ0028-01); and the 'Tianfu Granary' Digital Agriculture Sichuan-Chongqing Joint Innovation Key Laboratory Project—Intelligent Planting Strategy Model and Application for Facility Tomatoes (TFLCSZ-JB3).

**Institutional Review Board Statement:** Not applicable.

**Data Availability Statement:** The data presented in this study are available on request from the corresponding author due to our image datasets being self-built.

**Conflicts of Interest:** The authors declare no conflicts of interest.

## References


1. Alajrami, M.A.; Abu-Naser, S.S. Type of Tomato Classification Using Deep Learning. *Int. J. Acad. Pedagog. Res.* **2020**, *3*, 21–25.
2. Wei, Y.; Qin, R.; Ding, D.; Li, Y.; Xie, Y.; Qu, D.; Zhao, T.; Yang, S. The impact of the digital economy on high-quality agricultural development--Based on the regulatory effects of financial development. *J. Huazhong Agric. Univ.* **2023**, *43*, 9–21.
3. Hinton, G.E.; Salakhutdinov, R.R. Reducing the dimensionality of data with neural networks. *Science* **2006**, *313*, 504–507. [CrossRef] [PubMed]
4. Hinton, G.E.; Osindero, S.; Teh, Y.-W. A fast learning algorithm for deep belief nets. *Neuralcomputation* **2006**, *18*, 1527–1554. [CrossRef] [PubMed]
5. Tomas, M.; Kai, C.; Greg, C.; Jeffrey, D. Efficient Estimation of Word Representations in Vector Space. *arXiv* **2013**, arXiv:1301.3781.
6. Jacob, D.; Ming, W.; Kenton, L.; Kristina, T. BERT: Pre-training of Deep Bidirectional Transformers for Language Understanding. *arXiv* **2019**, arXiv:1810.04805.
7. Mahesh, G.; Sweta, J. Stacked Convolutional Neural Network for Diagnosis of COVID-19 Disease from X-ray Images. *arXiv* **2020**, arXiv:2006.13871.
8. Mahesh, G.; Sweta, J.; Raghav, A. DeepRNNetSeg: Deep Residual Neural Network for Nuclei Segmentation on Breast Cancer Histopathological Images. In Proceedings of the International Conference on Computer Vision and Image Processing (CVIP), Singapore, 27–29 September 2019; pp. 243–253.
9. Xu, Y.S.; Fu, T.J.; Yang, H.K.; Lee, C.Y. Dynamic video segmentation network. In Proceedings of the 2018 IEEE/CVF Conference on Computer Vision and Pattern Recognition (CVPR), Salt Lake City, UT, USA, 18–23 June 2018; pp. 6556–6565.
10. Tang, H.; Ding, L.; Wu, S.; Ren, B.; Sebe, N.; Rota, P. Deep Unsupervised Key Frame Extraction for Efficient Video Classification. *ACM Trans. Multimedia Comput. Commun. Appl.* **2023**, *119*, 1–17. [CrossRef]
11. Amreen, A.; Sweta, J.; Mahesh, G.; Swetha, V. Tomato plant disease detection using transfer learning with C-GAN synthetic images. *Comput. Electron. Agric.* **2021**, *187*, 106279.
12. Xu, J.; Wang, J.; Xu, X.; Ju, S. Image recognition for different developmental stages of rice by RAdam deep convolutional neural networks. *Trans. Chin. Soc. Agric. Eng.* **2021**, *37*, 143–150.
13. Zhang, X.; Hou, T.; Hao, Y.; Shangguan, H.; Wang, A.; Peng, S. Surface Defect Detection of Solar Cells Based on Multiscale Region Proposal Fusion Network. *IEEE Access* **2021**, *9*, 62093–62101. [CrossRef]
14. Krizhevsky, A.; Sutskever, I.; Hinton, G.E. ImageNet classification with deep convolutional neural networks. *Commun. ACM* **2017**, *60*, 84–90. [CrossRef]
15. Al-Qizwini, M.; Barjasteh, I.; Al-Qassab, H.; Radha, H. Deep learning algorithm for autonomous driving using GoogLeNet. In Proceedings of the 2017 IEEE Intelligent Vehicles Symposium(IV), Los Angeles, CA, USA, 11–14 June 2017; pp. 89–96.
16. Alexey, D.; Lucas, B.; Alexander, K.; Dirk, W.; Zhai, X.; Thomas, U.; Mostafa, D.; Matthias, M.; Georg, H.; Sylvain, G.; et al. An Image is Worth 16x16 Words: Transformers for Image Recognition at Scale. *arXiv* **2020**, arXiv:2010.11929.
17. Bai, P.; Feng, Y.; Li, G.; Zhao, M.; Zhou, H.; Hou, Z. Algorithm of wheat disease image identification based on Vision Transformer. *J. Chin. Agric. Mech.* **2024**, *45*, 267–274.
18. Karen, S.; Andrew, Z. Very Deep Convolutional Networks for Large-Scale Image Recognition. *arXiv* **2014**, arXiv:1409.1556.
19. Wang, Y.; Li, Y.; Xu, J.; Wang, A.; Ma, C.; Song, S.; Xie, F.; Zhao, C.; Hu, M. Crop Disease Recognition Method Based on Improved Vision Transformer Network. *J. Chin. Comput. Syst.* **2024**, *45*, 887–893.
20. He, K.; Zhang, X.; Ren, S.; Sun, J. Deep Residual Learning for Image Recognition. *arXiv* **2015**, arXiv:1512.03385.
21. Rasti, S.; Bleakley, C.J.; Silvestre, G.C.M.; Holden, N.M.; Langton, D.; Gregory, M.P.; O'Hare, G.M. Crop growth stage estimation prior to canopy closure using deep learning algorithms. *Neural Comput. Appl.* **2021**, *33*, 1733–1743. [CrossRef]
22. Tan, S.; Liu, J.; Lu, H.; Lan, M.; Yu, J.; Liao, G.; Wang, Y.; Li, Z.; Qi, L.; Ma, X. Machine Learning Approaches for Rice Seedling Growth Stages Detection. *Front. Plant Sci.* **2022**, *13*, 914771. [CrossRef] [PubMed]
23. Liu, W.; Lu, X. Research Progress of Transformer Based on Computer Vision. *Comput. Eng. Appl.* **2022**, *58*, 1–16.
24. Goodfellow, I.; Pouget-Abadie, J.; Mirza, M.; Xu, B.; Warde-Farly, D.; Ozair, S.; Courville, A.; Bengio, Y. Generative Adversarial Nets. *arXiv* **2014**, arXiv:1406.2661.
25. Han, X.; Li, Y.; Gao, A.; Ma, J.; Gong, Q.; Song, Y. Data Augmentation Method for Sweet Cherries Based on Improved Generative Adversarial Network. *Trans. Chin. Soc. Agric. Mach.* **2024**, *55*, 1–17.
26. Karras, T.; Aittala, A.; Laine, S.; Harkonen, E.; Hellsten, J.; Lehtinen, J.; Aila, T. Alias-Free Generative Adversarial Networks. *arXiv* **2021**, arXiv:2106.12423.
27. Karras, T.; Laine, S.; Aila, T. A Style-Based Generator Architecture for Generative Adversarial Networks. *arXiv* **2018**, arXiv:1812.04948.
28. Mirza, M.; Osindero, S. Conditional Generative Adversarial Nets. *arXiv* **2014**, arXiv:1411.1784.
29. Radford, A.; Metz, L.; Chintala, S. Unsupervised Representation Learning with Deep Convolutional Generative Adversarial Networks. *arXiv* **2015**, arXiv:1511.06434.

30. Brock, A.; Donahue, J.; Simonyan, K. Large Scale GAN Training for High Fidelity Natural Image Synthesis. *arXiv* **2018**, arXiv:1809.11096.
31. Huang, G.; Liu, Z.; Maaten, L.; Weinberger, K.Q. Densely Connected Convolutional Networks. *arXiv* **2016**, arXiv:1608.06993.

agriculture

MDPI

*Article*

# Pear Fruit Detection Model in Natural Environment Based on Lightweight Transformer Architecture

**Zheng Huang [1,2], Xiuhua Zhang [1,2,*], Hongsen Wang [1,2], Huajie Wei [1,2], Yi Zhang [1,2] and Guihong Zhou [2,3]**

[1] College of Mechanical and Electrical Engineering, Hebei Agricultural University, Baoding 071001, China; 20237091054@pgs.hebau.edu.cn (Z.H.); 20237091051@pgs.hebau.edu.cn (H.W.); 20227090993@pgs.hebau.edu.cn (H.W.); zhangyi@hebau.edu.cn (Y.Z.)

[2] Hebei Intelligent Agriculture Technology Innovation Center, Baoding 071001, China; zhouguihong@hebau.edu.cn

[3] College of Information Science and Technology, Hebei Agricultural University, Baoding 071001, China

* Correspondence: jdzhxh@hebau.edu.cn or zhang72xh@163.com; Tel.: +86-139-3089-3593

**Abstract:** Aiming at the problems of low precision, slow speed and difficult detection of small target pear fruit in a real environment, this paper designs a pear fruit detection model in a natural environment based on a lightweight Transformer architecture based on the RT-DETR model. Meanwhile, Xinli No. 7 fruit data set with different environmental conditions is established. First, based on the original model, the backbone was replaced with a lightweight FasterNet network. Secondly, HiLo, an improved and efficient attention mechanism with high and low-frequency information extraction, was used to make the model lightweight and improve the feature extraction ability of Xinli No. 7 in complex environments. The CCFM module is reconstructed based on the Slim-Neck method, and the loss function of the original model is replaced with the Shape-NWD small target detection mechanism loss function to enhance the feature extraction capability of the network. The comparison test between RT-DETR and YOLOv5m, YOLOv7, YOLOv8m and YOLOv10m, Deformable-DETR models shows that RT-DETR can achieve a good balance in terms of model lightweight and recognition accuracy compared with other models, and comprehensively exceed the detection accuracy of the current advanced YOLOv10 algorithm, which can realize the rapid detection of Xinli No. 7 fruit. In this paper, the accuracy rate, recall rate and average accuracy of the improved model reached 93.7%, 91.9% and 98%, respectively, and compared with the original model, the number of params, calculation amount and weight memory was reduced by 48.47%, 56.2% and 48.31%, respectively. This model provides technical support for Xinli No. 7 fruit detection and model deployment in complex environments.

**Keywords:** fruit detection; lightweight Transformer; RT-DETR; Xinli No. 7; FasterNet; Slim-Neck

## 1. Introduction

Pear is one of the main fruits in China. It has a cultivation history of more than 3000 years in China. Its cultivation area and output rank first in the world, and it is known as the king of fruits in China. China's pear production has a positive growth trend, and the export volume has always been above 500,000 tons, which plays an important role in China's agricultural economy. The mechanization level of the pear industry is low, and the picking process requires a large amount of labor, accounting for 35–45% of the total labor input [1], and the picking cost is 50–70% of all the links [2]. The overall process of intelligent and mechanized picking can be generally divided into two parts: one is to realize rapid identification of pear fruits, and the other is to accurately and losslessly grasp the identified fruits [3]. However, due to the influence of light changes, branches and leaves occlusion, fruit overlap and distance environment changes, the model in the actual picking environment is prone to problems such as identification difficulties and slow detection speed. Therefore, how to quickly and accurately identify pear fruits has become the primary problem of automatic picking.

Real-time target detection is an important technology field that is widely used in all walks of life. Integrating visual recognition technology into global agriculture can carry out real-time monitoring and accurate analysis of various factors in the agricultural environment, thus significantly improving agricultural production efficiency, quality and sustainability. In the agricultural field, it is embodied in forest fruit quality detection [4], animal breeding detection [5], plant pest detection [6], fruit target detection [7], fruit density classification [8] and orchard road example segmentation [9]. Existing real-time target detection is generally based on CNN architecture, the most famous of which is the YOLO model [10] because it reasonably balances the trade-off between speed and accuracy. Tan et al. [11] proposed a fragrant pear object detection method based on improved YOLOv8n. Taking YOLOv8n as the base model, he used residual convolution module to optimize C2 f for feature fusion, optimized Spatial Pyramid Pooling Fast (SPPF) to Simplified Spatial Pyramid Pooling Fast (simSPPF), introduced PConv convolution and used Inner-CIoU loss function. Weight parameter sharing is proposed to achieve a lightweight detection head. The average accuracy of the self-built Sweet pear dataset is 94.7%, and the reasoning time of the original model on the self-built dataset is 62.9 ms. Zheng et al. [12] designed a lightweight pear target detection M-YOLOv7-SCSN+F model. The data enhancement method based on the Fourier transform generated new image data by analyzing image frequency domain information and reconstructing image amplitude components, thereby improving the model generalization ability. Liu et al. [13] proposed a detection method based on MAE-YOLOv8 for small objects in a real complex orchard environment. By replacing the feature pyramid network, the authors improved the detection accuracy of small target objects. In order to alleviate the problem of missing detection and inaccurate positioning caused by overlapping occlusion, the minimum point distance intersection is introduced as a regression loss function. Chen et al. [14] proposed an improved YOLOv8-based multi-objective segmentation method for the apple tree at the emerging stage and combined advanced convolutional network modules (ConvNeXt V2, Multi-Scale Dilated Attention (MSDA) and Distribution Shifting Convolution (DSConv)) to enhance YOLOv8 and improve the accuracy of organ segmentation in complex natural environments. The above models show certain advantages in accuracy and speed, but YOLO models usually require Non-Maximum Suppression (NMS) for post-processing, and the introduction of hyperparameters makes the accuracy and speed of the model unstable, slowing down the reasoning speed of the model [15]. In addition, the need to select reasonable NMS thresholds has hindered the development of real-time object detection.

DETR (Detection Transformer) [16] was first proposed by Facebook, and Transformer architecture [17] was introduced into the target detection network, treating detection as a collection prediction problem without the need to generate candidate regions and post-processing steps. In recent years, the application of Transformers in real-time target detection has become an important research direction in the field of computer vision. The introduction of the Transformer provides a new way of thinking for traditional Convolutional Neural Networks (CNN). It is particularly good when it comes to handling long-term dependencies, global context information, and enhancing the expressiveness of the model. Although DETR is very successful, its training convergence speed is slow, and its detection performance for small targets is poor. Therefore, RT-DETR [18], an end-to-end real-time target detector based on Transformer architecture, provides a method to solve such problems. The hybrid encoder of DETR introduces multi-scale features to accelerate training convergence and improve performance [19], but the sharply increased sequence length still causes the encoder to become a computational bottleneck. An efficient hybrid encoder is designed in RT-DETR to replace the original Transformer encoder. Multi-scale features can be rapidly processed by cross-scale fusion and decoupling intra-scale interaction. The encoder can effectively process features of different scales, greatly reduce the computational load of the encoder and significantly improve the reasoning speed. In order to reduce the difficulty of the object query, DETR uses the confidence score to select the best feature in the encoder to initialize the object query [20,21]. However, the current query selection leads to uncertainty in the selection feature, which affects the model performance. Therefore, RT-DETR (The full name of RT-DETR is in the Appendix A Table A1) proposes the IoU-aware query selection in the decoder. By providing IoU constraints in the training process, higher quality initial object queries are provided for the decoder and the detection accuracy is improved. Zhao et al. [22] proposed a lightweight cherry tomato ear state recognition model based on an improved Transformer. By replacing the trunk structure and adding an adaptive detail fusion module, the calculation and model parameters are significantly reduced, and the average accuracy of 90% is guaranteed, while low calculation and fast detection are realized. Hu et al. [23] proposed an improved RT-DETR detection model, RIC-DETR, in which ResNet18 was selected as the backbone feature network, and the reverse residual mobile module was introduced while the second innovation and improvement were carried out. Under the condition that the average accuracy of 97.2% was maintained, the computation, parameter count and memory footprint are greatly reduced. Li et al. [24] proposed a chicken target detection model with high precision and strong generalization based on improved Real-Time Detection Transformer (RT-DETR) Efficient Multi-Scale-Conv Detection Transformer (EMSC-DETR). In order to solve the problem of small target features being easily lost, a Study Data Tabulation Model (SDTM) was introduced. The module significantly improves the computational efficiency of the converter, with $mAP_{0.5}$ being 98.6%. Li et al. [25] proposed an end-to-end semi-supervised object detection method based on DEtection TRansformer (DETR), which simplified the post-processing process without the need for Non-Maximum Suppression (NMS) and adopted a more advanced binary matching allocation strategy. The proposed method only used 5% of the total data to achieve 74.1% of the mAP. Although the performance of RT-DETR is slightly inferior to that of YOLO in small target detection, it is superior in speed and accuracy to current real-time detectors of similar scale.

Aiming at the problems of low precision and slow detection speed of pear fruit in a real environment and difficult detection of small target fruit in a long-term environment, this study improved and designed a pear fruit detection model in a natural environment based on RT-DETR. Firstly, ResNet-r18 was replaced with a lightweight FasterNet network. Secondly, HiLo, an efficient attention mechanism for extracting high and low-frequency information, is used to improve the Attention-based Intrascale Feature Interaction module (AIFI), which achieves higher performance and faster speed. A new convolutional GSConv is introduced into the Compact Convolutional Feature Fusion Module (CCFM) to reduce the complexity of the model and maintain high recognition accuracy. Finally, the loss function in RT-DETR is replaced with Shape-IoU and used in combination with the Normalized Wasserstein Distance (NWD) small target detection mechanism to further improve the detection performance of small target pears.

## 2. Materials and Methods

### 2.1. Construction of Data Sets

#### 2.1.1. Image Acquisition

The image data was collected from the modern pear garden demonstration base of Hebei Wokang Agricultural Science and Technology Co., Ltd., Xingtai City, Hebei Province, China. The collected pear variety was Xinli No. 7, and the collection time was from July to August 2024, during the fruit ripening and picking period. Fruit characteristic parameters of Xinli No. 7; The shape is oval, the fruit is medium to large, the average single fruit weight is 185.85 g, the longitudinal diameter is 72.61 mm, and the transverse diameter is 71.73 mm. This variety has the characteristics of precocious maturity and a long natural harvesting period, as well as resistance to disease and insect pests and storage resistance [26]. It is rare for Chinese pear crossbreeding to combine these four fine traits in a single fruiting individual at the same time, which is a germplasm innovation in Chinese pear breeding.

Through field investigation, the plant spacing of a modern pear garden is 1~2 m, row spacing is 4 m, tree height is 3~4 m, and the main cultivation mode is wide row close planting. Taking the JAKA C12 robotic arm as the picking robotic arm, the picking working radius of the robotic arm is 1327 mm, so two shooting distances are designed for close-range shooting (100~500 mm) and long-range shooting (700~1350 mm). In order to make the time distribution of the collected data set closer to the actual picking time of fruit farmers, the shooting time of this data set was selected in two periods: 7:00–12:00 and 13:00–19:00. The image shooting equipment was iPhone 13, Honor Magic 3 pro and ZED 2i binocular cameras, which were saved as jpg format. A total of 7468 images of Xinli No. 7 fruit in different environments and scenes were collected in the experiment. The data set was classified according to near-vision conditions, including 4001 close-range images and 3467 long-range images. According to lighting conditions, 3357 low-light images, 1890 high-light images and 2221 backlight images were classified. In the data set, the factors that influenced the collection, such as branch and leaf occlusion, fruit overlap, different distances, and lighting conditions, were comprehensively considered, and part of the collected images are shown in Figure 1.

**Figure 1.** Fruit image of Xinli No. 7 under different environment. (**a**) Branches and leaves cover. (**b**) Frontlight close view. (**c**) Backlight close view. (**d**) Fruit overlap. (**e**) Frontlight distant view. (**f**) Backlight distant view.

2.1.2. Data Set Creation

In order to improve the annotation efficiency, YOLO semi-automatic annotation was adopted [27]. First, labelImg was used to annotate 2000 fruit images of Xinli No. 7 in different scenes, and the YOLOv7 model was applied to train the annotated images to obtain the optimal weight file, and then semi-automatic annotation was performed on the remaining fruit images of Xinli No. 7. Finally, labelImg was used to manually adjust the wrong label and missing label of the annotation result.

After manual and semi-automatic annotation, a total of 7468 XML files of Xinli No. 7 were obtained and randomly divided according to the ratio of 8:1:1 to form a training set, verification set and test set, in which the training set contained 5974 images, the verification set and the test set each contained 747 images.

*2.2. Experimental Method*

2.2.1. Target Detection Selection Index

In this paper, Precision (P, %), Recall (R, %), mean Average Precision (mAP, %), Frames Per Second (FPS, frames/s), Floating-point Operations Per second (FLOPs, G), parameter number (Params, M) and Model size (MB) are used as model evaluation indicators, and P, R and mAP are used to measure the detection accuracy of the model [28], FPS is used to measure the detection speed of the model, model lightweight is measured by the number of parameters and the amount of computation, and deployment cost is assessed using the space occupied by the model.

2.2.2. Training of Network Models

In this study, the Intel core i7-14700 KF Win11 operating system is used, the main frequency is 3.4 GHZ, the running memory is 32 G, and the Nvidia GeForce RTX 4070 Ti SUPER graphics card is installed. Experiments were carried out on the Pycharm platform, the PyTorch deep learning framework was configured for environment construction, and Python language was used for algorithm writing. Model hyperparameters were set as follows: the default image is 640 $\times$ 640 pixels, the number of model training iteration cycles was set to 200, the number of samples processed in each batch was set to 8, the number of worker threads during data loading was set to 4, and the initial learning rate was set to 0.0001. All other training hyperparameters were used as default values, and all tests of the model were carried out under the same environment.

## 3. Network Model and Improvement

This paper improves a pear detection model under a natural environment based on the end-to-end real-time target detector RT-DETR based on Transformer architecture. First, the backbone of the original model is replaced with a lightweight FasterNet network. Secondly, HiLo, an efficient attention mechanism that can extract high and low-frequency information, is used to improve the Attention-based Intrascale Feature Interaction (AIFI), and a GSConv convolution is introduced into the Cross-Scale Feature Fusion Module (CCFM). Finally, the loss function in the original model is replaced with the loss function of the Shape-NWD small target detection mechanism. The model performance is improved. The network structure diagram of the improved model is shown in Figure 2.

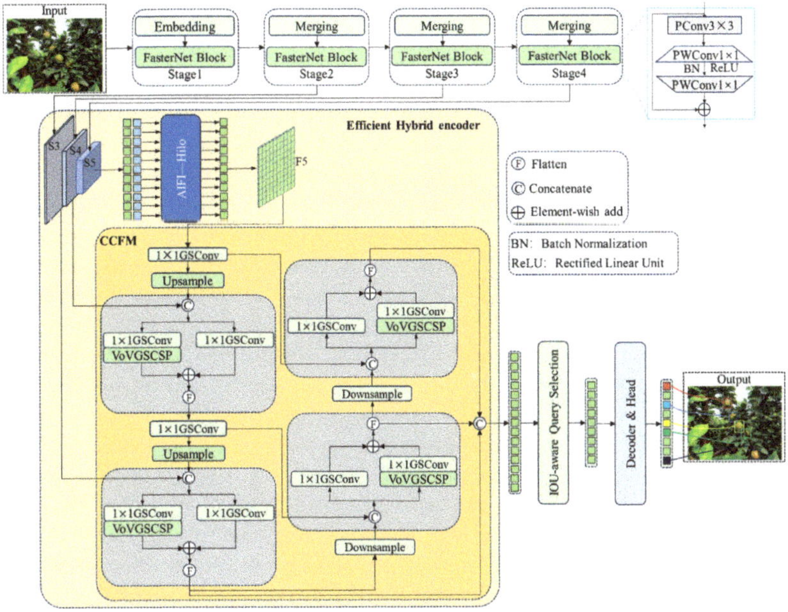


**Figure 2.** Improved network model structure diagram.

### *3.1. Detect Model Backbone Network Replacement*

In order to make the improved RT-DETR model closer to practical applications, this paper re-studies the commonly used backbone network model and finds that most models focus on reducing Floating point operations (FLOPs), but the reduction of FLOPs does not necessarily mean the reduction of the same horizontal delay, as shown in Formula (1). The main reason is that Floating point operations per second (FLOPS) need to be optimized at the same time to achieve truly low latency. Therefore, this paper introduces a lightweight FasterNet network as the backbone network for feature extraction and introduces a simple but fast and effective convolutional PConv, which can extract spatial features more efficiently while reducing redundant computation and memory access. The FasterNet Block is shown in the blue dashed box in Figure 2.

$$\text{Latency} = \frac{\text{FLOPs}}{\text{FLOPS}} \tag{1}$$

Compared with the traditional ResNet network, the FasterNet network can greatly reduce the parameter number and calculation amount of the model when the precision rate, recall rate, and mean average precision are slightly reduced [29], realizing the light weight of the model, which is more conducive to deployment on the pear fruit picking robot. There are four hierarchical stages in the FasterNet backbone network, each of which is preceded by an embedding layer (step size 4, Conv4×4) or a merging layer (step size 2, Conv2×2) for spatial downsampling and channel number expansion, respectively [30]. In order to fully and efficiently utilize the information from all channels, each directional residual block consists of a PConv layer at the front end and two PWConv layers at the back end, in which a Batch Normalization (BN) is placed in the middle of the two PWConv layers. The. Rectified Linear Unit (ReLU) activation function is used to enhance the input features of the Xinli No. 7 fruit image, thereby improving model performance and training speed.

### 3.2. Introduction of Improved Efficient Attention Mechanism HiLo

In order to solve the problem of the huge computational cost of throughput in high-resolution images in the original RT-DETR model, especially in this task, a new efficient attention mechanism, HiLo, was introduced, and the scaled dot product attention of the attention branch was re-improved to adapt the two high and low-frequency attention branches. One path encodes high-frequency interactions by scaling dot-product attention and relatively high-resolution feature maps, while the other encodes low-frequency interactions by dot-product attention and downsampled features and finally incorporates improved attention mechanisms into the in-scale feature interaction module. The HiLo attention mechanism uses two kinds of effective attention to decoupling the high and low frequencies in the feature graph [31], eliminating the huge computational burden of different lower frequencies in the standard Multi-head Self-Attention layer (MSA) features.

Compared to existing standard attention mechanisms, such as self-attention mechanisms, CBAM or Transformer models, HiLo attention mechanisms improve the representation and computational efficiency of the model through more sophisticated processing and fusion of high-level and low-level information. As shown in Figure 3, the HiLo attention mechanism divides the MSA layer into two components: One is used to encode high-frequency attention branch Hi-Fi with local self-attention mechanism and high-resolution feature map, and the other is used to globally encode low-frequency attention branch Lo-Fi by subsampling features. The output of each HiLo attention mechanism is a series of high-frequency attention branches and low-frequency attention branches, as shown in formula (2). Thus, the information extraction efficiency [32] is effectively improved, which is more effective than standard MSA.

$$\text{HiLo}(X) = [\text{Hi-Fi}(X);\ \text{Lo-Fi}(X)] \tag{2}$$

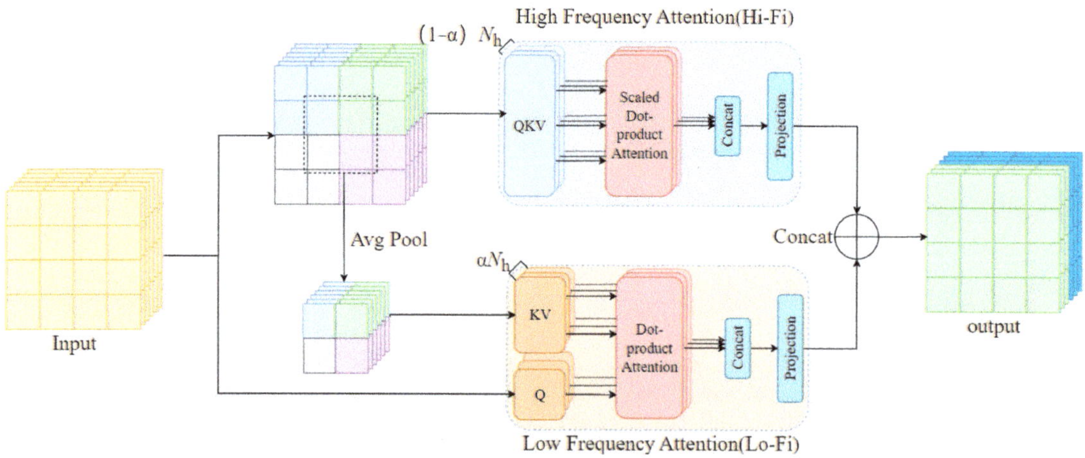


**Figure 3.** High-efficiency attention mechanism HiLo framework diagram.

HiLo allocates Hi-Fi and Lo-Fi with the same structure as the standard Multi-head Self-Attention layer (MSA). In order to make the allocation scheme more favorable, the same number of magnetic heads in the MSA are divided into two groups, as shown in Formula (3), and the division ratio is: where $(1 - \alpha)N_h$ heads are used for Hi-Fi, Other $\alpha N_h$ heads are used for Lo-Fi, and each attentional architecture is less complex than a standard MSA, so the overall framework of the HiLo attentional mechanism guarantees a low complexity model with high throughput for high-resolution images.

$$\text{HiLo} = \alpha\text{Hi-Fi} + (1 - \alpha)\text{Lo-Fi} \quad (3)$$

### 3.3. *Reconstruction of Cross-Scale Feature Fusion Module Based on Slim-Neck Method*

In order to further meet the real-time detection requirements of the pear fruit-picking robot, this study introduced lightweight convolutional GSConv into the Cross-Scale Feature Fusion Module (CCFM), which can utilize local information and global information at the same time and realize the fusion of these different scale information. It can further reduce the complexity of the model while maintaining accuracy and solve the problem of the speed of prediction calculation in convolutional neural networks. As shown in Figure 4 below, GSConv first inputs an ordinary convolution undersampling, then uses DWConv deep convolution to concatenate the output results of the two CONVs, and finally performs data distribution shuffle operation to concatenate the corresponding channel numbers of the previous two convolution. Therefore, when the spatial information of the input image is gradually transferred to the channel, GSConv convolution avoids the phenomenon of partial loss of semantic information caused by spatial compression and channel expansion of each feature image [33].

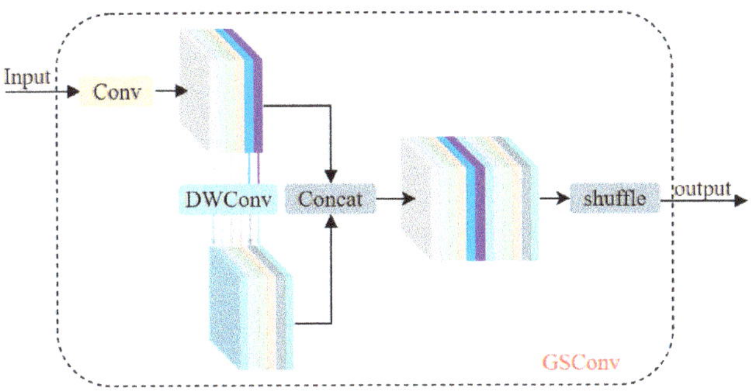


**Figure 4.** GSConv module framework diagram.

After introducing lightweight GSConv convolution, we continue to introduce GS bottleneck and interstage partial network modules VoV-GSCSP, which are designed to further improve feature utilization efficiency and network performance. An improved transformation structure of the VoV-GSCSP module is proposed, as shown in Figure 5. Due to its simple structure, this module only consists of lightweight GSConv convolution and GS bottleneck, which requires less training hardware while ensuring performance.

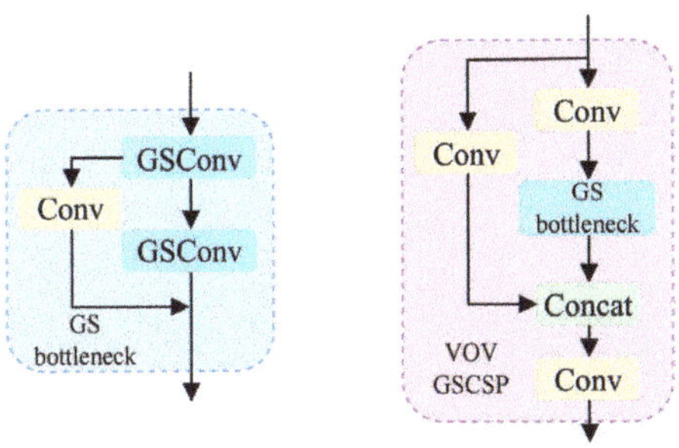


**Figure 5.** GS bottleneck and VoV-GSCSP cross-stage network modules.

Therefore, this cross-scale feature Fusion Module based on Slim-Neck architecture is more in line with the requirements of this paper for a lightweight model and low computing cost and can achieve significant accuracy improvement while meeting the deployment conditions of mobile terminals.

### 3.4. *Shape-NWD Small Target Detection Mechanism Loss Function*

In the RT-DETR algorithm, the prediction frame regression loss function uses GIoU. When the two prediction frames have the same height and width and are in the same horizontal plane, GIoU degenerates into IoU loss function, which leads to the problem of slow convergence and inaccurate regression. To solve the above problems, this study adopts the shape-IoU loss function to replace the GIoU loss function used by RT-DETR [34]. This method can calculate the loss by focusing on the shape and scale of the bounding box itself so as to make bounding box regression more accurate. Figure 6 shows the schematic diagram of Shape-IoU parameters.

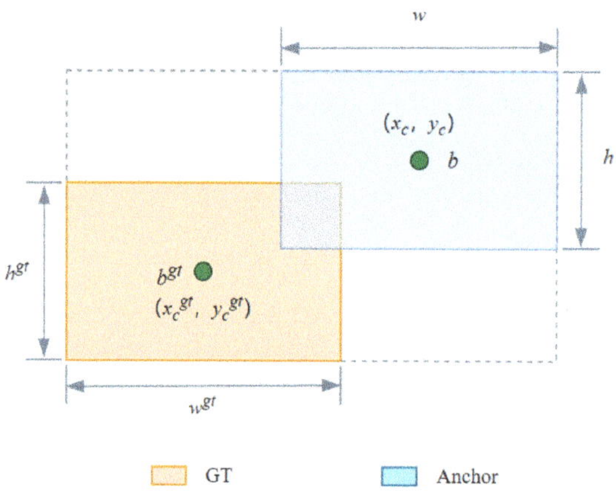


**Figure 6.** Schematic diagram of Shape-IoU parameters.

The formula of Shape-IoU can be derived from Figure 6 and Formulas (4)–(6).

$$\text{IoU} = \frac{|B \cap B^{gt}|}{|B \cup B^{gt}|} \tag{4}$$

$$distance^{shape} = hh \times (x_c - x_c^{gt})^2 / c^2 + ww \times (y_c - y_c^{gt})^2 / c^2 \tag{5}$$

$$\Omega^{shape} = \sum_{t=w,h} (1 - e^{-wt})^{\theta}, \theta = 4 \tag{6}$$

where IoU is the actual crossover ratio, $ww$ and $hh$ represent the weight coefficients in the horizontal direction and vertical direction respectively, and their values are related to the shape of the GT frame. Shape-IoU loss is defined by Formula (7).

$$L_{\text{Shape-IoU}} = 1 - \text{IoU} + distance^{shape} + 0.5 \times \Omega^{shape} \tag{7}$$

The NWD [35] small target detection method was introduced into the target detection of Xinli No. 7 fruit in the test, and the NWD small target detection mechanism and Shape-IoU loss function were combined. In this experiment, the strategy with 50% of each is selected. As shown in Formula (8), the loss function of the Shape-NWD small target detection mechanism can not only improve the target detection accuracy but also ensure the model detection speed.

$$\text{Shape-NWD} = (1 - \text{IoU})(1 - 0.5\text{NWD}) + \text{IoU}(1 - 0.5\text{Shape-IoU}) \quad (8)$$

# 4. Test Results and Analysis

## *4.1. Performance Comparison Test of RT-DETR Model*

In order to select the optimal model file, the performance of ResNet-r18, ResNet-r34 and ResNet-r50 backbone networks commonly used in the ResNet series were compared under the same conditions, taking Precision, Recall, $mAP_{0.5}$, Params, FLOPs, Model size and FPS as the basic evaluation indexes. The test results are shown in Table 1 below. Under the condition that the accuracy rate, recall rate and average accuracy mean are not different, the RT-DETR-r18 model file meets the requirements of rapid detection of pear fruits while meeting the lightweight requirements, and the generated weight model is 38.5 MB, which is more conducive to deployment on the mobile end of picking equipment.

**Table 1.** Performance comparison of RT-DETR models.

| Model | Params/M | FLOPs/G | P/% | R/% | mAP/% | Weights Size/MB | FPS·(f s$^{-1}$) |
|---|---|---|---|---|---|---|---|
| RT-DETR-r18 | 19.87 | 57.3 | 89.7 | 90.1 | 96.6 | 38.5 | 75.7 |
| RT-DETR-r34 | 31.1 | 88.8 | 89.5 | 90.6 | 96.7 | 60 | 61.6 |
| RT-DETR-r50 | 41.96 | 129.5 | 90.4 | 89.8 | 96.6 | 164 | 49.8 |

## *4.2. Comparative Analysis of Different Algorithms*

YOLOv5m, YOLOv7, YOLOv8m, YOLOv10m, Deformable-DETR and RT-DETR-r18 models were used to perform performance comparison on Xinli No. 7 data set, and the test results were shown in Table 2. According to the results in Table 2 below, it can be seen that the number of parameters and calculation amount displayed by YOLOv5m are small, but they do not meet the requirements for rapid detection of pear fruit. Although YOLOv7 maintains high recognition accuracy, the model volume is large, which does not meet the requirements of a lightweight model. Although YOLOv8m and YOLOv10m have high detection accuracy, they do not meet the requirements of lightweight models due to the large amount of calculation. Compared with the YOLOv8m model, RT-DETR-r18 has a 5.97 M and 21.4 G reduction in parameter number and computation amount, a 0.8% increase in average accuracy, and an 11.1 MB reduction in weight memory. Compared with the YOLOv10m model, RT-DETR-r18 is superior to YOLOv10m in precision, recall and average accuracy, although the params and weight memory are increased by 3.42 M and 6.6 MB, the computation amount is decreased by 6.1 G and the FPS is increased by 5.28 f/s.

**Table 2.** Analysis of comparison results of different algorithms.

| Model | Params/M | FLOPs/G | P/% | R/% | mAP/% | Weights Size/MB | FPS·(f·s$^{-1}$) |
|---|---|---|---|---|---|---|---|
| YOLOv5m | 21.04 | 50.2 | 87.7 | 91 | 96.4 | 40.4 | 15.9 |
| YOLOv7 | 36.48 | 103.2 | 91.8 | 89.5 | 97.1. | 71.3 | 24.04 |
| YOLOv8m | 25.84 | 78.7 | 89.1 | 89.7 | 95.8 | 49.6 | 75.19 |
| YOLOv10m | 16.45 | 63.4 | 89.3 | 90 | 96.3 | 31.9 | 70.42 |
| Deformable-DETR | 40 | 196 | 88.6 | 88.3 | 95.7 | 86 | 29.5 |
| RT-DETR-r18 | 19.87 | 57.3 | 89.7 | 90.1 | 96.6 | 38.5 | 75.7 |

Deformable-DETR has a significant gap with RT-DETR in all evaluation indexes in this paper, especially in terms of parameter number, computation amount and weight memory, 20.13M, 138.7 G and 47.5 MB higher than the RT-DETR model. However, RT-DETR can achieve a good balance in terms of model lightweight and recognition accuracy, and the transmission frame per second can reach 75.7 f/s, meeting the needs of practical applications and realizing the real-time accurate detection of Xinli No.7 fruit.

### 4.3. Ablation Study

The ablation test is an important means to evaluate the effectiveness of model improvement. In order to verify the validity of the modules and loss functions proposed in this paper, the ablation test was designed and carried out under the same experimental environment following the principle of the control variable method. As shown in Table 3, test 1 represents the performance indicators of RT-DETR before improvement. Through the following test improvements, the comparison of key evaluation indicators of the improvement points can be visually observed. The results show that the identification performance of the model is improved compared with that of the original model.

**Table 3.** RT-DETR ablation results based on Xinli No. 7 data set.

| Model | FN | HiLo | SN | Shape-NWD | Params/M | FLOPs/G | P/% | R/% | mAP/% | Weights Size/MB |
|---|---|---|---|---|---|---|---|---|---|---|
| Experiment1 | × | × | × | × | 19.87 | 57.3 | 89.7 | 90.1 | 96.6 | 38.5 |
| Experiment2 | √ | × | × | × | 10.81 | 28.5 | 89.2 | 89.1 | 96.1 | 21.2 |
| Experiment3 | √ | √ | × | × | 10.78 | 28.6 | 92 | 90.1 | 97.1 | 21.2 |
| Experiment4 | √ | √ | √ | × | 10.24 | 25.1 | 91.3 | 91.5 | 97.3 | 19.9 |
| Experiment5 | √ | √ | √ | √ | 10.24 | 25.1 | 93.7 | 91.9 | 98 | 19.9 |

It can be seen from experiment 2 in Table 3 that after introducing the lightweight FasterNet (FN) network as the backbone network for feature extraction, the number of params and calculation amount of the model is reduced by 45.6% and 50.26%, respectively, and the weight memory is also significantly reduced, which realizes the lightweight of the model. This improvement is attributed to the simple and efficient extraction of spatial features of the Xinli No. 7 fruit image after the introduction of PConv convolution, the reduction of computational redundancy and memory access, and the introduction of new convolution can maintain high FLOPs while reducing FLOPS. Secondly, after introducing HiLo, an improved and efficient attention mechanism for extracting high and low-frequency information, in experiment 3, the recognition accuracy of the algorithm was greatly improved, with the recall rate, accuracy rate and average accuracy increased by 1%, 2.8% and 1%, respectively, while the number of params, calculation amount and weight memory remained basically unchanged. This improvement is due to the attention mechanism. HiLo uses two efficient attention types to decouple the high/low frequencies in the feature map, refining the high/low-frequency features and connecting them by capturing the fine grain features of the high/low-frequency Xinli No.7 fruit image through high-frequency attention (Hi-Fi) and low-frequency attention (Lo-Fi). Moreover, in experiment 4, the Slim-Neck (SN) method was used to reconstruct the Cross-Scale Feature Fusion Module (CCFM). Although the accuracy was slightly reduced, the weight memory and computing cost were reduced to 19.9 MB and 25.1 G, respectively, because GSConv was used to replace Conv convolution. Because the structure of the VoV-GSCSP module is improved and transformed, the model structure is simple and efficient, and the simultaneous recall rate and average accuracy value are improved slightly.

Finally, in experiment 5, the loss function was replaced by the Shape-NWD small target detection mechanism loss function. Through the improvement of the above module, the model recognition accuracy is continuously improved. Finally, the accuracy rate, recall rate and average accuracy of the improved model are increased by 4%, 1.8% and 1.4%, respectively, compared with the original model in experiment 1.

After the ablation test, all four modules in this paper are effective, and the average accuracy of the improved model reaches 98%, which is 9.7% and 0.77% higher than the 88.3% and 97.23% in previous pear fruit detection model literature [11,12], respectively, showing the advanced and deployable of the improved model in this paper.

Figure 7 shows the effect of the original model and the improved model on the target detection of Xinli No. 7 fruit under different scenario conditions. Both the RT-DETR model and the improved model can accurately identify the fruit target of Xinli No. 7 in the image under the frontlight close view, but the improved model could detect the missed fruit to various degrees under branches and leaves cover, backlight close view, fruit overlap, frontlight distant view and backlight distant view, blue boxes missed due to fruit occlusion in Figure 7a,f and blue baskets missed due to branches and leaves occlusion in Figure 7c–e is shown. Moreover, the improved model has higher confidence in different environments. As shown in Figure 7b, the fruit confidence detected by the improved model is 17% higher than that of RT-DETR for Xinli No. 7 fruit, that is, fruit overlap. The test results show that the improved model has a better detection effect and higher confidence in the complex environment of branches and leaves cover, fruit overlap and reverse light near and far, and meets the recognition accuracy requirements of Xinli No. 7 fruit picking robot.

RT-DETR

Improve the model

(a) (b) (c) (d) (e) (f)

**Figure 7.** Fruit target detection results of Xinli No.7 under different environments. (**a**) Branches and leaves cover. (**b**) Frontlight close view. (**c**) Backlight close view. (**d**) Fruit overlap. (**e**) Frontlight distant view. (**f**) Backlight distant view.

### 4.4. Construction of Grading Detection Model

The improved model in this paper tested Xinli No. 7 fruit in 6 typical environments, and the test results are shown in Table 4 below. As can be seen from the results in the table, there is a large difference in recognition accuracy between recall rate and accuracy rate in frontlight close view. The reason for this result is that the data collected in frontlight close view is small, and the model cannot be fully learned and verified, and it is difficult to ensure generalization. Another reason is that the color and shape characteristics of the Xinli No. 7 fruit are not prominent enough in the close view of the frontlight, which makes it easy to confuse the appearance of branches and leaves, resulting in the missing some fruits in the environment. The model showed good detection ability in the environment of backlight near prospect, branches and leaves occlusion and fruit overlap. The detection effect of pear fruit in the frontlight distant view is especially the best, and the accuracy rate, recall rate, and average accuracy average have reached a good balance. This result can be said that the improved model is suitable for the recognition distance requirement of the robot arm picking pear fruit set in this paper. No matter what kind of typical pear fruit environment, the average accuracy of the improved model is the lowest, 98.4%, which realizes the accurate identification of Xinli No. 7 fruit in the natural environment.

**Table 4.** An improved model for pear fruit detection results in different environments.

| Typical Pear Fruit Environment | P/% | R/% | mAP/% |
|---|---|---|---|
| Frontlight close view | 98.8 | 91.6 | 98.4 |
| Frontlight distant view | 96 | 94.3 | 98.8 |
| Backlight close view | 94.3 | 94.8 | 98.4 |
| Backlight distant view | 95.7 | 93.1 | 98.7 |
| Branches and leaves cover | 94.7 | 93.9 | 98.5 |
| Fruit overlap | 95 | 93 | 98.6 |

## 5. Discussion

In terms of data set construction, although this paper comprehensively considered the influence factors such as branch and leaf occlusion, fruit overlap, different distances and light conditions, it did not build the diverse Xinli No. 7 fruit dataset under complex environmental conditions in different pear orchards in different regions, and it also lacked the construction of other pear fruit variety data sets. Therefore, the subsequent data set construction can comprehensively consider the above requirements. Continue to increase the stability and generalization of the model to detect the pear fruit. In terms of limitations, the model in this paper was only trained and applied in the experimental equipment and environment, and it can be adapted to other hardware devices such as laptops, smartphones or embedded devices in the later stage. It is also necessary to run stably under extreme environmental conditions with limited computing resources and declining data quality to verify the adaptive ability and delay of the improved model. Overall, the pear detection model, through the application of deep learning and computer vision technology, can greatly improve the efficiency of agriculture, food safety and supply chain management, promote the development of precision agriculture, and play an important role in the sustainability of the agricultural industry.

In the future practical application, we will continue to study how to deploy the model algorithm in the mobile end of the pear fruit picking equipment and integrate debugging with the robot arm, the chassis of the picking device and the end effector so as to achieve the long-term goal of automatic and non-destructive pear fruit picking [3].

## 6. Conclusions

(1) Based on the end-to-end real-time target detector RT-DETR model of Transformer architecture, this paper designs a pear fruit detection model in a natural environment based on lightweight Transformer architecture, aiming at the problems of low detection accuracy, slow speed and difficult detection of small target pears in a real environment. The accuracy rate, recall rate and average accuracy of the model reach 93.7%, 91.9% and 98%, respectively, and the number of parameters, calculation amount and weight memory reach 10.24 M, 25.1 G and 19.9 MB, respectively. Therefore, this model not only achieves high recognition accuracy but also has the requirements for deployment in automated pear-picking robots, fruit measurement devices and mobile terminals of automatic sorting systems.

(2) The performance comparison tests of YOLOv5m, YOLOv7, YOLOv8m and YOLOv10m, Deformable-DETR and RT-DETR-r18 models on Xinli No. 7 dataset were designed, and three comprehensive evaluation indexes of model lightweight, recognition accuracy and detection speed were used to evaluate the self-built Xinli No. 7 fruit dataset. The results show that the RT-DETR-r18 model can achieve a good balance in

terms of model lightweight and recognition accuracy compared with other models. The transmission frame per second is 75.7 f/s, which can realize rapid and accurate detection of Xinli No.7 fruit.

(3) The ablation experiment was divided into five groups. Based on the original model of the first group, ResNet-r18 was replaced with a lightweight FasterNet backbone network respectively. Secondly, the AIFI module was improved by using HiLo, an improved and efficient attention mechanism with high and low-frequency information extraction. A simple and efficient GSConv convolution is introduced into the CCFM module, and the loss function GIoU in RT-DETR is replaced by the Shape-NWD small target detection mechanism loss function. The results show that compared with the original model, the accuracy, recall and average accuracy of the improved model are increased by 4%, 1.8% and 1.4%, respectively, and the number of params, calculation and weight memory is reduced by 48.47%, 56.2% and 48.31%, respectively, so as to meet the requirements of model lightweight and accurate identification of pear fruits.

**Author Contributions:** Conceptualization, X.Z. and Z.H.; Methodology, Z.H. and Y.Z.; Software, Z.H., H.W. (Hongsen Wang) and H.W. (Huajie Wei); Validation, X.Z., Z.H. and H.W. (Hongsen Wang); Formal analysis, Z.H. and G.Z.; Investigation, H.W. (Hongsen Wang) and (Huajie Wei); Resources, X.Z. and G.Z.; Data curation, X.Z., Y.Z. and H.W. (Hongsen Wang); Writing—original draft preparation, X.Z., Z.H. and G.Z.; Writing—review and editing, Z.H., Y.Z. and H.W. (Hongsen Wang) and H.W. (Huajie Wei); Visualization, Z.H., Y.Z., G.Z. and H.W. (Huajie Wei); Supervision, X.Z.; Project administration, X.Z.; Funding acquisition, X.Z. All authors have read and agreed to the published version of the manuscript.

**Funding:** This study was funded by Department of Agriculture and Rural Affairs of Hebei Province (Hebei Agriculture Research System HBCT2024170207).

**Institutional Review Board Statement:** Not applicable.

**Data Availability Statement:** Data are contained within the article.

**Acknowledgments:** The authors gratefully acknowledge the editors and anonymous reviewers for their constructive comments on our manuscript.

**Conflicts of Interest:** The authors declare no conflicts of interest.

## Appendix A

**Table A1.** Abbreviation list.

| Abbreviation | Full Title |
|---|---|
| RT-DETR | Real-Time Detection Transformer |
| SPPF | Spatial Pyramid Pooling Fast |
| simSPPF | Simplified Spatial Pyramid Pooling Fast |
| MSDA | Multi-Scale Dilated Attention |
| NMS | Non Maximum Suppression |
| CNN | Convolutional Neural Networks |
| EMSC-DETR | Efficient Multi-Scale-Conv Detection Transformer |
| SDTM | Study Data Tabulation Model |
| FPS | Frames Per Second |
| AIFI | Attention-based Intrascale Feature Interaction |
| DSConv | Distribution Shifting Convolution |
| Shape-NWD | Shape-Normalized Wasserstein Distance |
| CCFM | Cross-Scale Feature Fusion Module |
| GSConv | Generalized-Sparse Convolution |
| FLOPs | Floating point operations |
| FLOPS | Floating point operations per second |
| BN | Batch Normalization |
| ReLU | Rectified Linear Unit |
| MSA | Multi-head Self-Attention |

## References

1. Yuan, Y.; Bai, S.; Niu, K.; Zhou, L.; Zhao, B.; Wei, L.; Xiong, S.; Liu, L. Research progress on mechanized harvesting technology and equipment for forest fruit. *Trans. Chin. Soc. Agric. Eng.* **2022**, *38*, 53–63.
2. Wang, B. Research on Key Technologies of Pear Fruit Picking Robot Based on ROS and YOLOv5. Master's Thesis, Hebei University, Hebei, China, 2024.
3. Li, M.; Liu, P. A bionic adaptive end-effector with rope-driven fingers for pear fruit harvesting. *Comput. Electron. Agric.* **2023**, *211*, 107952. [CrossRef]
4. Hai, T.; Zhang, N.; Lu, X.; Xu, J.; Wang, X.; Hu, J.; Ji, M.; Zhao, Z.; Wang, J.; Dong, M. Implementation and Evaluation of Attention Aggregation Technique for Pear Disease Detection. *Agriculture* **2024**, *14*, 1146. [CrossRef]
5. Liu, L.; Xu, S.; Chen, J.; Wang, H.; Zheng, X.; Shen, M.; Liu, L. Detection of Feeding Behavior in Lactating Sows Based on Improved You Only Look Once v5s and Image Segmentation. *Agriculture* **2024**, *14*, 1402. [CrossRef]
6. Dai, G.; Tian, Z.; Fan, J.; Sunil, C.; Dewi, C. DFN-PSAN: Multi-level deep information feature fusion extraction network for interpretable plant disease classification. *Comput. Electron. Agric.* **2024**, *216*, 108481. [CrossRef]
7. Zhao, Y.; Li, Y.; Xu, X. Object Detection in High-Resolution UAV Aerial Remote Sensing Images of Blueberry Canopy Fruits. *Agriculture* **2024**, *14*, 1842. [CrossRef]
8. Jiang, L.; Wang, Y.; Wu, C.; Wu, H. Fruit Distribution Density Estimation in YOLO-Detected Strawberry Images: A Kernel Density and Nearest Neighbor Analysis Approach. *Agriculture* **2024**, *14*, 1848. [CrossRef]
9. Wu, W.; He, Z.; Li, J.; Chen, T.; Luo, Q.; Luo, Y.; Wu, W.; Zhang, Z. Instance Segmentation of Tea Garden Roads Based on an Improved YOLOv8n-seg Model. *Agriculture* **2024**, *14*, 1163. [CrossRef]
10. Wang, A.; Chen, H.; Liu, L.; Chen, K.; Lin, Z.; Han, J.; Ding, G. YOLOv10: Real-Time End-to-End Object Detection. *arXiv* **2024**, arXiv:2405.14458. [CrossRef]
11. Tan, H.; Ma, W.; Tian, Y.; Zhang, Q.; Li, M.; Li, M.; Yang, X. Improved YOLOv8n object detection of fragrant pears. *Trans. Chin. Soc. Agric. Eng.* **2024**, *40*, 178–185.
12. Zheng, W.; Yang, Y. Mature Pear Target Detection Method Based on Frequency Domain Data Enhancement and Lightweight YOLO v7 Model. *J. Agric. Mach.* **2024**, *55*, 244–253.
13. Liu, Q.; Lv, J.; Zhang, C. MAE-YOLOv8-based small object detection of green crisp plum in real complex orchard environments. *Comput. Electron. Agric.* **2024**, *226*, 109458. [CrossRef]
14. Chen, J.; Ji, C.; Zhang, J.; Feng, Q.; Li, Y.; Ma, B. A method for multi-target segmentation of bud-stage apple trees based on improved YOLOv8. *Comput. Electron. Agric.* **2024**, *220*, 108876. [CrossRef]
15. Chen, F.; Chen, C.; Zhu, X.; Shen, D.; Zhang, X. Detection of Camellia oleifera fruit maturity based on improved YOLOv7. *Trans. Chin. Soc. Agric. Eng.* **2024**, *40*, 177–186.
16. Carion, N.; Massa, F.; Synnaeve, G.; Usunier, N.; Kirillov, A.; Zagoruyko, S. End-to-End Object Detection with Transformers. *arXiv* **2020**, arXiv:2005.12872. [CrossRef]
17. Vaswani, A.; Shazeer, N.; Parmar, N.; Uszkoreit, J.; Jones, L.; Gomez, A.N.; Kaiser, L.; Polosukhin, I. Attention is All you Need. In Proceedings of the 31st Conference on Neural Information Processing Systems (NIPS 2017), Long Beach, CA, USA, 4–9 December 2017; pp. 6000–6010.
18. Lv, W.; Xu, S.; Zhao, Y.; Wang, G.; Wei, J.; Cui, C.; Du, Y.; Dang, Q.; Liu, Y. DETRs Beat YOLOs on Real-time Object Detection. In Proceedings of the 2024 IEEE/CVF Conference on Computer Vision and Pattern Recognition (CVPR), Seattle, WA, USA, 16–22 June 2024; pp. 16965–16974.
19. Zhu, X.; Su, W.; Lu, L.; Li, B.; Wang, X.; Dai, J. Deformable DETR: Deformable Transformers for End-to-End Object Detection. *arXiv* **2020**, arXiv:2010.04159. [CrossRef]
20. Yao, Z.; Ai, J.; Li, B.; Zhang, C. Efficient DETR: Improving End-to-End Object Detector with Dense Prior. *arXiv* **2021**, arXiv:2104.01318. [CrossRef]
21. Zhang, H.; Li, F.; Liu, S.; Zhang, L.; Su, H.; Zhu, J.; Ni, L.M.; Shum, H. DINO: DETR with Improved DeNoising Anchor Boxes for End-to-End Object Detection. *arXiv* **2022**, arXiv:2203.03605. [CrossRef]
22. Zhao, B.; Liu, S.; Zhang, W.; Zhu, L.; Han, Z.; Feng, X.; Wang, R. Research on Performance Optimization of Lightweight Transformer Architecture for Cherry Tomato Picking. *J. Agric. Mach.* **2024**, *55*, 1–13.
23. Hu, J.; Zhang, G.; Shen, M.; Li, W. Detecting surface defects of pine wood using an improved RT-DETR model. *Trans. Chin. Soc. Agric. Eng.* **2024**, *40*, 210–218. [CrossRef]
24. Li, X.; Cai, M.; Tan, X.; Yin, C.; Chen, W.; Liu, Z.; Wen, J.; Han, Y. An efficient transformer network for detecting multi-scale chicken in complex free-range farming environments via improved RT-DETR. *Comput. Electron. Agric.* **2024**, *224*, 109160. [CrossRef]
25. Li, H.; Shi, F. A DETR-like detector-based semi-supervised object detection method for Brassica Chinensis growth monitoring. *Comput. Electron. Agric.* **2024**, *219*, 108788. [CrossRef]
26. Wang, R.; Zhang, B.; Guo, T.; He, T.; Cui, H.; Wang, Z.; Man, S. Physiological response and cold resistance evaluation of 5 pear varieties under low temperature stress. *Shandong Agric. Sci.* **2023**, *55*, 57–63.

27. Qiao, C.; Han, M.; Gao, W.; Gao, W.; Li, K.; Zhu, X.; Zhang, L. Quantitative Detection of Cucumber Downy Mildew Spores at Multi-scale Based on Faster-NAM-YOLO. *J. Agric. Mach.* **2023**, *54*, 288–299.
28. Fu, C.; Ren, L.; Wang, F. Recognizing beef cattle behavior under automatic scene distinction using lightweight FABF-YOLOv8s. *Trans. Chin. Soc. Agric. Eng.* **2024**, *40*, 152–163.
29. Yang, F.; Li, X.; Cheng, H.; Guo, Y.; Chen, L.; Li, J. MSB-FCN: Multi-Scale Bidirectional FCN for Object Skeleton Extraction. *IEEE Trans. Image Process.* **2020**, *30*, 2301–2312. [CrossRef]
30. Chen, J.; Kao, S.; He, H.; Zhuo, W.; Wen, S.; Lee, C.; Chan, S.G. Run, Don't Walk: Chasing Higher FLOPS for Faster Neural Networks. In Proceedings of the 2023 IEEE/CVF Conference on Computer Vision and Pattern Recognition (CVPR), Vancouver, BC, Canada, 17–24 June 2023; pp. 12021–12031.
31. Pan, Z.; Cai, J.; Zhuang, B. Fast Vision Transformers with HiLo Attention. *arXiv* **2020**, arXiv:2205.13213. [CrossRef]
32. Wang, S.; Jiang, H.; Li, Z.; Yang, J.; Ma, X.; Chen, J.; Tang, X. PHSI-RTDETR: A Lightweight Infrared Small Target Detection Algorithm Based on UAV Aerial Photography. *Drones* **2024**, *8*, 240. [CrossRef]
33. Li, H.; Li, J.; Wei, H.; Liu, Z.; Zhan, Z.; Ren, Q. Slim-neck by GSConv: A lightweight-design for real-time detector architectures. *J. Real-Time Image Process.* **2022**, *21*, 62. [CrossRef]
34. Zhang, H.; Zhang, S. Shape-IoU: More Accurate Metric considering Bounding Box Shape and Scale. *arXiv* **2023**, arXiv:2312.17663. [CrossRef]
35. Wang, J.; Xu, C.; Yang, W.; Yu, L. A Normalized Gaussian Wasserstein Distance for Tiny Object Detection. *arXiv* **2021**, arXiv:2110.13389. [CrossRef]

agriculture

MDPI

*Article*

# Mastitis Classification in Dairy Cows Using Weakly Supervised Representation Learning

**Soo-Hyun Cho [1], Mingyung Lee [2], Wang-Hee Lee [1], Seongwon Seo [2,*] and Dae-Hyun Lee [1,*]**

[1] Department of Biosystems Machinery Engineering, Chungnam National University, Daejeon 34134, Republic of Korea; kuaiche12@o.cnu.ac.kr (S.-H.C.); wanghee@cnu.ac.kr (W.-H.L.)

[2] Division of Animal and Dairy Sciences, Chungnam National University, Daejeon 34134, Republic of Korea; mingyung1203@cnu.ac.kr

* Correspondence: swseo@cnu.ac.kr (S.S.); leedh7@cnu.ac.kr (D.-H.L.)

**Abstract:** Detecting mastitis on time in dairy cows is crucial for maintaining milk production and preventing significant economic losses, and machine learning has recently gained significant attention as a promising solution to address this issue. Most studies have detected mastitis on time series data using a supervised learning model, which requires the scale of labeled data; however, annotating the onset of mastitis in milking data from dairy cows is very difficult and costly, while supervised learning relies on accurate labels for ensuring the performance. Therefore, this study proposed a mastitis classification based on weakly supervised representation learning using an autoencoder on time series milking data, which allows for concurrent milking representation learning and weakly supervision with low-cost labels. The proposed method employed a structure where the classifier branches from the latent space of a 1D-convolutional autoencoder, enabling representation learning of milking data to be conducted from the perspective of reconstructing the original information and detecting mastitis. The branched classifier backpropagate the mastitis symptoms, which are less costly than mastitis diagnosis, during the encoder's representation learning. The results showed that the proposed method achieved an F1-score of 0.6 that demonstrates performance comparable to previous studies despite using low-cost labels. Our method has the advantage of being easily reproducible across various data domains through low-cost annotation for supervised learning and is practical as it can be implemented with just milking data and weak labels, which can be collected in the field.

**Keywords:** dairy cow; mastitis; representation learning; autoencoder; weakly supervised learning; anomaly detection; 1D-CNN

## 1. Introduction

Dairy cow mastitis, which is caused by bacteria or fungi invading the udder, is the most prevalent and economically damaging of the dairy cow diseases [1]. It leads to a decrease in both milk production and quality, which results in significant economic losses for the dairy industry, with an average reduction of about 30% in milk production and approximately 20% in dairy income per cow [2]. The early detection of mastitis and intervention can improve both dairy farm income and cow welfare [3,4]. The increase in the number of dairy cows has necessitated the use of a significant amount of time and labor for disease detection, which was traditionally performed manually [5,6]. Therefore, monitoring for early disease response and reducing labor time are essential, and recent advancements in automation technologies, such as automated milking systems and robotic milkers, have enabled real-time health monitoring [4,7–9].

Disease detection using milking data has been extensively studied over the past 30 years. Initially, manually collected data were predominantly used, and linear regression methods were used to predict the linear relationships between milking data and diseases.

De Mol et al. [10] developed a time series analysis-based mastitis detection model by combining manually constructed time series data with a Kalman filter. The model showed high sensitivity and specificity, demonstrating potential for disease detection. However, linear models such as the Kalman filter have limitations when used to analyze the nonlinear relationships between variables associated with mastitis; thus, they fail to guarantee high performance [11–13]. It has been confirmed that mastitis in dairy cows is associated with nonlinear relationships between various milking parameters, such as milk yield (MY) and milk electrical conductivity (MEC) [14]. Therefore, the effective detection of diseases in dairy cows requires a nonlinear approach to the data.

Recent advancements in ML, such as deep learning (DL), have significantly improved the extraction of features from the data; they have shown high performance in nonlinear tasks, such as mastitis detection, and are gaining attention in the automation research of the livestock industry. Over the past decade, the livestock industry has been using conventional ML for mastitis detection, continuously improving performance. Miekley et al. [15] analyzed the MY, MEC, lactation stage, and mastitis history of 215 cows using a support vector machine (SVM) model and reported on its usefulness for disease detection. Mansour et al. [16] collected approximately 360,000 milking data samples using an automated livestock monitoring system, and they used six parameters, including MY and MEC, to find the optimal predictive model. Seven models and statistical techniques were employed to predict mastitis, and the decision tree showed high performance in predicting subclinical mastitis. Fuyang et al. [17] collected data on milk's electrical properties and resistance from commercial farms, using a learning vector quantization (LVQ) model to classify mastitis into positive and negative categories, showing high accuracy for clinical mastitis detection and demonstrating practical performance in automated disease detection. The studies mentioned above were conducted using supervised learning, which requires labels paired with training data. The accuracy of the labels significantly influences model performance as the model is trained to output predetermined labels based on inputs. Therefore, high performance in supervised learning for mastitis detection requires accurately labeled data that distinguish between healthy and diseased states. However, determining the precise onset of mastitis involves high expertise and extensive analysis efforts (in terms of time and cost), which can pose challenges. Additionally, disease detection generally suffers from label imbalance, where positive data are significantly lower compared to the overall dataset [18–20]. Label imbalance can result in a detection bias towards the majority class [21]. To overcome such issues, studies using unsupervised learning methods have been conducted. Franceschini et al. [22] used Ward's agglomerative hierarchical clustering method on collected data points to detect milk components related to mastitis, achieving high accuracy. Liu et al. [23] employed the unsupervised learning-based cluster algorithm to segment data, followed by hierarchical clustering to group similar behavioral patterns. Kiouvrekis et al. [24] developed a predictive model for mastitis prevalence using the unsupervised learning method K-means clustering. By analyzing data from 111 sheep farms, they effectively categorized farms with high and low mastitis prevalence rates. However, this method requires the user to predefine hyperparameters, such as the number of clusters and their structure, and the formation of clusters is based on predetermined rules and distance calculations rather than being optimized through data-driven learning. This approach has limitations as it may lead to cumulative errors during the hierarchical clustering process, and since it does not perform global optimization across the entire dataset in an end-to-end manner, it may negatively impact the results [25,26]. An end-to-end learning approach that facilitates training using input data in a single, unified process should be considered to overcome these challenges.

Autoencoders, which are unsupervised methods capable of end-to-end learning, have been increasingly utilized for various tasks involving the encoding of the most salient information from input data through self-reconstruction [27,28]. In livestock-related applications, autoencoders have recently been used to predict milk production in cows, showing comparable performance to supervised-learning-based multi-layer perceptron in unsupervised learning [29]. Notably, autoencoder-based anomaly detection techniques have been extensively explored for the detection of anomalies in milk production [30] and cow behavior [31]. However, the anomaly detection method using autoencoders relies on the differences in the reconstruction performances of the input data, which are dependent on the frequency of class-specific training data; this makes it not only passive but also highly sensitive to the threshold used for reconstruction error, which can significantly affect model performance [32]. To address these issues, incorporating labels with approximate information can guide the model in learning robust representations for anomaly detection [33,34].

Therefore, this study proposes an autoencoder model using weakly supervised learning (WSL) for mastitis detection in dairy cows. This approach used an autoencoder as a baseline for performing the feature representation of milking data, and a classifier was added to the latent space to use mastitis labels in the training of feature representation. The autoencoder performs two tasks simultaneously: input reconstruction and mastitis classification. These two tasks can provide feature representations that can be separated between classes. In this study, the labels were defined as fixed time intervals relative to the mastitis onset date, providing approximate information about the mastitis and enabling WSL. In addition, our framework can be implemented in an end-to-end manner and can be expanded to other practical domains.

## 2. Materials and Methods

### 2.1. Overview of the Proposed Model

Anomaly detection using autoencoders leverages the distributional differences between normal and abnormal data during the process of reconstructing input data. The autoencoder primarily learns from normal data, forming a manifold that reflects the patterns of the normal data. Abnormal data, which deviate from these patterns, do not fit well within this manifold, leading to reconstruction errors [35]. In this study, to enable more separable distributions between the normal and abnormal data in the latent space, we connected a classifier to the latent space and trained the encoder in a supervised method. The proposed architecture, illustrated in Figure 1, introduces a method for detecting mastitis by utilizing the trained classifier instead of relying solely on relative reconstruction error differences.

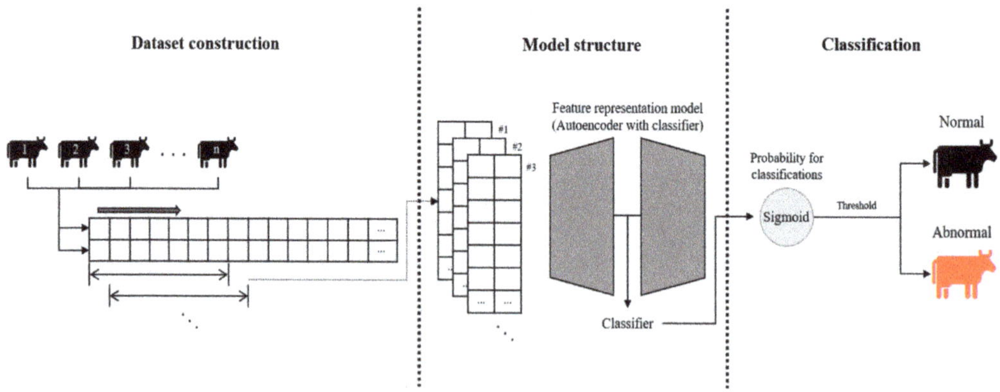


**Figure 1.** Overview of the proposed pipeline for mastitis based on autoencoder with classifier.

The input data consists of multichannel time series data related to milking information; these are fed into the encoder after undergoing preprocessing. The multichannel data are transformed by the encoder into feature vectors within the latent space. These feature vectors are then used by both the classifier and decoder to learn the dual tasks of classification and reconstruction. Notably, in this study, the labels correspond to the duration of mastitis occurrence rather than the exact onset time, enabling the model to actively learn representations in the latent space that can effectively classify mastitis. WSL typically involves training a model using incomplete, weak, or no labels, thereby reducing the labor required for label data collection. In this study, we leveraged this advantage by using approximate labels for the mastitis data that are challenging to define [36].

The proposed framework trains the autoencoder and the classifier simultaneously using an end-to-end learning method, where the data reconstruction and mastitis classification tasks are jointly optimized. This approach enables the model to learn representations from the milking data so that the representation can facilitate effective mastitis classification. Unlike conventional approaches that rely on the reconstruction error differences that are inherent to autoencoders, this study integrates a classifier for final mastitis detection. During the training phase, the autoencoder minimizes reconstruction error to optimize feature representation learning in the latent space, while the classifier is concurrently trained on these representations. In the testing phase, the classifier leverages the latent features to perform mastitis classification, resulting in a more proactive detection mechanism compared to those of unsupervised methods. The mastitis is finally determined using the classifier by thresholding the score (probability) of the mastitis class; this is referred to as the mastitis score (MS) in this study.

## *2.2. Dataset Description*

### 2.2.1. Data Collection

The milking data used in this study were collected from January 2020 to August 2021 from four farms located in Cheonan-si, Dangjin-si, and Asan-si, Chungcheongnam-do, South Korea. This study was conducted in accordance with ethical standards and was approved by the Animal Care and Use Committee of Chungnam National University in 2020 (Approval Number: 202009A-CNU-121). All the dairy cows on the farms were of the Holstein breed, and milking was performed twice daily using a herringbone milking system equipped with electronic flow meters and individual identification systems. Data were collected using the automated milking management systems (Afimilk, Kibbutz Afikim, Israel; DeLaval system, Tumba, Botkyrka, Sweden; BouMatic system, Madison, WI, USA) installed on each farm. The collected data were automatically transmitted daily and synchronized to a central server via a gateway system (AgriRoboTech, Icheon, Republic of Korea).

The number of milking cows per farm and the total amount of collected data are presented in Table 1. Milking data from a total of 113 Holstein cows were collected across the four farms, with a 1-day sampling interval. The daily average milk yield ranged from 20 to 36 kg, with an overall average of 28.25 kg across all the farms. The milking data were composed of milk yield and milk electrical conductivity, which are parameters that are easily obtainable on farms and are highly correlated with mastitis; milk yield (MY; kg/cow per day) and milk electrical conductivity (MEC; mS/cm) were used. Labels were assigned using three mastitis diagnostic methods: the clot filter, the California mastitis test (CMT), and visual observation (VO) [37–39]. The clot filter is effective for detecting clinical mastitis by identifying the presence of clots in the milk, indicating an intramammary infection. After milking, an inline filter attached to the milking line is visually inspected by the examiner for clots, and if any are found, the filter is replaced and the case is recorded as mastitis. The CMT is effective for detecting subclinical mastitis by observing the somatic cell count (SCC) through the addition of a reagent to milk samples; this reagent reacts with leukocyte proteins to form a gel-like substance [40]. The examiner manually assesses the thickness of the mixture and records the case as positive if mastitis is confirmed. VO is a traditional method of mastitis diagnosis, where a trained operator visually inspects the cow's udder to determine the presence of clinical mastitis, making it effective for diagnosing advanced stages of the disease [41]. All four farms utilized these three diagnostic methods, and this study employed all three to maximize label acquisition. If any of the three methods yielded a positive result, the cow was recorded as positive for mastitis. All mastitis diagnoses were made based on the examiner's observations, and the labels were recorded as '1' for positive cases and '0' for negative cases and were synchronized with the milking parameters.

**Table 1.** Number of data points collected from four farms.

| Farm Code | Number of Cows | Number of Data Points | | | | |
|---|---|---|---|---|---|---|
| | | Normal (Health) | Abnormal (Mastitis) | | | |
| | | | Clot | CMT | VO | Total |
| A (Cheonan-si) | 26 | 9272 | 23 | 35 | - | 58 |
| B (Dangjin-si) | 36 | 9751 | 2 | - | 41 | 43 |
| C (Asan-si) | 15 | 2179 | 11 | 33 | 16 | 60 |
| D (Asan-si) | 19 | 3892 | 24 | - | 6 | 30 |

2.2.2. Data Preprocessing

Milking data from dairy cows require preprocessing due to potential issues such as missing data caused by sensor malfunctions, variations arising from measurement environments, and individual differences [42]. The collected raw data contained instances of missing data and zero outputs due to sensor errors, as well as biases associated with individual cows, time, and spatial factors. These biases may stem from genetic variations, physiological states, and environmental conditions. Temporal factors include diurnal variability, seasonal changes, and cyclical behavioral patterns, while spatial factors are related to barn location, environmental conditions, and stocking density [43–45]. To address these issues, two filtering processes were applied during preprocessing. First, any data missing due to sensor errors during the collection period were treated as missing values and subsequently removed. Second, to verify data stability, an augmented Dickey–Fuller (ADF) test was conducted for each cow to assess the stationarity of the MY and MEC data, allowing the identification and removal of non-stationary data. The ADF test normalizes time series data by removing trends and volatility, ensuring stationary statistical properties, which enhance the model's capacity to capture temporal patterns and improve predictive accuracy [46].

Lactation refers to the period during which a dairy cow produces milk, and it is a parameter closely associated with mastitis detection [47]. In this study, to incorporate the characteristic patterns of MY, which typically increase or decrease during the lactation cycle, Accumulative MY (AMY) was added as a predictive variable [48]. The AMY is calculated by adding the daily MY to the cumulative total from the previous days within the same lactation cycle, as shown in Equation (1). This approach effectively mitigates daily fluctuations and outliers, providing a clearer view of long-term production patterns. Moreover, it ensures data continuity, even when some daily data points are missing, by preserving the overall trend in MY [49]. After preprocessing, the data were filtered down to 8505 records from 38 cows. The prevalence rates for each farm were approximately 0.5% for Farms A and B, and approximately 0.6% for Farms C and D. The average prevalence rate across Farms A, B, C, and D was approximately 0.55%. Each record included the farm ID, cow ID, collection date, MY, MEC, AMY, and label data.

$$AMY_t = \sum_{i=1}^{t} MY_i \quad (1)$$

where $AMY_t$ is the AMY on day $t$ of the lactation cycle, and $MY_i$ is the daily MY on day $i$ for $i = 1, 2, 3 \ldots, t$.

2.2.3. Dataset Construction

The dataset was constructed by extracting time series segments and training examples from the collected data of each cow sample. Each training example was generated using a sliding window method, which extracts consecutive time series data from the preprocessed data points, resulting in samples of size n (window size) × 3 (number of channels). The window slides took place at 1-day intervals, with samples extracted separately for each cow and added to the dataset; if a time gap greater than 1 day resulted in discontinuities, the dataset was split and stored accordingly.

The labels for the extracted training examples were generated based on the onset interval, as illustrated in Figure 2. The onset interval refers to the time surrounding the mastitis detection when the cow's health status undergoes significant changes. Given the difficulty in pinpointing the exact date of mastitis onset due to the potential gap between the actual onset and the detection, the onset interval is determined relative to the detection date; thus, it is typically defined as being up to 14 days before and after the detection [10,13,50]. Early warning signs of subclinical mastitis can be detected up to two months before diagnosis [51]; however, setting the onset interval more than one week before the mastitis may not offer practical utility for farm management [52,53]. Therefore, this study defined the onset interval as being a maximum of 7 days before and after detection. In this study, the entire onset interval was used as the label, and mastitis detection was performed using the MS output by the classifier; the performances were compared based on the size of the time series window and the onset interval. The time series window size was set at three levels, 3, 5, and 7, by adjusting the window size for data extraction. Here, to ensure the timeliness and operational relevance of the mastitis detection system, the maximum window size was restricted to 7, divided into three intervals [54]. The window length was matched to the onset interval to determine whether the data from these periods influenced mastitis detection, with window sizes set to 3, 5, and 7 days accordingly.

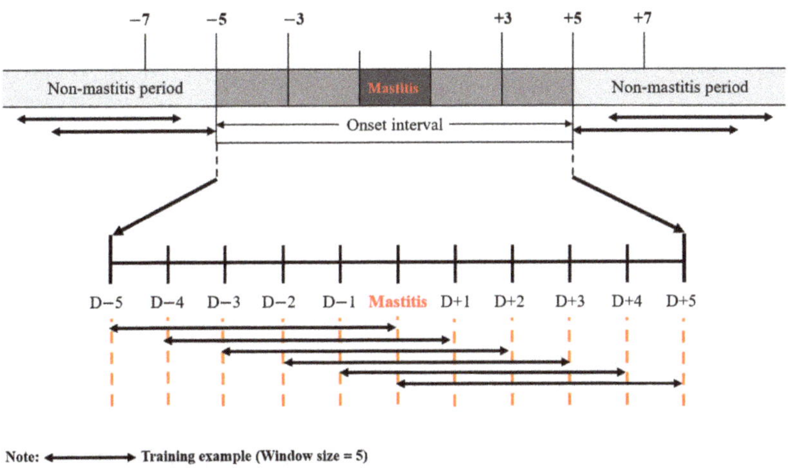


**Figure 2.** Extraction of training examples using the window sliding method with label annotation based on mastitis onset period. The figure shows an example with a 5-day size window sliding.

### 2.3. Weakly Supervised Autoencoder-Based Mastitis Detection

An autoencoder is a neural network in which the encoder and decoder are connected; the encoder maps input data to a lower-dimensional space to extract essential feature vectors (Z vectors), and the decoder reconstructs the input from these features. In this study, the encoder and decoder were implemented using 1D-CNNs. The 1D-CNN is particularly effective at capturing long-term dependencies in time series data and is better suited for studying dynamic changes and patterns within data samples [55]. The classifier in this model is constructed with fully connected layers and obtains one-dimensional outputs that provide a binary classification result.

Figure 3 illustrates the proposed autoencoder-based mastitis detection model, where both the feature representation and the classification tasks are performed concurrently. The representation learning process in the autoencoder aims to transform the data collected from various cows into a new latent space that captures common features. During this process, the autoencoder compresses the input data to extract essential features and seeks to find characteristic representations for reconstructing the original data. As mastitis data, which occur at a relatively lower frequency in the input data, typically result in higher reconstruction loss compared to normal data, they can be detected as anomalies based on the lower reconstruction accuracy. This study optimizes the general autoencoder-based anomaly detection by incorporating label data to enable more proactive mastitis detection. To achieve this, the encoder transforms the input data into a feature vector, which is then input into the latent space. A classifier was connected to the latent space to learn the mastitis label during feature representation learning; this encourages the encoder to represent the features and to distinguish between those that are distributed between the normal and abnormal data in the latent space during autoencoder training. Additionally, the features extracted from the latent space are fed into the classifier, which outputs binary classification results through a sigmoid activation function.

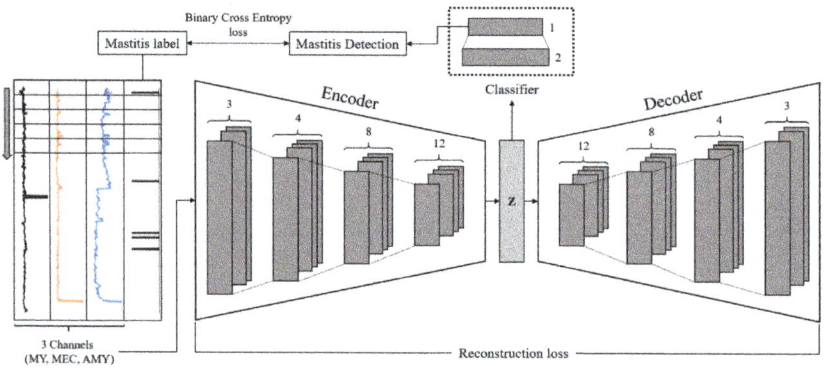


**Figure 3.** The proposed supervised autoencoder-based mastitis detection for dairy cows. The model comprises an autoencoder structure, utilizing a 1D-CNN as the backbone network. In addition, a classifier is included to differentiate between the normal and mastitis feature vectors.

The encoder consists of three convolutional layers, each reducing the sequence length of the input data while increasing the number of channels from 3 to 4, 4 to 8, and finally 8 to 12. This architecture facilitates the learning of local patterns and the extraction of relevant features from the input data. The dimensional expansion allows the learned feature vectors to represent more information, with the final layer of the encoder outputting a feature vector that represents the mastitis-related characteristics of the cow. The decoder operates in the reverse order of the encoder, receiving the latent vector Z and reconstructing the sequence length back to its original form while sequentially reducing the number of channels from 12 to 8, 8 to 4, and 4 to the original 3 input channels. The classifier connected to the latent space Z consists of two fully connected layers and determines the presence of mastitis. It transforms the feature vector extracted by the encoder into a 2-dimensional vector, compressing the essential features related to mastitis into a lower-dimensional space, thus enabling the extraction of information that distinguishes between mastitis and non-mastitis states. The output is then reduced into a 1-dimensional vector that provides a probability value indicating the presence of mastitis.

### 2.4. *Training and Evaluation*

#### 2.4.1. Model Training

In this study, leave-one-out cross-validation (LOOCV) was employed as the validation method as it is effective for small datasets. Given that our dataset is imbalanced, with a mastitis prevalence of 2%, evaluating the model using a single train–test split may result in performance fluctuations depending on whether the mastitis label data are included in the training or test set, thereby increasing the likelihood of overfitting [56,57]. This approach minimizes bias introduced by the mastitis data by evaluating the model's predictive performance on each data point individually; the data of every cow are used once as a test set, thereby preserving the model's generalization ability [58]. In this validation setup, the dataset was constructed by designating one cow's data as the test set, while the remaining data were split into training and validation sets in a 7:3 ratio. The model's performance was estimated by averaging the results obtained from the evaluations of all the cows.

The input data were fed into the model in a time-ordered sequence according to the specified window size and were normalized using min–max scaling. The training loss function comprised two components: L1 loss for reconstruction loss and cross-entropy loss for classification loss. The final loss function was the weighted sum of these two losses, with specific weights assigned to the reconstruction loss and classification loss to update the model parameters. The parameters were optimized using the Adam optimizer, with a learning rate set to $4 \times 10^{-3}$. The model was trained for a total of 1000 epochs, with early stopping implemented to prevent overfitting; training was halted when the validation loss reached its minimum, and the corresponding parameters were saved for performance evaluation [59]. The model was implemented in Python 3.8 and developed using PyTorch 1.0. The hardware used for the experiments included an Intel Core i7-11,700 K CPU and an NVIDIA RTX 3070Ti GPU with 6144 CUDA cores and a 1770 MHz boost clock.

2.4.2. Evaluation Metrics

The performance of the proposed approach was evaluated from two perspectives: the representation learning performance, measured through the distribution of reconstruction errors based on the learning method, and the detection of mastitis. The distribution of reconstruction errors was observed using a visualization technique which compared the differences between a general autoencoder using unsupervised learning and the model employed in this study, which utilized a weakly supervised learning approach. The comparison focused on examining the loss distribution in the non-mastitis intervals and mastitis intervals for each model. This approach allowed a visual assessment of how the reconstruction loss between the input data and the reconstructed data was distributed after training, depending on the learning method. Additionally, to quantify the magnitude of differences between the models in the mastitis and non-mastitis intervals, Cohen's d was calculated. Cohen's d is a standardized measure of effect size that assesses the strength of the relationship between two groups [60]. A d value of 0.2 indicates a small effect, 0.5 indicates a medium effect, and 0.8 indicates a large effect; values greater than 0.8 suggest a very large effect. These values provide benchmarks for the interpretation of effect size.

The performance of mastitis detection was evaluated using six metrics: accuracy, sensitivity, specificity, F1-score, and receiver operating characteristic area under the curve (ROC AUC). The performance was compared across different onset intervals and window sizes, as defined by Equations (2)–(5), by analyzing and comparing the evaluation metrics while varying the threshold from 0.5 to 0.9 in increments of 0.1 [61]. The F1-score emphasizes the model's ability to correctly predict true positives, making it suitable for minimizing missed cases in mastitis detection. ROC AUC assesses the model's discriminative power across various thresholds, providing a comprehensive evaluation of overall classification performance [62,63]. Therefore, these metrics were employed to comprehensively evaluate the performance of the mastitis detection model. In this study, the results of the binary classifier were utilized for mastitis diagnosis by applying a threshold to the probability values to determine the final mastitis status. To assess the similarity of performance across different threshold conditions, a one-way analysis of variance (ANOVA) was performed for each evaluation metric [64]. This statistical test was chosen to determine whether there were any significant differences in the performance metrics (accuracy, sensitivity, specificity,

F1-score, and ROC AUC) across the different threshold levels (0.5, 0.6, 0.7, 0.8, and 0.9). The ANOVA results indicated whether the variations in threshold settings led to statistically significant changes in model performance. Following the ANOVA, Tukey's honest significant difference (HSD) method was applied as a post hoc test to identify specific pairs of thresholds that exhibited significant differences in performance metrics [65]. This post hoc analysis provided detailed insights into which threshold levels were significantly different from each other, ensuring that the observed differences were not due to random variation.

$$\text{Accuracy} = \frac{TP + TN}{TP + TN + FP + FN} \tag{2}$$

$$\text{Sensitivity} = \frac{TP}{TP + FN} \tag{3}$$

$$\text{Specificity} = \frac{TN}{TN + FP} \tag{4}$$

$$\text{F1-score} = 2 \times \frac{\text{Precision} \times \text{Sensitivity}}{\text{Precision} + \text{Sensitivity}} \tag{5}$$

where the true positives (*TPs*) are the cases where the model correctly identifies mastitis within the label period; the true negatives (*TNs*) are the cases where the model correctly identifies the absence of mastitis outside the label period; the false positives (*FPs*) are the cases where the model incorrectly identifies mastitis outside the label period; and the false negatives (*FNs*) are the cases where the model fails to identify mastitis within the label period.

## 3. Results

### 3.1. Cow Mastitis Estimation

Figure 4 illustrates the loss progression during model training; the figure displays training loss in the graph (a) and validation loss in the graph (b). Each graph is plotted at 100 epoch intervals, representing the mean and standard deviation of the total loss across all the models. In the training loss curve, the mean loss consistently decreases with each epoch, indicating that the proposed model effectively learns from the data. However, the standard deviation of the loss begins to increase after 300 epochs, suggesting variability in the model performance across different training iterations. In contrast to the training loss, the validation loss reaches its minimum at approximately 300 epochs and then gradually increases. Additionally, the standard deviation of the validation loss continues to rise beyond this point. These observations imply that after 300 epochs, the model begins to overfit individual specific features rather than capturing common characteristics across the dataset. To minimize overfitting and ensure the model's generalization capability, the parameters corresponding to the lowest loss at around 300 epochs were selected for the final model.

Figure 5 presents the input and output results for two representative samples processed by the mastitis detection model. Each graph is aligned with the days in milk (DIM) and comprises an upper row showing the input values and a lower row displaying the output results. The input values consist of daily MY and MEC, both of which are normalized between 0 and 1. The output values represent the MS, which is generated by the classifier. The dataset used label intervals set at 5 days before and after the mastitis diagnosis date. In the left graph, significant fluctuations are observed in both MY and MEC, whereas the right graph shows relatively stable MEC and a trend in MY influenced by individual variability. Mastitis occurrences predominantly coincide with periods of abrupt changes in MY. The MS effectively detects these intervals, including the mastitis periods. Although some false positives are present, the MS predominantly spikes near the onset period and approximately a month prior, suggesting that the model may be capturing subclinical mastitis in the cow's milking data.

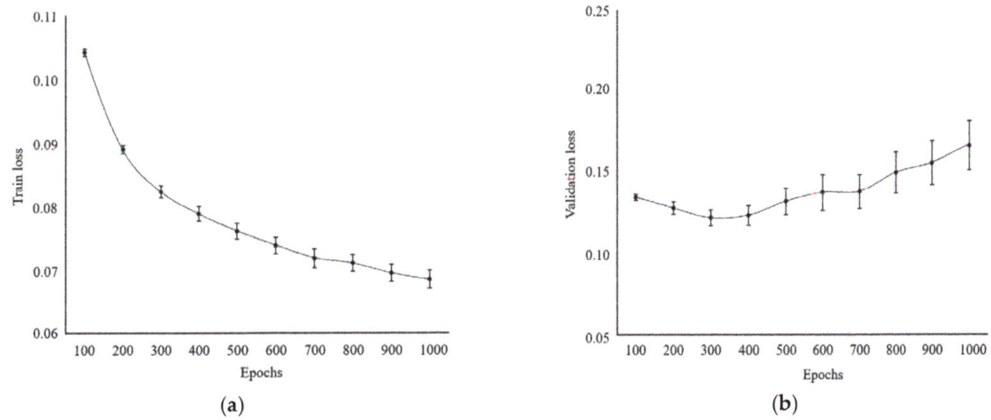


**Figure 4.** Loss curves for training (**a**) and validation (**b**) sets during model training; each loss value was expressed as mean and standard deviation within 100 epoch intervals.

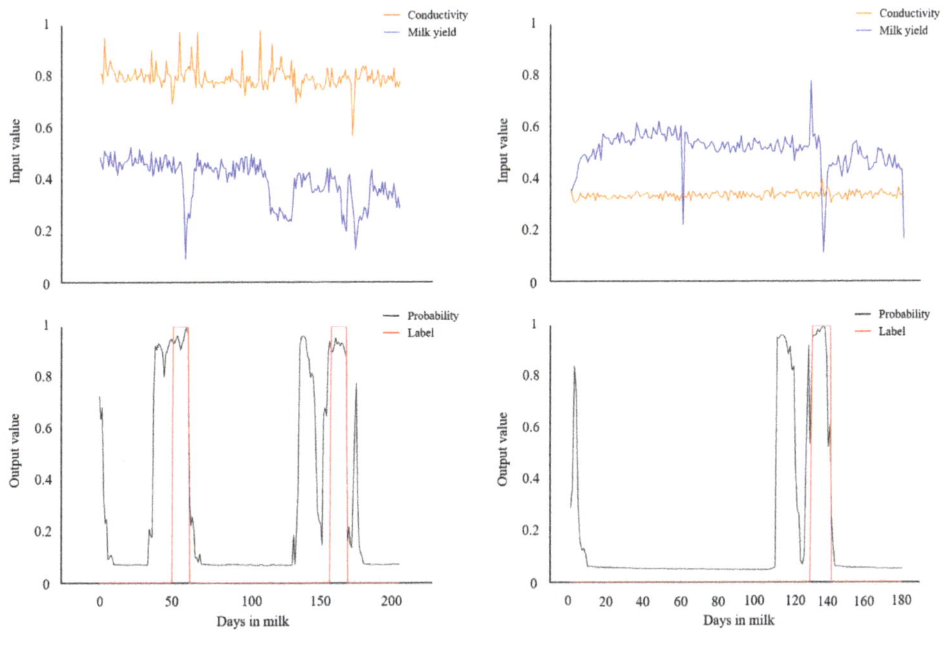


**Figure 5.** Results of representative cow samples for mastitis detection using deep learning model.

### *3.2. Performance Evaluation*

Figure 6 compares the reconstruction loss of the MY data between the proposed model (ours) and those trained using a typical autoencoder (AE). The mastitis intervals are highlighted by red boxes. In the autoencoder, higher reconstruction loss is observed during mastitis periods and lower loss during non-mastitis periods, which is favorable for anomaly detection. However, in many cases, the reconstruction loss remains similar across all the intervals or is higher in non-mastitis periods, reducing the model's effectiveness. In contrast, the proposed model shows a noticeable increase in reconstruction loss, particularly within the mastitis intervals, while also exhibiting a slight increase in non-mastitis periods.

Notably, the peaks in reconstruction loss predominantly occur within the mastitis intervals, indicating robust performance even with the stringent threshold settings. This suggests that the representation learned through supervised learning effectively separates anomalies from the main data distribution in the latent space, making it less reliant on the frequency of training data and more reliant on the learned features.

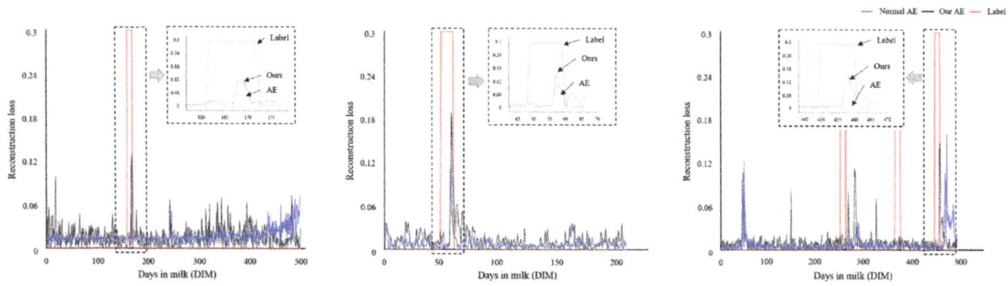


**Figure 6.** Comparison of reconstruction loss between AE and our model with label period.

Table 2 compares the anomaly detection performances of the AE and our model using mean loss and peak loss metrics. For the AE, the mean reconstruction loss in both the mastitis and non-mastitis intervals was approximately 0.01, showing negligible differences based on label presence. In contrast, the proposed model exhibited a mean loss of 0.02 in the non-mastitis intervals and 0.03 in the mastitis intervals, reflecting a roughly 50% increase in loss for the mastitis intervals. The Cohen's d effect sizes were 2.04 for the mastitis intervals and 1.44 for the non-mastitis intervals, indicating substantial differences between the models. Regarding peak loss, the AE recorded 0.06 in the mastitis intervals and 0.08 in the non-mastitis intervals, showing a 25% lower loss in the mastitis intervals compared to the non-mastitis intervals. This suggests that the AE exhibits limited capability to distinguish between mastitis and non-mastitis intervals. Conversely, the proposed model reported peak losses of 0.10 in the non-mastitis intervals and 0.12 in the mastitis intervals, representing an increase of approximately 20%. The Cohen's d value of 1.03 for the mastitis intervals further confirms the significant enhancement in the ability to distinguish between non-mastitis and mastitis conditions compared to the AE.

**Table 2.** Comparison of mean reconstruction loss and peak loss between AE and our model.

| | Mean Loss | | | Peak Loss | | |
|---|---|---|---|---|---|---|
| | **AE** | **Ours** | **Cohen's d** | **AE** | **Ours** | **Cohen's d** |
| Mastitis | 0.01 ± 0.007 | 0.03 ± 0.012 | 2.04 | 0.06 ± 0.055 | 0.12 ± 0.061 | 1.03 |
| Non-mastitis | 0.01 ± 0.004 | 0.02 ± 0.009 | 1.44 | 0.08 ± 0.049 | 0.10 ± 0.047 | 0.42 |

### 3.3. Model Performance Analysis

Table 3 presents the detection performance across different threshold levels for mastitis estimation, which is evaluated using various onset periods and window sizes, resulting in nine different conditions. The performance metrics included accuracy, sensitivity, specificity, and F1-score. A one-way ANOVA was conducted to assess the significant differences between the threshold conditions, followed by Tukey's HSD post hoc tests to identify specific significant differences. Accuracy showed a consistent upward trend with increasing thresholds, starting at 0.93 for a threshold of 0.5 and rising to 0.97 at a threshold of 0.9. Sensitivity was the highest at a threshold of 0.5 with a value of 0.67 but decreased by up to 5% as the threshold increased to 0.8, and by over 10% at a threshold of 0.9. Specificity exhibited an inverse relationship with sensitivity, increasing from 0.94 at a threshold of 0.5 to 0.98 at a threshold of 0.9. The F1-score, representing the harmonic mean of sensitivity and specificity, peaked at a threshold of 0.8 with a value of 0.41, indicating the best balance between sensitivity and specificity.

**Table 3.** Summary of mastitis estimation performance across multiple thresholds, onset periods, and window sizes.

| Threshold | Accuracy | Sensitivity | Specificity | F1-Score |
|---|---|---|---|---|
| 0.5 | 0.93 ± 0.024 [a*] | 0.65 ± 0.015 [a] | 0.94 ± 0.026 [a] | 0.35 ± 0.110 [a] |
| 0.6 | 0.94 ± 0.021 [b] | 0.62 ± 0.015 [b] | 0.95 ± 0.025 [a] | 0.37 ± 0.114 [a] |
| 0.7 | 0.95 ± 0.019 [c] | 0.59 ± 0.017 [c] | 0.96 ± 0.022 [b] | 0.39 ± 0.121 [a] |
| 0.8 | 0.96 ± 0.017 [d] | 0.55 ± 0.017 [d] | 0.97 ± 0.020 [c] | 0.41 ± 0.115 [b] |
| 0.9 | 0.97 ± 0.014 [d] | 0.44 ± 0.014 [e] | 0.98 ± 0.012 [d] | 0.40 ± 0.108 [a,b] |

* Means with different superscripts (a, b, c, d, e) in each column are significantly different at $p < 0.05$ using Tukey's range tests.

The results of the statistical analysis confirmed significant differences across all the metrics. Specifically, accuracy at a threshold of 0.9 was significantly higher than at the other thresholds, while sensitivity at the same threshold was significantly lower. Specificity and F1-score at the thresholds of 0.8 and 0.9 were significantly higher compared to the lower thresholds. These findings suggest that a threshold of 0.8 offers an optimal balance between correctly identifying mastitis cases and minimizing false positives.

Figure 7 compares the receiver operating characteristic (ROC) curves of the proposed mastitis detection model under varying time series lengths and onset periods. For the onset periods, extending the interval from ±3 days to ±5 days improved the area under the curve (AUC) from 0.84 to 0.90, a 7% increase. However, further extending the onset period to ±7 days reduced the AUC to 0.87, a 3% decrease. This indicates that while a longer onset period allows the model to capture subclinical mastitis and increases the proportion of true positives, excessively long periods may introduce normal intervals into the mastitis class, compromising label accuracy. Similarly, increasing the window size from ±3 days to ±5 days enhanced the AUC from 0.79 to 0.93, an 18% improvement. Extending the window size to ±7 days resulted in a slight decrease in AUC to 0.87, a 6% reduction. This suggests that a window size of 5 days effectively balances the inclusion of sufficient temporal information for accurate detection while avoiding the complexity and increased false positives associated with larger window sizes. Therefore, setting both the onset period and window size to 5 days was determined to be the most effective configuration for mastitis detection.

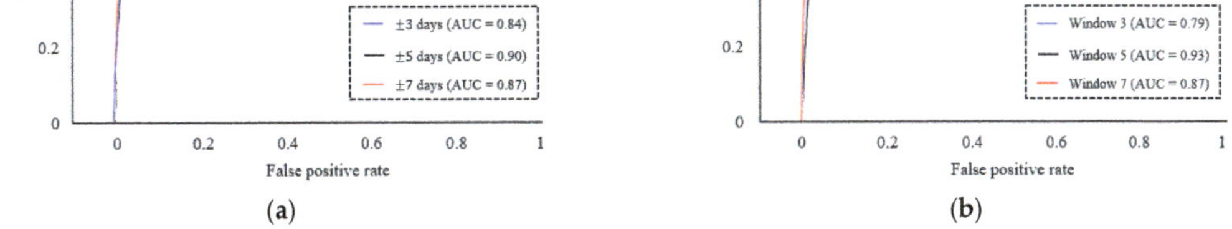


**Figure 7.** The ROC curve with AUC by onset period (**a**) and window size (**b**).

## 4. Discussion

This study proposed a weakly supervised autoencoder model to effectively detect mastitis in dairy cows. By integrating a classifier into the latent space of the autoencoder, the model learns to reconstruct input data and classify mastitis events concurrently, facilitating a more pronounced separation between normal and abnormal data distributions. This dual-task framework alleviates the dependency on reconstruction error differences alone, which is a common limitation in traditional unsupervised anomaly detection methods.

The proposed model demonstrated the ability to detect early signs of mastitis approximately two weeks prior to clinical diagnosis. This early detection capability is critical for timely intervention, which can significantly reduce economic losses and enhance animal welfare. The increase in reconstruction loss during mastitis intervals, coupled with the clear separation in the latent space, indicates that the model effectively learns the subtle changes in milking data associated with mastitis. Statistical analysis using Cohen's d effect size confirmed that the proposed model provides a substantial distinction between normal and mastitis data compared to the normal autoencoder. The higher reconstruction loss in mastitis periods and the consistent separation in the latent space underscore the model's enhanced capability to identify anomalies beyond mere frequency-based detection.

Previous studies on mastitis detection predominantly utilized supervised learning methods, which require precise labeling of mastitis onset. Methods such as support vector machines (SVMs) and decision trees have achieved high accuracy; for instance, some research using decision trees reported an accuracy of approximately 98%, which is comparable to that of the proposed model [66]. However, these approaches are constrained by the necessity for accurately labeled datasets and the challenges posed by label imbalance. Furthermore, studies employing machine learning models with fewer parameters, such as k-nearest neighbors (k-NNs), logistic regression, and gradient boosting machines, have demonstrated difficulties in distinguishing parameter distributions, yielding an area under the curve (AUC) score of 0.793, which is lower than the AUC of approximately 0.9 achieved by our proposed method [67].

The implementation of the proposed model on dairy farms has significant practical benefits. Utilizing readily available milking parameters such as milk yield and electrical conductivity, the model can be seamlessly integrated into existing automated milking systems without the need for additional sensors or substantial modifications. This facilitates the monitoring and detection of mastitis, enabling farmers to respond swiftly to the disease, thereby reducing the severity of outbreaks and the associated economic losses. Moreover, the weakly supervised approach diminishes the burden of data labeling by reducing the need for extensive manual annotation; it leverages limited or less precise labels to train the model effectively, thereby saving time and resources in the data preparation process.

Despite the promising results, this study has several limitations that warrant further investigation. Firstly, the dataset used was relatively small, with mastitis cases constituting approximately 2% of the total data. This imbalance may affect the model's potential to be generalized to larger and more diverse populations. Future studies should aim to collect more extensive datasets from various farms and cow breeds to enhance the model's robustness and applicability. Secondly, the model inputs include only two milking parameters that are highly correlated with mastitis in cows. In this study, the purpose was to develop a generalized model that minimizes reliance on the physiological parameters of cows during training, thereby lessening the need for manual feature selection. The input data were restricted to milk yield and electrical conductivity, which are parameters that can be automatically collected, and have demonstrated a high correlation with mastitis in studies focused on mastitis detection, prediction, and its association with physiological characteristics, thereby excluding additional data fields to prioritize a generalized representation learning of the milking information itself. In future work, improvements in representation learning outcomes are anticipated through post-processing methods based on factors such as seasonality, DIM, and other relevant parameters. Additionally, performance improvements are planned through the expansion of data and augmentation of parameters.

## 5. Conclusions

This study presents a novel approach to mastitis detection by employing a weakly supervised autoencoder framework. By simultaneously learning data reconstruction and mastitis classification tasks, the model achieves an effective separation of normal and abnormal patterns within the latent space. The proposed method outperforms traditional unsupervised models in anomaly detection, demonstrating superior detection capabilities for mastitis in dairy cows. Implementing this model in dairy farming practices can facilitate early interventions, reduce economic losses, and improve animal welfare. We believe that this study establishes a foundation for future research aimed at refining and validating the approach with larger and more diverse datasets.

**Author Contributions:** Conceptualization, S.-H.C. and D.-H.L.; methodology, S.-H.C. and D.-H.L.; software, S.-H.C.; validation, S.-H.C., M.L., W.-H.L. and D.-H.L.; formal analysis, S.-H.C., W.-H.L., S.S. and D.-H.L.; investigation, S.-H.C.; resources, M.L.; data curation, M.L.; writing—original draft preparation, S.-H.C.; writing—review and editing, S.-H.C., M.L., W.-H.L. and D.-H.L.; visualization, S.-H.C.; supervision, W.-H.L., S.S. and D.-H.L.; project administration, S.S.; funding acquisition, S.S. All authors have read and agreed to the published version of the manuscript.

**Funding:** This work was supported by Korea Institute of Planning and Evaluation for Technology in Food, Agriculture and Forestry (IPET) and Korea Smart Farm R&D Foundation (KosFarm) through Smart Farm Innovation Technology Development Program, funded by Ministry of Agriculture, Food and Rural Affairs (MAFRA) and Ministry of Science and ICT (MSIT), Rural Development Administration (RDA) (Project No. 421022–04).

**Institutional Review Board Statement:** This study was conducted in accordance with ethical standards and was approved by the Animal Care and Use Committee of Chungnam National University in 2020 (Approval Number: 202009A-CNU-121).

**Data Availability Statement:** The data presented in this study are available on request from the corresponding author due to confidentiality.

**Conflicts of Interest:** The authors declare no conflicts of interest.

## References

1. Seegers, H.; Fourichon, C.; Beaudeau, F. Production effects related to mastitis and mastitis economics in dairy cattle herds. *Vet. Res.* **2003**, *34*, 475–491. [CrossRef] [PubMed]
2. Ibrahim, N. Review on mastitis and its economic effect. *Can. J. Sci. Res.* **2017**, *6*, 13–22.
3. Leslie, K.E.; Petersson-Wolfe, C.S. Assessment and Management of Pain in Dairy Cows with Clinical Mastitis. *Vet. Clin. N. Am. Food Anim. Pract.* **2012**, *28*, 289–305. [CrossRef]
4. Fadul-Pacheco, L.; Delgado, H.; Cabrera, V.E. Exploring machine learning algorithms for early prediction of clinical mastitis. *Int. Dairy J.* **2021**, *119*, 105051. [CrossRef]
5. Reynolds, L.; Nierenberg, D. Disease and Drought Curb Meat Production and Consumption. In *Vital Signs*, 20th ed.; Island Press: Washington, DC, USA, 2013. [CrossRef]
6. Hurt, C. Weekly Outlook: Beef Herd Expanding Quickly, But Will It Continue? *Farmdoc Daily* **2015**, *5*, 140. Available online: http://farmdocdaily.illinois.edu/2015/08/weekly-outlook-beef-herd-expanding-continue.html (accessed on 22 October 2024).
7. De Koning, K.; Rodenburg, J. Automatic milking: State of the art in Europe and North America. In *A Better Understanding, Automatic Milking*; Wageningen Academic: Leiden, The Netherlands, 2004. [CrossRef]
8. Heikkila, A.M.; Vanninen, L.; Manninen, E. Economics of small-scale dairy farms having robotic milking. In Proceedings of the First North American Conference on Precision Dairy Management, Toronto, ON, Canada, 2–5 March 2010.
9. Schewe, R.L.; Stuart, D. Diversity in agricultural technology adoption: How are automatic milking systems used and to what end? *Agric. Hum. Values* **2015**, *32*, 199–213. [CrossRef]
10. De Mol, R.M.; Kroeze, G.H.; Achten, J.M.F.H.; Maatje, K.; Rossing, W. Results of a multivariate approach to automated oestrus and mastitis detection. *Livest. Prod. Sci.* **1997**, *48*, 219–227. [CrossRef]
11. Heald, C.W.; Kim, T.; Sischo, W.M.; Cooper, J.B.; Wolfgang, D.R. A Computerized Mastitis Decision Aid Using Farm-Based Records: An Artificial Neural Network Approach. *J. Dairy Sci.* **2000**, *83*, 711–720. [CrossRef] [PubMed]

12. Bausewein, M.; Mansfeld, R.; Doherr, M.G.; Harms, J.; Sorge, U.S. Sensitivity and Specificity for the Detection of Clinical Mastitis by Automatic Milking Systems in Bavarian Dairy Herds. *Animals* **2022**, *12*, 2131. [CrossRef]
13. Haxhiaj, K.; Wishart, D.S.; Ametaj, B.N. Mastitis: What It Is, Current Diagnostics, and the Potential of Metabolomics to Identify New Predictive Biomarkers. *Dairy* **2022**, *3*, 722–746. [CrossRef]
14. Castillo-Gallegos, E.; Cordero-Montoya, M.; Alonso-Díaz, M.Á. Use of a nonlinear model to estimate milk losses due to subclinical mastitis in Holstein-Zebu cows. *Trop. Anim. Health Prod.* **2022**, *54*, 299. [CrossRef] [PubMed]
15. Miekley, B.; Traulsen, I.; Krieter, J. Mastitis detection in dairy cows: The application of support vector machines. *J. Agric. Sci.* **2013**, *151*, 889–897. [CrossRef]
16. Mansour, E.; Mohammadi-Dehcheshmeh, M.; Ebrahimie, E.; Petrovski, K.R. Comprehensive analysis of machine learning models for prediction of sub-clinical mastitis: Deep Learning and Gradient-Boosted Trees outperform other models. *Comput. Biol. Med.* **2019**, *114*, 103456. [CrossRef]
17. Fuyang, T.; Zhonghua, W.; Sufang, Y.; Benhai, X.; Shunxi, W. An Automated On-Line Clinical Mastitis Detection System Using Measurement of Electrical Parameters and Milk Production Efficiency. *J. Phys. Conf. Ser.* **2020**, *1676*, 012190. [CrossRef]
18. Lemaître, G.; Nogueira, F.; Aridas, C.K. Imbalanced-learn: A python toolbox to tackle the curse of imbalanced datasets in machine learning. *J. Mach. Learn. Res.* **2017**, *18*, 1–5.
19. Kibebew, K. Bovine mastitis: A review of causes and epidemiological point of view. *J. Biol. Agric. Healthc.* **2017**, *7*, 1–14.
20. Zhdanovskaya, A.; Baidakova, D.; Ustalov, D. Data Labeling for Machine Learning Engineers: Project-Based Curriculum and Data-Centric Competitions. *Proc. AAAI Conf. Artif. Intell.* **2024**, *37*, 15886–15893. [CrossRef]
21. Yang, Y.; Xu, Z. Rethinking the value of labels for improving class-imbalanced learning. *Adv. Neural Inf. Process. Syst.* **2020**, *33*, 19290–19301. [CrossRef]
22. Franceschini, S.; Grelet, C.; Leblois, J.; Gengler, N.; Soyeurt, H. Can unsupervised learning methods applied to milk recording big data provide new insights into dairy cow health? *J. Dairy Sci.* **2022**, *105*, 6760–6772. [CrossRef]
23. Liu, J.; Bailey, D.W.; Cao, H.; Son, T.C.; Tobin, C.T. Development of a Novel Classification Approach for Cow Behavior Analysis Using Tracking Data and Unsupervised Machine Learning Techniques. *Sensors* **2024**, *24*, 4067. [CrossRef]
24. Kiouvrekis, Y.; Vasileiou, N.G.C.; Katsarou, E.I.; Lianou, D.T.; Michael, C.K.; Zikas, S.; Katsafadou, A.I.; Bourganou, M.V.; Liagka, D.V.; Chatzopoulos, D.C.; et al. The Use of Machine Learning to Predict Prevalence of Subclinical Mastitis in Dairy Sheep Farms. *Animals* **2024**, *14*, 2295. [CrossRef] [PubMed]
25. Großwendt, A.; Röglin, H.; Schmidt, M. Analysis of Ward's Method. In Proceedings of the Thirtieth Annual ACM-SIAM Symposium on Discrete Algorithms, San Diego, CA, USA, 6–9 January 2019; Society for Industrial and Applied Mathematics: Philadelphia, PA, USA, 2019; pp. 2939–2957. [CrossRef]
26. Wu, W.; Zhang, C. Towards Understanding End-to-End Learning in the Context of Data: Machine Learning Dancing over Semirings & Codd's Table. In Proceedings of the Fifth Workshop on Data Management for End-To-End Machine Learning, Virtual, 20–25 June 2021; Association for Computing Machinery: New York, NY, USA, 2021; pp. 1–4.
27. Bank, D.; Koenigstein, N.; Giryes, R. Autoencoders. In *Machine Learning for Data Science Handbook*; Rokach, L., Maimon, O., Shmueli, E., Eds.; Springer: Cham, Switzerland, 2023. [CrossRef]
28. Davaslioglu, K.; Erpek, T.; Sagduyu, Y.E. End-to-end autoencoder communications with optimized interference suppression. *arXiv* **2021**, arXiv:2201.01388. [CrossRef]
29. Liseune, A.; Salamone, M.; Van den Poel, D.; Van Ranst, B.; Hostens, M. Leveraging latent representations for milk yield prediction and interpolation using deep learning. *Comput. Electron. Agric.* **2020**, *175*, 105600. [CrossRef]
30. Vasafi, P.S.; Paquet-Durand, O.; Brettschneider, K.; Hinrichs, J.; Hitzmann, B. Anomaly detection during milk processing by autoencoder neural network based on near-infrared spectroscopy. *J. Food Eng.* **2021**, *299*, 110510. [CrossRef]
31. Zhang, K.; Han, S.; Wu, J.; Cheng, G.; Wang, Y.; Wu, S.; Liu, J. Early lameness detection in dairy cattle based on wearable gait analysis using semi-supervised LSTM-Autoencoder. *Comput. Electron. Agric.* **2023**, *213*, 108252. [CrossRef]
32. Ndubuaku, M.U.; Anjum, A.; Liotta, A. Unsupervised Anomaly Thresholding from Reconstruction Errors. In *Internet and Distributed Computing Systems*; IDCS, 2019; Montella, R., Ciaramella, A., Fortino, G., Guerrieri, A., Liotta, A., Eds.; Lecture Notes in Computer Science; Springer: Cham, Switzerland, 2019; Volume 11874. [CrossRef]
33. Guo, L.Z.; Li, Y.F. Class-Imbalanced Semi-Supervised Learning with Adaptive Thresholding. In Proceedings of the 39th International Conference on Machine Learning, Baltimore, MD, USA, 17–23 July 2022; Volume 162, pp. 8082–8094.
34. Cai, J.; Hao, J.; Yang, H.; Zhao, X.; Yang, Y. A Review on Semi-Supervised Clustering. *Inf. Sci.* **2023**, *632*, 164–200. [CrossRef]
35. Torabi, H.; Mirtaheri, S.L.; Greco, S. Practical Autoencoder Based Anomaly Detection by Using Vector Reconstruction Error. *Cybersecurity* **2023**, *6*, 1. [CrossRef]
36. Grandvalet, Y.; Bengio, Y. Semi-Supervised Learning by Entropy Minimization. *Adv. Neural Inf. Process. Syst.* **2004**, *17*, 529–536.
37. Nielen, M.; Schukken, Y.H.; Van De Broek, J.; Brand, A.; Deluyker, H.A.; Maatje, K. Relations between On-Line Electrical Conductivity and Daily Milk Production on a Low Somatic Cell Count Farm. *J. Dairy Sci.* **1993**, *76*, 2589–2596. [CrossRef]
38. Ebrahimie, E.; Ebrahimi, F.; Ebrahimi, M.; Tomlinson, S.; Petrovski, K.R. A Large-Scale Study of Indicators of Sub-Clinical Mastitis in Dairy Cattle by Attribute Weighting Analysis of Milk Composition Features: Highlighting the Predictive Power of Lactose and Electrical Conductivity. *J. Dairy Res.* **2018**, *85*, 193–200. [CrossRef]

39. Tian, H.; Zhou, X.; Wang, H.; Xu, C.; Zhao, Z.; Xu, W.; Deng, Z. The Prediction of Clinical Mastitis in Dairy Cows Based on Milk Yield, Rumination Time, and Milk Electrical Conductivity Using Machine Learning Algorithms. *Animals* **2024**, *14*, 427. [CrossRef] [PubMed]
40. Sharma, D.; Kaniamuthan, S.; Manimaran, A.; Kumaresan, A.; Sivaram, M.; Rajendran, D.; Wankhade, P.R.; Sejian, V.; Banu, S. Seasonal, Physiological and Bacteriological Risk Factors for Subclinical Mastitis in Dairy Cows Maintained under Different Farming Conditions. *J. Dairy Res.* **2023**, *90*, 164–172. [CrossRef] [PubMed]
41. Malcata, F.B.; Pepler, P.T.; O'Reilly, E.L.; Brady, N.; Eckersall, P.D.; Zadoks, R.N.; Viora, L. Point-of-Care Tests for Bovine Clinical Mastitis: What Do We Have and What Do We Need? *J. Dairy Res.* **2020**, *87*, 60–66. [CrossRef] [PubMed]
42. Tedeschi, L.O.; Greenwood, P.L.; Halachmi, I. Advancements in Sensor Technology and Decision Support Intelligent Tools to Assist Smart Livestock Farming. *J. Anim. Sci.* **2021**, *99*, skab038. [CrossRef] [PubMed]
43. Vázquez Diosdado, J.A.; Barker, Z.E.; Hodges, H.R.; Amory, J.R.; Croft, D.P.; Bell, N.J.; Codling, E.A. Space-Use Patterns Highlight Behavioural Differences Linked to Lameness, Parity, and Days in Milk in Barn-Housed Dairy Cows. *PLoS ONE* **2018**, *13*, e0208424. [CrossRef]
44. Oehm, A.W.; Knubben-Schweizer, G.; Rieger, A.; Stoll, A.; Hartnack, S. A Systematic Review and Meta-Analyses of Risk Factors Associated with Lameness in Dairy Cows. *BMC Vet. Res.* **2019**, *15*, 346. [CrossRef]
45. Yang, X.; Daraz, U.; Ma, J.; Lu, X.; Feng, Q.; Zhu, H.; Wang, X.-B. Temporal–Spatial Variability of Grazing Behaviors of Yaks and the Drivers of Their Intake on the Eastern Qinghai-Tibetan Plateau. *Front. Vet. Sci.* **2024**, *11*, 1393136. [CrossRef]
46. Showalter, S.; Gropp, J. Validating Weak-Form Market Efficiency in United States Stock Markets with Trend Deterministic Price Data and Machine Learning. *arXiv* **2019**, arXiv:1909.05151. [CrossRef]
47. Græsbøll, K.; Kirkeby, C.; Nielsen, S.S.; Halasa, T.; Toft, N.; Christiansen, L.E. Models to Estimate Lactation Curves of Milk Yield and Somatic Cell Count in Dairy Cows at the Herd Level for the Use in Simulations and Predictive Models. *Front. Vet. Sci.* **2016**, *3*, 115. [CrossRef]
48. Kim, H.; Lee, J.T.; Fong, K.C.; Bell, M.L. Alternative Adjustment for Seasonality and Long-Term Time-Trend in Time-Series Analysis for Long-Term Environmental Exposures and Disease Counts. *BMC Med. Res. Methodol.* **2021**, *21*, 2. [CrossRef]
49. Hossein-Zadeh, N.G.; Kuhi, H.D.; France, J.; López, S. Introducing a Sinusoidal Equation to Describe Lactation Curves for Cumulative Milk Yield and Composition in Holstein Cows. *J. Dairy Res.* **2020**, *87*, 220–225. [CrossRef] [PubMed]
50. Maatje, K.; Huijsmans, P.J.M.; Rossing, W.; Hogewerf, P.H. The Efficacy of In-Line Measurement of Quarter Milk Electrical Conductivity, Milk Yield, and Milk Temperature for the Detection of Clinical and Subclinical Mastitis. *Livest. Prod. Sci.* **1992**, *30*, 239–249. [CrossRef]
51. Argaw, A. Review on Epidemiology of Clinical and Subclinical Mastitis on Dairy Cows. *Food Sci. Qual. Manag.* **2016**, *52*, 56–65.
52. Kamphuis, C.; Sherlock, R.; Jago, J.; Mein, G.; Hogeveen, H. Automatic Detection of Clinical Mastitis Is Improved by In-Line Monitoring of Somatic Cell Count. *J. Dairy Sci.* **2008**, *91*, 4560–4570. [CrossRef]
53. Hogeveen, H.; Kamphuis, C.; Steeneveld, W.; Mollenhorst, H. Sensors and Clinical Mastitis—The Quest for the Perfect Alert. *Sensors* **2010**, *10*, 7991–8009. [CrossRef]
54. Kamphuis, C.; Rue, B.D.; Eastwood, C.R. Field Validation of Protocols Developed to Evaluate In-Line Mastitis Detection Systems. *J. Dairy Sci.* **2016**, *99*, 1619–1631. [CrossRef]
55. Chengolova, Z.; Ivanov, Y.; Grigorova, G. The Relationship of Bovine Milk Somatic Cell Count to Neutrophil Level in Samples of Cow's Milk Assessed by an Automatic Cell Counter. *J. Dairy Res.* **2021**, *88*, 330–333. [CrossRef]
56. Johnson, J.M.; Khoshgoftaar, T.M. Survey on Deep Learning with Class Imbalance. *J. Big Data* **2019**, *6*, 27. [CrossRef]
57. Ghosh, K.; Bellinger, C.; Corizzo, R.; Branco, P.; Krawczyk, B.; Japkowicz, N. The Class Imbalance Problem in Deep Learning. *Mach. Learn.* **2024**, *113*, 4845–4901. [CrossRef]
58. Bürkner, P.C.; Gabry, J.; Vehtari, A. Efficient Leave-One-Out Cross-Validation for Bayesian Non-Factorized Normal and Student-t Models. *Comput. Stat.* **2021**, *36*, 1243–1261. [CrossRef]
59. Caruana, R.; Lawrence, S.; Giles, C. Overfitting in Neural Nets: Backpropagation, Conjugate Gradient, and Early Stopping. *Adv. Neural Inf. Process. Syst.* **2000**, *13*, 381–387.
60. Cohen, J. *Statistical Power Analysis for the Behavioral Sciences*, 2nd ed.; Routledge: New York, NY, USA, 1988. [CrossRef]
61. Kotlarz, K.; Mielczarek, M.; Biecek, P.; Wojdak-Maksymiec, K.; Suchocki, T.; Topolski, P.; Jagusiak, W.; Szyda, J. An Explainable Deep Learning Classifier of Bovine Mastitis Based on Whole-Genome Sequence Data—Circumventing the p >> n Problem. *Int. J. Mol. Sci.* **2024**, *25*, 4715. [CrossRef]
62. Fawcett, T. An Introduction to ROC Analysis. *Pattern Recognit. Lett.* **2006**, *27*, 861–874. [CrossRef]
63. Yacouby, R.; Axman, D. Probabilistic Extension of Precision, Recall, and F1 Score for More Thorough Evaluation of Classification Models. In Proceedings of the First Workshop on Evaluation and Comparison of NLP Systems, Virtual, 20 November 2020; pp. 79–91. [CrossRef]
64. Kim, T.K. Understanding One-Way ANOVA Using Conceptual Figures. *Korean J. Anesthesiol.* **2017**, *70*, 22–26. [CrossRef] [PubMed]
65. Abdi, H.; Williams, L.J. Tukey's Honestly Significant Difference (HSD) Test. *Encycl. Res. Des.* **2010**, *3*, 1–5.

66. Hyde, R.M.; Down, P.M.; Bradley, A.J.; Breen, J.E.; Hudson, C.; Leach, K.A.; Green, M.J. Automated Prediction of Mastitis Infection Patterns in Dairy Herds Using Machine Learning. *Sci. Rep.* **2020**, *10*, 4289. [CrossRef]
67. Jin, C.; Upton, J.; Bazargani, M.H.Z.; Mac Namee, B. Using Milk Flow Profiles for Subclinical Mastitis Detection. *Smart Agric. Technol.* **2024**, *9*, 100537. [CrossRef]

agriculture

MDPI

Article

# An Effective Yak Behavior Classification Model with Improved YOLO-Pose Network Using Yak Skeleton Key Points Images

**Yuxiang Yang †, Yifan Deng †, Jiazhou Li, Meiqi Liu, Yao Yao, Zhaoyuan Peng, Luhui Gu and Yingqi Peng ***

College of Mechanical and Electrical Engineering, Sichuan Agricultural University, Ya'an 625014, China; yangyx@stu.sicau.edu.cn (Y.Y.); dengyifan@stu.sicau.edu.cn (Y.D.); 202004728@stu.sicau.edu.cn (J.L.); 79073@sicau.edu.cn (M.L.); 201800772@stu.sicau.edu.cn (Y.Y.); pengzhaoyuan@stu.sicau.edu.cn (Z.P.); 2023217016@stu.sicau.edu.cn (L.G.)

* Correspondence: pengyingqi@sicau.edu.cn

† These authors contributed equally to this work.

**Abstract:** Yak behavior is a valuable indicator of their welfare and health. Information about important statuses, including fattening, reproductive health, and diseases, can be reflected and monitored through several indicative behavior patterns. In this study, an improved YOLOv7-pose model was developed to detect six yak behavior patterns in real time using labeled yak key-point images. The model was trained using labeled key-point image data of six behavior patterns including walking, feeding, standing, lying, mounting, and eliminative behaviors collected from seventeen 18-month-old yaks for two weeks. There were another four YOLOv7-pose series models trained as comparison methods for yak behavior pattern detection. The improved YOLOv7-pose model achieved the best detection performance with precision, recall, mAP0.5, and mAP0.5:0.95 of 89.9%, 87.7%, 90.4%, and 76.7%, respectively. The limitation of this study is that the YOLOv7-pose model detected behaviors under complex conditions, such as scene variation, subtle leg postures, and different light conditions, with relatively lower precision, which impacts its detection performance. Future developments in yak behavior pattern detection will amplify the simple size of the dataset and will utilize data streams like optical and video streams for real-time yak monitoring. Additionally, the model will be deployed on edge computing devices for large-scale agricultural applications.

**Keywords:** improved YOLOv7-pose; yak; image processing; deep learning

## 1. Introduction

The yak (*Bos grunniens*) is an important species that lives on the Qinghai–Tibetan Plateau and in high alpine locations. Around 95% of the yaks in the world are found in China, where they are distributed throughout the ecological conditions of low atmospheric oxygen, a lack of grass in the winter, and persistent cold temperatures all year long [1,2]. In the Qinghai–Tibetan regions, yaks are one of the most important economic resources, giving local farmers fuel, milk, meat, and hair. Yaks, through feeding, trampling, and defecation, significantly influence above- and belowground ecological processes such as plant succession and nutrient cycling. They adapt their grazing patterns seasonally to respond to variations in plant community composition, biomass, and abiotic environmental factors [3,4]. To improve the welfare of the animals and the effectiveness of breeding and reproduction, some livestock farms have raised yaks in pens in recent decades by improving their living conditions and feeds.

The behavior of cattle including yaks is an important welfare and health indicator. The critical status of livestock farming, including fattening, reproductive health, and diseases or infections, can be monitored through several indicative behavior patterns [5–8]. However, yaks' behavior habits are more varied as a result of their long-term grazing lifestyle. Yaks have evolved more aggressive behavior and stronger locomotor skills. This has helped them defend against predators and adapt to their surroundings. Studies have shown that yaks exhibit more active protein catabolism and glycolysis for energy production [9]. Based on prior studies and our own observations, yaks move more actively and with greater amplitude than normal cattle species, including dairy cows [10].

As a result of the habits of yaks mentioned above, it is challenging to directly use techniques which have been successful in cattle and dairy cows. For yaks, there are primarily two types of behavior- and route-tracking techniques used in various agricultural conditions. GPS monitoring can be used to track and locate yaks that are grazing across the pasture [11]. However, since indoor yak cultivation has a limited history, monitoring research and accompanying equipment rarely take this into account. Under conditions of continuous and real-time animal monitoring, dynamic analysis allows for object tracking and the recognition of various cow behavioral patterns [12]. With the aid of a monitoring and detection system, it is possible to identify self-protective behavior, estrus, and behavior associated with hoof illness [13–15]. Multi-sensors and deep learning models currently offer dependable solutions for precise livestock management [16,17]. By observing changes in the behavior and physical characteristics of the cattle, the corresponding important physiological and cultivation statuses can be tracked. Previous studies have fitted accelerometers, inertial measurement unit (IMU) sensors, and pressure sensors to the necks, ears, noses, and legs of cattle in order to track their behavior [18–22]. Usually, these studies achieve better monitoring results. However, the challenges with wearable devices have an impact or stress on livestock's daily activities [23,24]. Additionally, they usually use rechargeable batteries, which need regular replacement and cannot provide continuous behavior tracking in the long term. Acoustic sensors require specific settling locations and a quiet experimental environment. To obtain high-quality collection of individual cow vocalizations, additional processing is needed, including filtering out background noise and other cow vocalizations [25]. Problems with these devices can be addressed by employing methods of image processing to monitor the behavior of the livestock in the video. Therefore, pose estimation and behavior detection based on image processing techniques provide a potential method to understanding and analyzing animal behavior, as it enables the identification and tracking of key points on body of animal, thereby providing more precise data for behavior analysis [26,27]. To date, posture detection based on skeleton key points is one of the priority tasks for pose estimation. Human key-point detection can provide rich information about body structure, such as action recognition, behavior tracking, sports, and fitness [28–30]. Recently, animal pose estimation has set different key points for animals such as cows, pigs, sheep, and fish [31–34], providing new possibilities for animal behavior monitoring and analysis. The results showed the average precision (AP) of cattle and sheep pose estimation exceeding 87%. However, previous studies in livestock behavior detection were rarely found to recognize specific behaviors such as mounting and eliminative behaviors utilizing image processing techniques [35,36]. In particular, there are few behavior detection methods based on yak skeletons key points. Automated yak behavior monitoring using image-based key-point detection, while offering the potential for non-invasive, continuous observation with increased accuracy, is accompanied by technical challenges. Yaks' wide-ranging movements and significant activity result in large variations in their feature scales; on the other hand, there is variation in the image quality under different lighting conditions, which increases the complexity of the recognition algorithm.

For livestock detection, the YOLO (You Only Look Once) series models are a notable example of a single-stage detection algorithm, and YOLO series version 7 (YOLOv7) offers a compelling balance of speed and accuracy, making it suitable for real-time monitoring applications. The YOLOv7 architecture is designed to scale, offering a range of model sizes (e.g., YOLOv7-pose-tiny and YOLOv7-pose-w6) to meet the diverse memory and precision demands of various applications [37]. Therefore, in this study, a yak behavior detection model was developed based on the YOLOv7-pose architecture. To enhance the capability of subtle and diverse leg movement detection in a variety of environmental conditions, we developed an improved YOLOv7-pose model by using the mish activation function, SPPFCSPC (Spatial Pyramid Pooling, Feature Concatenate, and Spatial Convolution) extraction modules, and a dynamic head based on the original YOLOv7-w6-pose model.

This study aims to develop an effective, improved YOLOv7-pose model for detecting yak behavior patterns in a complex environment. Six yak behavior patterns including walking, feeding, standing, lying, mounting, and eliminative behaviors were detected with several YOLOv7-pose models using image datasets. To compare its detection performance, the model was also trained using the original YOLOv7-pose model on the same dataset. The effectiveness of the detection models was evaluated by comparing its detection precision, recall, mean average precision 0.5 (mAP0.5), and mAP0.5:0.95 across six behavioral patterns with those of other models. In summary, the main contributions of this paper are as follows:

(1) Six yak behavior patterns include walking, feeding, standing, lying, mounting and eliminative behaviors were detected with YOLOv7-pose models using yak skeleton key-point images.
(2) The yak behavior pattern detection model was improved by making three modifications to the original YOLOv7 architecture: replacing SiLU with the mish activation function, switching from the SPPCSPC module to the SPPFCSPC module, and replacing the original YOLO pose head with a dynamic head.
(3) The detection performance of the improved YOLOv7-pose model was compared with other five YOLOv7-pose models based on precision, recall, mAP0.5, and mAP0.5:0.95.

## 2. Materials and Methods

### 2.1. Location and Animals

The experiment was conducted from 15 July to 28 July 2022 on a livestock farm in Sichuan, China. Continuous behavior videos were collected from seventeen 18-month-old yaks raised in six pens. A guiding corral was situated outside the pens to transfer the yaks. A weighbridge was located adjacent to the pens for individual yak weight measurement. As shown in Figure 1, the yaks were raised in groups of 5, 6, and 6 in three pens (5000 mm in width and 8000 mm in length), separately. All yak was grown with unlimited access to water in concrete feed troughs. Soft sand and wood dust were used to set up the yaks' resting areas. The feed standards for the yaks were based on the Ministry of Agriculture of China's NYT/815/2004 dietary standard [38] for dry matter intake, net energy, crude protein, and minerals for beef cattle feeding. Total mixed diets, primarily composed of soybean meal, corn, rapeseed meal, wheat bran, and oat grass, were the feeding strategy. The feed was delivered to the troughs at 09:00 and 15:00 daily, followed by cleaning of the excrement and urine from each pen. The experiment was assessed by Sichuan Agricultural University Animal Ethical and Welfare committee (protocol No. 20220211).

### 2.2. Camera Setting and Data Acquisition

Two weeks' videos of yaks' daily behavior were recorded using DS-2CD3T47EWDV3-L cameras with a 4 mm focal length (Hikvision Inc., Hangzhou, China). The camera featured a frame rate of 25 frames per second (fps) with a resolution of 2560 × 1440 pixels. To monitor the yaks' behavior, the cameras were positioned as follows, camera1 was settled into the middle of the beam above the pen's geometric center at 4 m, camera2 was fixed in the corner close to the manger at 3 m, and camera3 was in the corner far from the manger at 3 m. The layout of the pen and the camera's placement are shown in Figure 1.

#### 2.2.1. Yak Behavior Pattern Definition

Six target behavior patterns of yak were defined based on previous research [24–26] and manual observation. Previous research has shown a connection between yaks' mounting activity and their social and reproductive activities [39,40]. Furthermore, basic behaviors like feeding, walking, lying, standing, and eliminative behavior are important indicators of welfare, productivity, and health. Each behavior pattern and its corresponding descriptions are shown in Table 1.

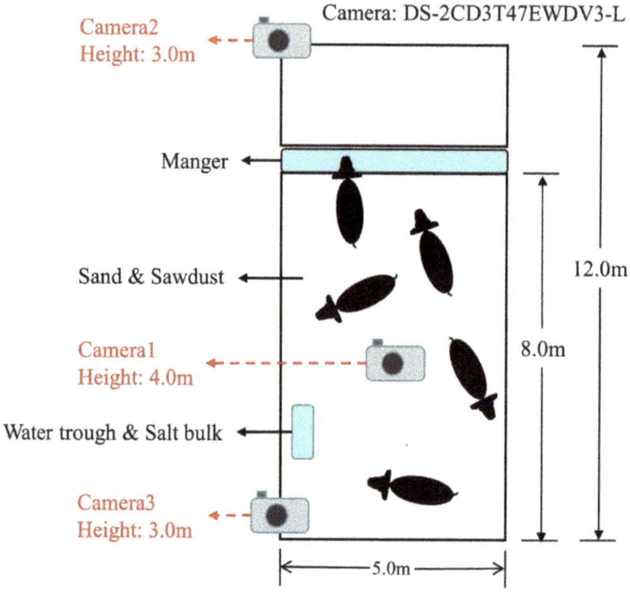


**Figure 1.** Layout of the pen and camera setting position.

**Table 1.** Definitions of yak behaviors.

| Behavior Class | Description |
|---|---|
| Standing | The yak stands on four legs with its head and neck moving |
| Lying | The yak lies on the ground without ruminating |
| Feeding | The yak puts its head into the feed trough and eats |
| Walking | The yak walks with its front and rear legs alternately |
| Mounting | The yak mounts another yak by placing its forelegs on its back |
| Eliminative behaviors | The yak defecates and urinates while standing with its hind legs spread apart, its tail up, and evidence of manure |

2.2.2. Yak Behavior Datasets

Labeled image datasets of yaks were prepared for the yak behavior pattern detection. Yak video segments with well-defined behavior patterns were manually edited, ensuring that the video clips were extracted under different times and lighting conditions and from multiple camera angles. A total of 187 videos were captured at a frame rate of 25 frames per second, with durations ranging from 5 to 30 s. The yak behavior pattern dataset was created using 2043 labels that were observed from these videos at two-second intervals. Then, each behavior pattern, as shown in Figure 2. was manually labeled with Labelme software according to the description in Table 1 (Labelme, 5.5.0, Heartex Inc., San Francisco, CA, USA). We selected nineteen key points on the yak's body for behavior detection. The key areas include the head, limbs, back, and tail (1—Yak Mouth, 2—Head, 3—Neck, 4—Back, 5—Hip, 6—Middle of Tail, 7—Tail Tip, 8—Front Left Leg Root, 9—Front Left Knee, 10—Front Left Hoof, 11—Front Right Leg Root, 12—Front Right Knee, 13—Front Right Hoof, 14—Rear Left Leg Root, 15—Rear Left Knee, 16—Rear Left Hoof, 17—Rear Right Leg Root, 18—Rear Right Knee, and 19—Rear Right Hoof), as shown in Figure 3. These key points highlight significant structural body features of the yak [41], which enable more efficient extraction of behavioral features. This study extends previous research by incorporating tail key-point identification for the recognition of eliminative behaviors.

We randomly selected 80% of the labels as the training dataset, and the remaining 20% as the testing data. The particular label numbers of each behavior are displayed in Table 2.

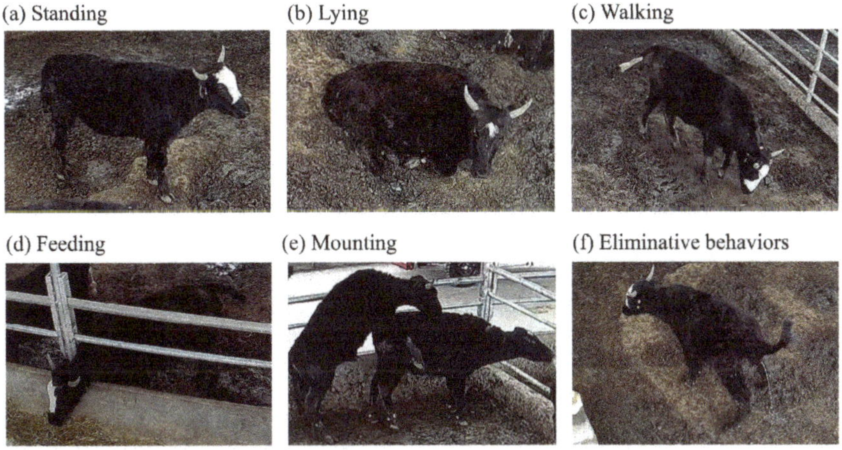


**Figure 2.** Sample image of each behavior.

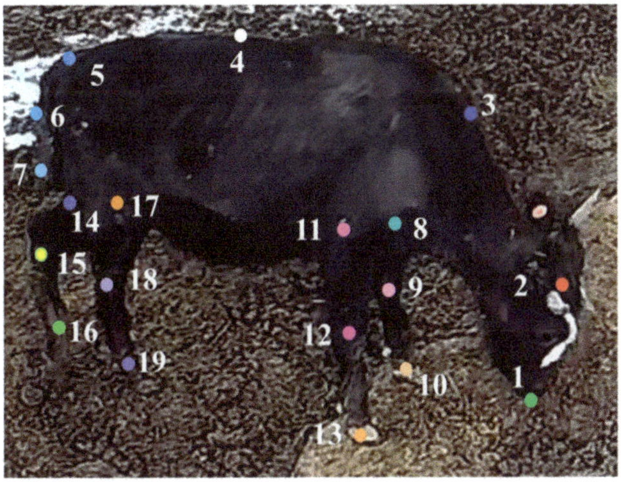


**Figure 3.** Nineteen key points of yak.

**Table 2.** Basic behavior image information on the datasets.

| Behavior Class | No. of Behavior Label for Training | No. of Behavior Label for Testing |
|---|---|---|
| Standing | 382 | 95 |
| Lying | 342 | 86 |
| Feeding | 475 | 119 |
| Walking | 188 | 47 |
| Mounting | 123 | 31 |
| Eliminative behaviors | 124 | 31 |
| Total | 1634 | 409 |

### *2.3. Behavior Detection Using Six Detection Models*

#### 2.3.1. Mish Activation Function

The original YOLOv7-w6-pose model uses SiLU as the activation function. The improved YOLOv7-pose model utilizes the mish activation function to detect the yak's behavior. The mish activation function is a self-regularizing non-monotonic activation function and provides a smooth and constant output. Compared with SiLU, the unbounded and non-saturating nature of the mish function effectively mitigates the vanishing gradient problem, while its lower bound and smooth transition in the negative half-axis prevent neuron death and enhance the generalization ability and optimization efficiency of the yak behavior detection network. The mish activation function can be represented as follows:

$$f(x) = xtanh(softplus(x)) \tag{1}$$

#### 2.3.2. SPPFCSPC Module

The SPPFCSPC module is a variation of the Spatial Pyramid Pooling, Cross-Stage Partial Channel (SPPCSPC) module by providing an effective feature extraction module [42], particularly in YOLOv7. The SPPFCSPC module improved the model efficiency by swapping out the SPP module with the SPPF module to maintain the performance of SPPCSPC while further reducing computational costs. The two main components of SPPFCSPC include SPPF (Spatial Pyramid Pooling-Fast) and CSP (Cross-Stage Partial Connections) [43]. To capture multi-scale characteristics, SPPF uses multiple max-pooling operations that use various kernel sizes and strides. These pooled features are then concatenated to provide a fixed-size representation, independent of the input size. CSP separates the input feature map into two branches in an effort to enhance gradient flow and reduce computations [44]. One branch is convolutional, while the other is connected by a shortcut. Subsequently, the outputs from both branches are combined, enabling more comprehensive feature representation and effective training.

Generally, by processing one branch of the CSP structure in the SPPFCSPC module, the SPPF component improves the multi-scale feature extraction capabilities substantially. A detailed configuration of the structure of the SPPFCSPC module is shown in Figure 4.

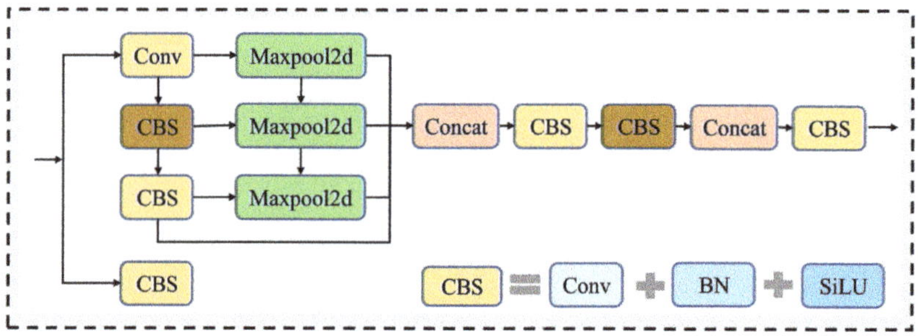


**Figure 4.** The structure of the SPPFCSPC module.

#### 2.3.3. Dynamic Head

Dynamic head is a framework that aims to integrate object detection heads with attention mechanisms. By dynamically adjusting the model structure and incorporating these mechanisms, it can more effectively process the subtle and diverse pose variations of yaks [45]. Dynamic head leverages three different self-attention mechanisms including scale-aware attention, space-aware attention, and task-aware attention to enhance object detection as follows:

Given the feature tensor $F \in R^{L \times S \times C}$, the general formulation fir applying self-attention is as follows:

$$w(F) = \pi(F) - F \tag{2}$$

where $\pi(\cdot)$ is an attention function. A naive solution to this attention function is implemented by fully connected layers. The attention function was converted into the three sequential attentions, each focusing on only one perspective:

$$w(F) = \pi_C(\pi_S(\pi_L(F) \cdot F) \cdot F) \cdot F \tag{3}$$

where $\pi_L(\cdot)$, $\pi_S(\cdot)$, and $\pi_C(\cdot)$ are three different attention functions applied on dimensions $L$, $S$, and $C$, respectively.

Scale-Aware Attention $\pi_L(\cdot)$. Scale-aware attention is introduced to dynamically combine features from different scales based on their semantic significance.

$$\pi_L(F)\cdot F = \sigma(f(\frac{1}{SC}\sum_{S,C} F))\cdot F \tag{4}$$

where $f(\cdot)$ is a linear function approximated by a $1 \times 1$ convolutional layer and $\sigma(x) = \max(0, \min(0, \min(1, \frac{x+1}{2})))$ is a hard-sigmoid function.

Space-Aware Attention $\pi_S$. Space-aware attention enables the model to consider the spatial relationships between objects and their surroundings. This focus helps the model concentrate on relevant areas of the input image, thereby improving predictions. Considering the high dimensionality in $S$, the module operates into two steps: it learns sparse attention using deformable convolution, and then, it aggregates features across different levels at the same spatial locations:

$$\pi_s(F)\cdot F = \frac{1}{L}(\sum_{l=1}^{L}\sum_{k=1}^{K} W_{l,k}\cdot F(l; p_k + \Delta p_k; c)\Delta m_k) \tag{5}$$

where $K$ is the number of sparse sampling locations, $p_k + \Delta p_k$ is a location shifted by the self-learned spatial offset $\Delta p_k$ to focus on a discriminative region, and $\Delta m_k$ is a self-learned importance scalar at location $p_k$. Both are learned from the input feature from the median level of $F$.

Task-Aware Attention $\pi_C$. The task-aware self-attention mechanism enables the model to adjust the importance of different channels for detection and localization tasks. The model can enhance its capacity to differentiate between object categories and precisely predict bounding boxes by focusing on relevant channels. It dynamically switches ON and OFF channels of features to favor different tasks:

$$\pi_C(F)\cdot F = max(\alpha^1(F)F_C + \alpha^2(F)F_C + \beta^2(F)) \tag{6}$$

where $F_C$ is the feature slice at the c-th channel and $\alpha^1, \alpha^2, \beta^1, \beta^2 = \theta(\cdot)$ is a hyperfunction that learns to control the activation threshold. $\theta(\cdot)$ is implemented similar to dynamic relu [46], which first conducts a global average pooling on L $\times$ S dimensions to reduce the dimensionality, then uses two fully connected layers and a normalization layer, and finally applies a shifted sigmoid function to normalize the output to $[-1, 1]$.

Finally, Equation (2) is nested multiple times to effectively stack multiple $\pi_L$, $\pi_s$, and $\pi_C$ blocks together. A detailed configuration of the dynamic head block is shown in Figure 5.

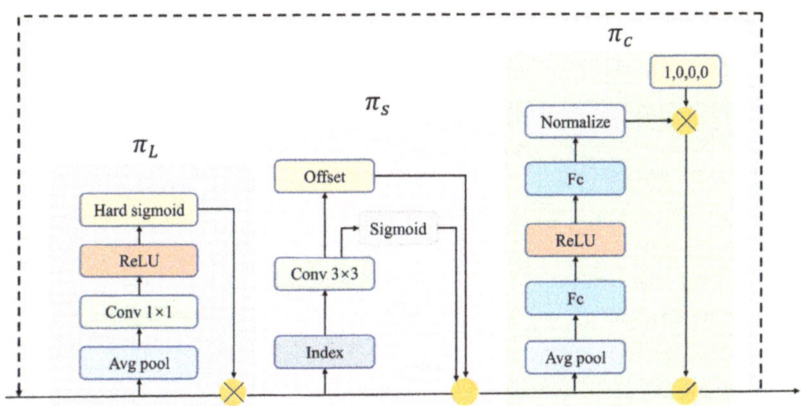


**Figure 5.** The structure of the dynamic head block.

#### 2.3.4. Original YOLOv7-Pose Model

The real-time detection of yak's behavior patterns is important for the development of smart livestock management systems. In order to improve the detection performance of multiple yak behaviors, especially in behavior patterns with physically similar movements such as walking and standing, the YOLOv7-Pose model was utilized to detect the target behavior patterns. YOLOv7-Pose is an extension of the YOLOv7 object detection model [47], primarily designed for human pose estimation. YOLOv7-Pose models can be categorized into YOLOv7-tiny-pose and YOLOv7-w6-pose, distinguished by their network depth and parameter count, as shown in Figures 6 and 7. Specifically, the YOLOv7-pose-w6 model uses 4 different scales of detection heads, while the YOLOv7-pose-tiny model simplifies the FPN structure and uses 4 different scales of detection heads, as shown in Figure 7. The YOLOv7-pose network model consists of three main parts: the backbone, neck, and head. In this study, the objective of the original YOLOv7-pose was to identify seventeen critical places on the yak body. The elements of YOLOv7-pose are described below:

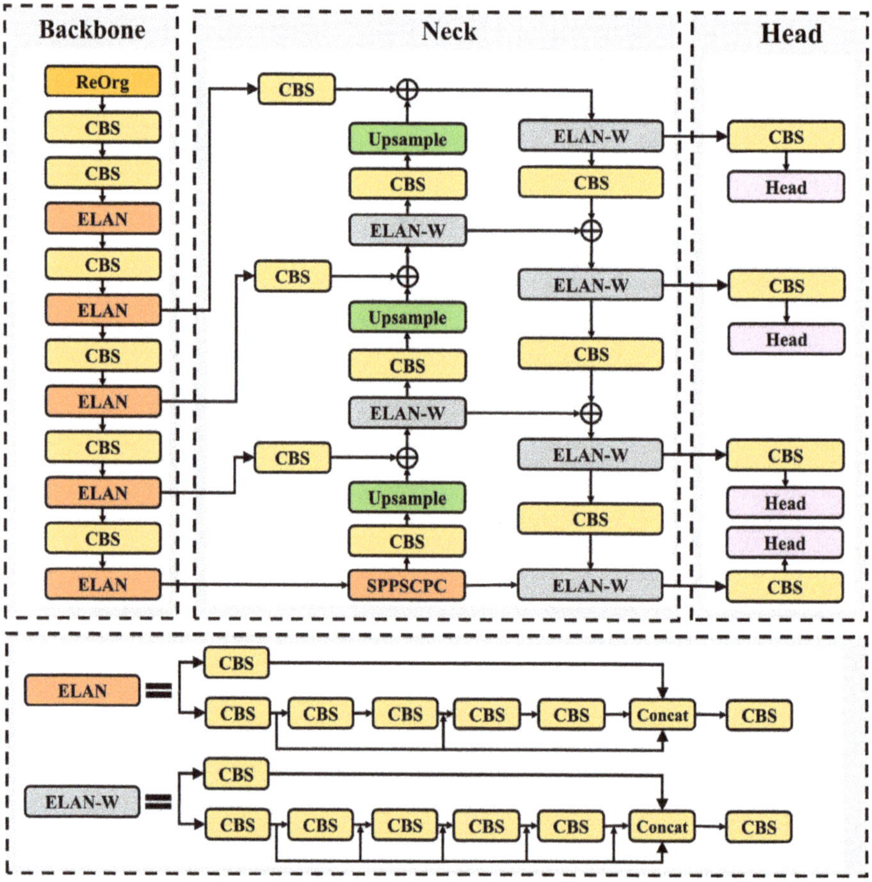


**Figure 6.** The structure of the YOLOv7-w6-pose model.

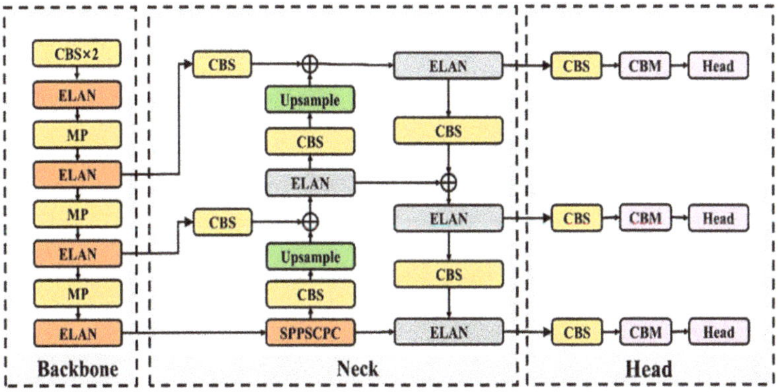


**Figure 7.** The structure of the YOLOv7-tiny-pose model.

Backbone: The function of the backbone is to extract features from the input image. In order to enhance the network's learning capability, it makes use of E-ELAN (Extended Efficient Layer Aggregation Network), an enhanced version of ELAN [48], to assesses variables such as memory access cost, I/O channel ratio, element-wise operations, activation functions, and gradient paths. E-ELAN aims to maximize the utilization of computational resources while enhancing feature representation through the application of innovative connection strategies.

Neck: YOLOv7-pose's neck uses an innovative framework that combines the Cross-Stage Partial Connections (CSPs) [43], Spatial Pyramid Pooling (SPP) [49], and Path Aggregation Network (PAN) [50] approaches to improve feature representation. While the SPP structure extracts characteristics from multiple scales and the PAN structure effectively combines features from different scales, the CSP structure efficiently minimizes processing costs.

Head: Two separate heads—one for object detection and one for key-point detection—make up the head network of YOLOv7-Pose. These heads independently forecast objects of multi-scale objects and key points by extracting data from four feature maps obtained from the neck network. While the key-point identification head predicts the coordinates of the yak's 17 key points, the object detection head predicts the class and location of objects to generate bounding boxes.

#### 2.3.5. YOLOv7-w6-Pose-Mish-SPPFCSPC-Dynamic Head Model

To improve the model performance of yak behavior detection, in this study, an improved YOLOv7-pose model was proposed to enhance the detection performance of yak behavior patterns. Based on the original YOLOv7-w6-pose model, we utilized the mish activation function instead of the SiLU activation function in the backbone, neck, and head; replaced SPPCSPC with the SPPFCSPC in the neck network; and modified the detection head with a dynamic head. The complete structure of the improved YOLOv7-pose is shown in Figure 8.

### *2.4. Experimental Platform*

The experiment was run on a hardware platform with the following specifications: Intel (R) Core (TM) i9-14900HX, NVIDIA GeForce RTX 4060 GPU, and 16 GB RAM. The PyTorch framework (version 2.0.0) and Python programming language (version 3.9.19) were utilized. To accelerate the model training speed, CUDA (version 11.8) was used. The detection models in this work were trained using a batch size of 8 for 200 epochs. The initial learning rate was set to 0.01, with a weight decay of 0.0005 and a momentum of 0.937. The optimizer used was SGD.

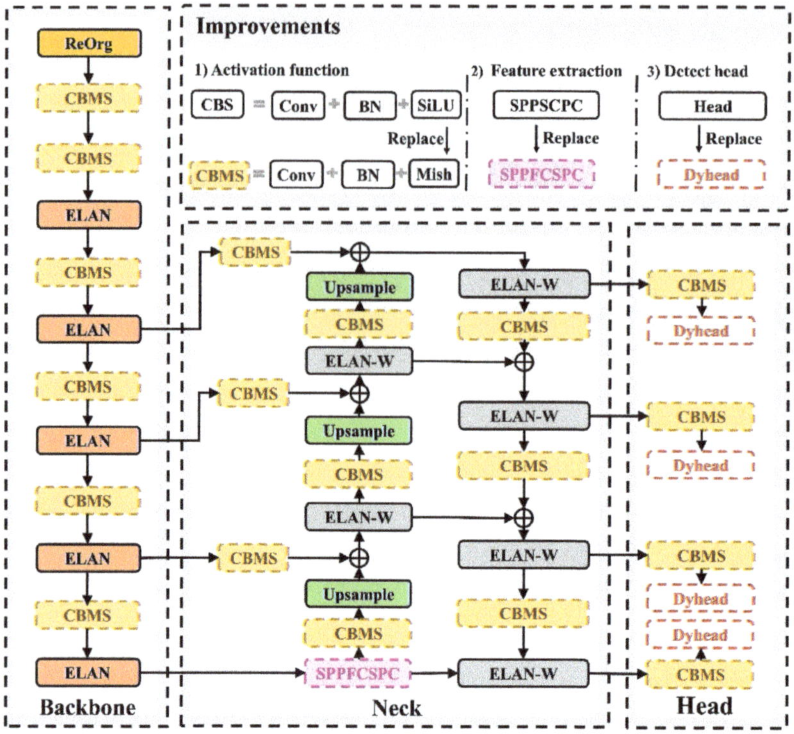


**Figure 8.** The structure of the improved YOLOv7-pose model.

### *2.5. Model Evaluation*

To validate the efficiency of the models, precision, recall, mAP0.5, and mAP0.5:0.95 were used to measure accuracy. Accuracy, precision, and recall were calculated as follows:

$$Accuracy = \frac{TP + TN}{TP + TN + FP + FN} \tag{7}$$

$$Precision = \frac{TP}{TP + FP} \tag{8}$$

$$Recall = \frac{TP}{TP + FN} \tag{9}$$

where $TP$ is true positive and $TN$ is true negative. Incorrect model predictions of the positive and negative classes are defined as $FP$ (false positive) and $FN$ (false negative).

The definition of mAP0.5 is an average precision at an IoU (Intersection over Union) threshold of 0.5. Moreover, mAP0.5:0.95 is the average precision averaged over IoU thresholds from 0.5 to 0.95, with a step size of 0.05. The formulas are shown as follows:

$$mAP0.5 = \frac{1}{C}\sum_{j=1}^{C} AP_j \tag{10}$$

$$mAP0.5{:}0.95 = \frac{1}{10}\sum_{t=0.5}^{0.95} AP(t) \tag{11}$$

where $AP_j$ reflects the model's ability to predict objects of class j accurately at 0.5 IoU thresholds and $AP(t)$ is the average precision at a specific IoU threshold.

## 3. Results

### 3.1. Comparison of Behavior Detection Performance of Different Models

This study aims to detect yak behavior using key-point detection. The detection results of the six yak behavior patterns including walking, feeding, standing, lying, mounting, and eliminative behaviors, using six skeleton key-point detection models are shown in Table 3. In order to compare the effectiveness of the improved YOLOv7-pose model in yak behavior patterns, the YOLOv7-tiny-pose, YOLOv7-w6-pose, YOLOv7-w6-pose-mish, YOLOv7-w6-pose-SPPFCSPC, and YOLOv7-w6-pose-dynamic head models were trained with the same dataset as the comparison model. The models' performance was evaluated using precision, recall, mAP0.5, and mAP0.5:0.95.

**Table 3.** Comparison of the detection results of the six models.

| Model | Precision | Recall | mAP0.5 | mAP0.5:0.95 |
|---|---|---|---|---|
| YOLOv7-tiny-pose | 0.758 | 0.778 | 0.799 | 0.603 |
| YOLOv7-w6-pose | 0.836 | 0.830 | 0.864 | 0.745 |
| YOLOv7-w6-pose-mish | 0.879 | 0.853 | 0.893 | 0.756 |
| YOLOv7-w6-pose-SPPFCSPC | 0.855 | 0.872 | 0.897 | 0.767 |
| YOLOv7-w6-pose-dynamic head | 0.896 | 0.839 | 0.895 | 0.756 |
| YOLOv7-w6-pose-mish SPPFCSPC-dynamic head (improved YOLOv7-pose) | 0.899 | 0.877 | 0.904 | 0.767 |

The improved YOLOv7-pose model exceeded the other models in its detection of six different yak behavior patterns using skeleton key-point images. By employing the YOLOv7-w6-pose model rather than the YOLOv7-tiny-pose model, the detection performance was clearly enhanced overall; the precision increased from 75.8% to 83.6% along with the other indicators. Based on the YOLOv7-w6-pose model, the utilization of the mish activation function, SPPFCSPC, and the dynamic head module improved the detection precision to 87.9%, 85.5%, and 89.6%, respectively. The best detection performance was attained by the improved YOLOv7-pose model, which used the mish activation function, SPPFCSPC, and the dynamic head module on the original YOLOv7-w6-pose model. Its precision, recall, mAP0.5, and mAP0.5:0.95 were 89.9%, 87.7%, 90.4%, and 76.7%, respectively. These results demonstrate that the improved YOLOv7-pose model enhances the performance of yak behavior detection by adding the mish activation function, SPPFCSPC extraction modules, and the dynamic head in complex environments.

### 3.2. Behavior Pattern Detection Results

The detection accuracy for particular behavior patterns of all models is shown in Figure 9. In all yak behavior pattern detection models, the behavior patterns feeding, lying, and mounting were recognized with high precision which are all over 89%. In particular, the precision of lying and mounting both exceeded 97% in the five models' results, except the YOLOv7-tiny-pose model. Mutual misidentification of standing and walking contributed to the majority of mistakes in detecting all yak behavior patterns. Especially in the results of the YOLOv7-tiny-pose model, 42% of standing behavior images were misdetected as walking and 15% of them were recognized as eliminative behaviors. The detection performance was improved by using YOLOv7-w6-pose and combining with SPPFCSPC, mish, and dynamic head modules. The detection precision of standing and walking was finally enhanced to 69% and 81% with the improved YOLOv7-pose model. Feeding cases was mainly misclassified as standing while using all detection models, with a ratio from 3% to 8%.

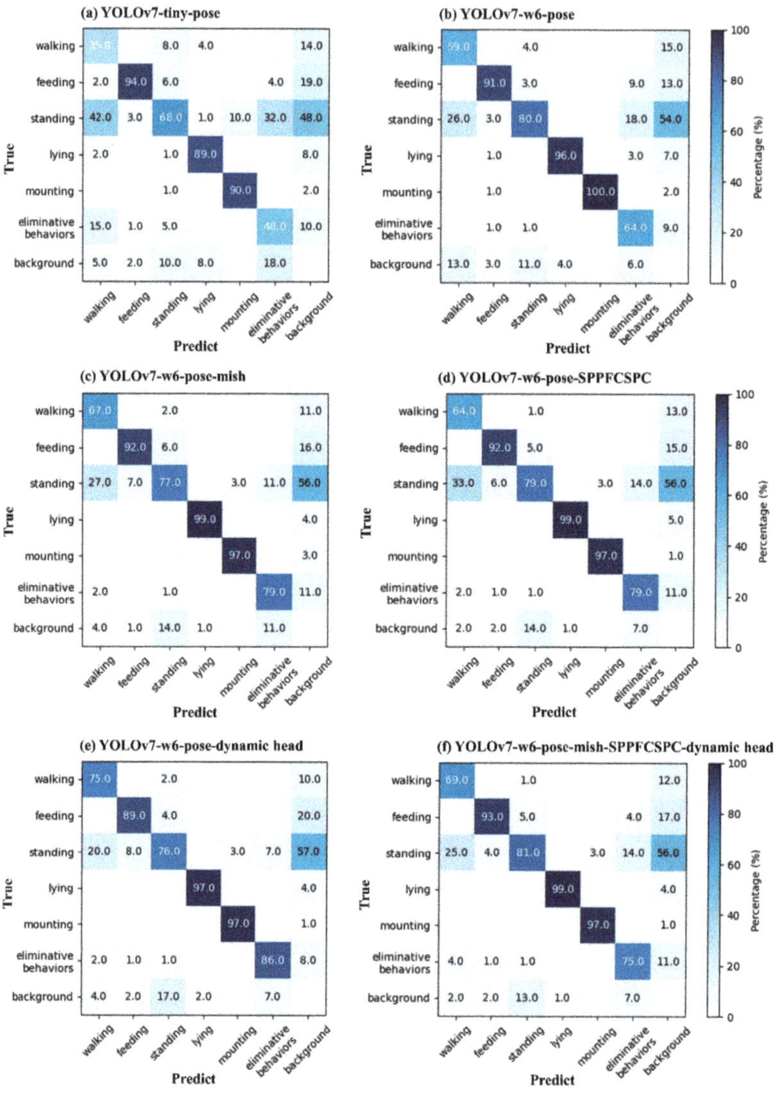


**Figure 9.** Confusion matrix of the detection accuracy of six behavior patterns of yak. The diagonal represents the detection accuracy for each behavior. The color is darker for higher accuracies.

For eliminative behaviors, the detection performance was obviously improved by using the YOLOv7-w6-pose-mish, YOLOv7-w6-pose-SPPFCSPC, YOLOv7-w6-pose-dynamic head, and improved YOLOv7-pose models. The amount of eliminative behaviors misclassified as standing behavior ranged from 1% to 5%. Moreover, there was a misdetection between standing and the background due to suboptimal lighting conditions at night and their black hair color being similar to the background color. Among all the detection models' results, 10% to 17% of background images were misrecognized as standing.

As shown in Figure 10, the detection performance of this study showed improvement compared with the original YOLOv7-pose model. Particularly, the detection performance of the three behavior patterns with physically similar movements significantly improved.

Walking increased from 35% to 69%. Standing increased from 68% to 81%, and eliminative behaviors increased from 48% to 75%.

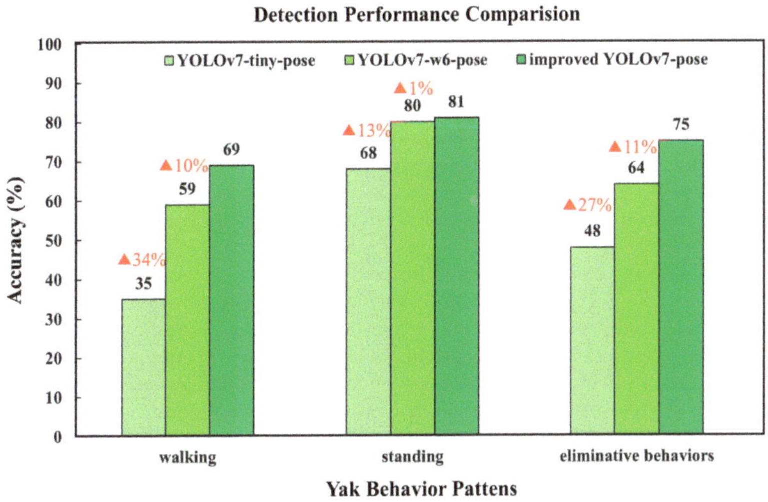


**Figure 10.** Detection performance comparison of the YOLOv7-pose and improved YOLOv7-pose models.

## 4. Discussion

### *4.1. Comparison with the State of the Art*

It is not straightforward to directly compare the performance of our behavior pattern detection model with related work, as there are significant differences in the materials, methods, and evaluation metrics used in different papers, even when the task being addressed is the same (e.g., detecting yak behavior from images). This study developed a yak behavior detection model using an improved YOLOv7-pose model with yak key-point-labeled images.

IMU, pressure sensors, and microphones have been used in livestock behavior pattern classification in recent years. These devices may cause discomfort, potentially altering natural behaviors and even eliciting rejection responses, leading to sensor detachment and subsequent data loss. This issue is exacerbated by frequent head-butting behaviors among yaks [51]. Furthermore, as compared to visual feature-based animal behavior detection methods, posture estimation-based methods that use the key points and skeletons of body part detection methods may identify finer-grained limb motions and behavioral changes while being more robust to interference. While pose estimation methods offer advantages in yak behavior detection, their performance is still limited by the difficulties of visual feature extraction. The similarity in color between yaks and the ground, especially under poor lighting conditions, makes it difficult to accurately extract their contours and small targets like yak leg features, thus impacting the accuracy of pose estimation and leading to difficulties in distinguishing between similar postures like walking, standing, and defecation. Moreover, current studies were rarely found to monitor specific behaviors such as mounting and eliminative behaviors utilizing image processing techniques. In this study, six behavior patterns of yak were detected using an improved YOLOv7-pose model and achieved better performance.

### *4.2. General Discussion*

Among all the models, the improved YOLOv7-pose yak behavior pattern detection model enhanced the detection performance. This model with the mish activation function, SPPFCSPC, dynamic head modules achieved superior performance in all evaluation indicators including precision, recall, mAP0.5, and mAP0.5:0.95.

Overall, the yaks in the camera footage exhibited obvious scale variations due to different distances and angles from the camera, as well as scale shifts between different action transitions, providing difficulty in detecting yak behavior across different scales. The dynamic head mitigates the challenge of scale variations in yak behavior detection. Compared to the YOLOv7-pose-w6 model, precision improved by 6% and mAP0.5 increased by 3.1%. This improvement is attributed to the scale-aware attention mechanism, which effectively handles multi-scale feature information. The spatial-aware attention mechanism flexibly combines multi-level features, enhancing the detection head's ability to distinguish different feature points and thus improving key-point localization accuracy. The results from the YOLOv7-w6-pose-mish network demonstrate improved performance using the mish activation function, which enhances adaptive deep learning capabilities compared to SiLU.

More specifically, eliminative, standing, and walking behaviors were difficult to differentiate in previous visual pose detection algorithms because of their similar leg postures. In this study, we labeled the key points of the yaks' leg skeleton to provide additional leg posture information for tracking these three behavior patterns. The standing and walking behavior cases of yak were commonly misclassified as each other in the detection process. Yak's habit to walk slowly most of the time led to incorrect detections because both the eliminative behavior and the walking postures in the images resembled standing positions. The SPPFCSPC module improves feature extraction for yak behavior detection by integrating multi-scale information, providing a network to capture contextual information over a wide spatial range. The multi-scale technique increases detection accuracy by effectively detecting both yak legs and yak bodies. To further increase the accuracy of the behavior patterns described above, a larger number of samples of similar movement patterns should be prepared in the future. In addition, a certain percentage, from 3% to 8%, of feeding behavior images were detected as standing due to a similar reason to that above.

All yak behavior patterns, including those that were accurately detected, were misidentified as the background. One reason is that yaks frequently linger near feeding troughs, which causes the trough to occasionally obstruct a portion of their bodies. In addition, the barn padding and the yaks' similar dark brown color cause the models to sometimes identify yaks as backgrounds. In this study, in order to enhance the detection performance of the YOLOv7-pose model on tasks involving capturing multi-scale features and subtle, diverse pose variations of yaks, the mish activation function, SPPFCSPC, and dynamic head modules were implemented in the original structure. The performance of the yak behavior pattern detection between the original YOLOv7-pose model and the improved YOLOv7 model is compared in Figure 11.

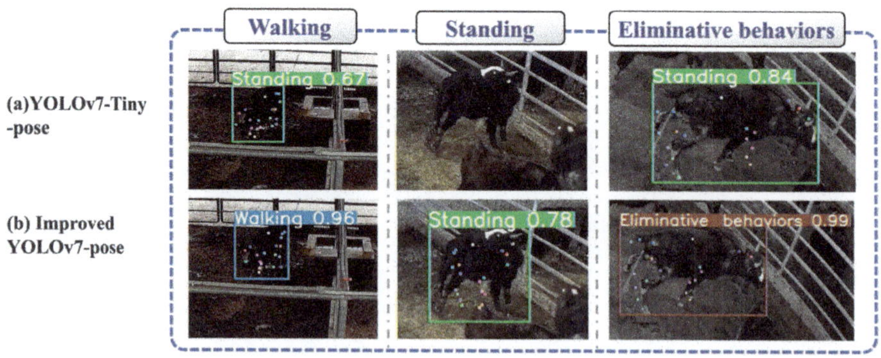


**Figure 11.** The detection performance of the improved behavior monitoring model based on the YOLOv7-pose and improved YOLOv7-pose models.

### 4.3. Limitations and Future Work

In this study, there exist several limitations that necessitate further investigation in future research. Firstly, the detection accuracy of physically similar movements such as walking, standing, and eliminative behavior is relatively lower even in the best performance model. While the dynamic head module effectively enhances the model's ability to capture spatial information, it might benefit from finer granularity when detecting small objects like legs and tails. This finer granularity could enable the detection of more subtle targets, such as the curvature features of legs and tails, which can be indicative of different behavior patterns. Secondly, while key-point-based behavior monitoring methods have made progress in improving the accuracy of behavior monitoring, their robustness to scene variations and different lighting conditions still needs further improvement. Based on these considerations, this paper proposes several future research directions:

- Expanding the database to include yaks of varying sexes, ages, and breeds and refining yak key-point definitions to enhance spatial information capture, improving the detection of subtle movements by providing a more comprehensive understanding of yak body structure and behavior patterns;
- Optimizing the capability of spatial information capture for subtle targets byb adding optical flow or depth information and employing finer granularity to improve the detection of slight movements.

## 5. Conclusions

In this study, six yak behavior patterns including walking, feeding, standing, lying, mounting, and eliminative behaviors were detected and classified using an improved YOLOv7-pose behavior detection model with yak skeleton key-point images. The model's detection performance was enhanced by combining the mish activation function, SPPFCSPC extraction modules, and dynamic head modules based on the original YOLOv7-w6-pose model. Compared to other YOLO series models, the improved YOLOv7-pose model achieved a superior performance with an overall precision of 89.9% and mAP0.5 of 90.4%. With the improved YOLOv7-pose model, there is an obvious improvement in the detection of the walking, standing, and eliminative behaviors of yak; these behavior patterns are difficult to distinguish as these behaviors have similar movements. Because the improved YOLOv7-pose model effectively extracts multi-scale features of yaks and identifies their subtle and diverse pose variations. The identification of key points on yak legs significantly improves the accuracy of recognizing physically similar movements such as walking, standing, and eliminative behaviors. In the future, the yak dataset's simple size will be expanded to include a wider range of ages, breeds, sexes, and breeding environments. The detection capability of small objects and slight movements will be improved by finer granularity. A multimodal signal including optical flow and video streaming will be added for more accurate yak behavior detection and monitoring. Additionally, we will further reduce the model size to enable its deployment on edge computing devices, allowing for the monitoring and detection of yak behavior patterns in livestock farm applications.

**Author Contributions:** Conceptualization, Y.P.; methodology, Y.Y. (Yuxiang Yang) and Y.D.; software, Y.Y. (Yuxiang Yang) and Y.D.; validation, Y.Y. (Yuxiang Yang) and Y.D.; formal analysis, Y.Y. (Yuxiang Yang) and Y.D.; investigation, Y.Y. (Yuxiang Yang) and Y.D.; resources, Y.Y. (Yuxiang Yang), Y.D., J.L., M.L. and Y.Y. (Yao Yao); data curation, J.L., M.L., Z.P. and L.G.; writing—original draft preparation, Y.Y. (Yuxiang Yang), Y.D. and Y.P.; writing—review and editing, Y.Y. (Yuxiang Yang), Y.D. and Y.P.; visualization, Y.Y. (Yuxiang Yang) and Y.D.; supervision, Y.P.; project administration, Y.P.; funding acquisition, Y.P. All authors have read and agreed to the published version of the manuscript.

**Funding:** This study was supported by the National Key R&D Program of China (grant number 2021YFD1600200) and by the subject double support program of Sichuan Agriculture University (grant number T202107).

**Institutional Review Board Statement:** The animal study protocol was approved by Sichuan Agricultural University Institutional Animal Care and Use Committee (protocol code No. 20220211, 2 July 2022).

**Informed Consent Statement:** Not applicable.

**Data Availability Statement:** The data presented in this study are available from the corresponding author on request. The data are not publicly available due to confidentiality requirements related to ongoing experiments.

**Conflicts of Interest:** The authors declare no conflicts of interest.

## References

1. Hu, R.; Zou, H.; Wang, H.; Wang, Z.; Wang, X.; Ma, J.; Shah, A.M.; Peng, Q.; Xue, B.; Wang, L.; et al. Dietary Energy Levels Affect Rumen Bacterial Populations That Influence the Intramuscular Fat Fatty Acids of Fattening Yaks (*Bos grunniens*). *Animals* **2020**, *10*, 1474. [CrossRef]
2. Zou, H.; Hu, R.; Wang, Z.; Shah, A.M.; Zeng, S.; Peng, Q.; Xue, B.; Wang, L.; Zhang, X.; Wang, X.; et al. Effects of Nutritional Deprivation and Re-Alimentation on the Feed Efficiency, Blood Biochemistry, and Rumen Microflora in Yaks (*Bos grunniens*). *Animals* **2019**, *9*, 807. [CrossRef]
3. Liu, P.; Ding, L.; Zhou, Y.; Jing, X.; Degen, A.A. Behavioural Characteristics of Yaks Grazing Summer and Winter Pastures on the Qinghai-Tibetan Plateau. *Appl. Anim. Behav. Sci.* **2019**, *218*, 104826. [CrossRef]
4. Yang, X.; Daraz, U.; Ma, J.; Lu, X.; Feng, Q.; Zhu, H.; Wang, X.B. Temporal–Spatial Variability of Grazing Behaviors of Yaks and the Drivers of Their Intake on the Eastern Qinghai-Tibetan Plateau. *Front. Vet. Sci.* **2024**, *11*, 1393136. [CrossRef]
5. Xiao, J.; Tian, J.; Peng, Y.; Rui, H.; Peng, Q.; Wang, L.; Bai, X.; Wang, Z. Effects of Forage Feeding to Calves on Performance, Rumen Fermentation and Nutrient Digestibility: A Meta-Analysis. *J. Dairy Sci.* **2024**, *100*, 1136–1150. [CrossRef]
6. Wu, Y.; Liu, M.; Peng, Z.; Liu, M.; Wang, M.; Peng, Y. Recognising Cattle Behaviour with Deep Residual Bidirectional LSTM Model Using a Wearable Movement Monitoring Collar. *Agriculture* **2022**, *12*, 1237. [CrossRef]
7. Peng, Y.; Kondo, N.; Fujiura, T.; Suzuki, T.; Ouma, S.; Wulandari; Yoshioka, H.; Itoyama, E. Dam Behavior Patterns in Japanese Black Beef Cattle Prior to Calving: Automated Detection Using LSTM-RNN. *Comput. Electron. Agric.* **2020**, *169*, 105178. [CrossRef]
8. Russello, H.; van der Tol, R.; Holzhauer, M.; van Henten, E.J.; Kootstra, G. Video-Based Automatic Lameness Detection of Dairy Cows Using Pose Estimation and Multiple Locomotion Traits. *Comput. Electron. Agric.* **2024**, *223*, 109040. [CrossRef]
9. Wang, H.B.; Zan, L.S.; Zhang, Y.Y. Profiling of the Yak Skeletal Muscle Tissue Gene Expression and Comparison with the Domestic Cattle by Genome Array. *Animal* **2014**, *8*, 28–35. [CrossRef]
10. Peng, Y.; Peng, Z.; Zou, H.; Liu, M.; Hu, R.; Xiao, J.; Liao, H.; Yang, Y.; Huo, L.; Wang, Z. A Dynamic Individual Yak Heifer Live Body Weight Estimation Method Using the YOLOv8 Network and Body Parameter Detection Algorithm. *J. Dairy Sci.* **2024**, *107*, 6178–6191. [CrossRef]
11. Yang, C.; Tsedan, G.; Fan, Q.; Wang, S.; Wang, Z.; Chang, S.; Hou, F. Behavioral Patterns of Yaks (*Bos grunniens*) Grazing on Alpine Shrub Meadows of the Qinghai-Tibetan Plateau. *Appl. Anim. Behav. Sci.* **2021**, *234*, 105182. [CrossRef]
12. Zheng, Z.; Qin, L. PrunedYOLO-Tracker: An Efficient Multi-Cows Basic Behavior Recognition and Tracking Technique. *Comput. Electron. Agric.* **2023**, *213*, 108172. [CrossRef]
13. Wang, Z.; Hua, Z.; Wen, Y.; Zhang, S.; Xu, X.; Song, H. E-YOLO: Recognition of Estrus Cow Based on Improved YOLOv8n Model. *Expert Syst. Appl.* **2024**, *238*, 122212. [CrossRef]
14. Li, Q.; He, Z.; Liu, X.; Chu, M.; Wang, Y.; Kang, X.; Liu, G. Lameness Detection System for Dairy Cows Based on Instance Segmentation. *Expert Syst. Appl.* **2024**, *249*, 123775. [CrossRef]
15. Chopra, K.; Hodges, H.R.; Barker, Z.E.; Vázquez Diosdado, J.A.; Amory, J.R.; Cameron, T.C.; Croft, D.P.; Bell, N.J.; Thurman, A.; Bartlett, D.; et al. Bunching Behavior in Housed Dairy Cows at Higher Ambient Temperatures. *J. Dairy Sci.* **2024**, *107*, 2406–2425. [CrossRef]
16. Chelotti, J.O.; Martinez-Rau, L.S.; Ferrero, M.; Vignolo, L.D.; Galli, J.R.; Planisich, A.M.; Rufiner, H.L.; Giovanini, L.L. Livestock Feeding Behaviour: A Review on Automated Systems for Ruminant Monitoring. *Biosyst. Eng.* **2024**, *246*, 150–177. [CrossRef]
17. Cockburn, M. Review: Application and Prospective Discussion of Machine Learning for the Management of Dairy Farms. *Animals* **2020**, *10*, 1690. [CrossRef]
18. Peng, Y.; Chen, Y.; Yang, Y.; Liu, M.; Hu, R.; Zou, H.; Xiao, J.; Jiang, Y.; Wang, Z.; Xu, L. A Multimodal Classification Method: Cow Behavior Pattern Classification with Improved EdgeNeXt Using an Inertial Measurement Unit. *Comput. Electron. Agric.* **2024**, *226*, 109453. [CrossRef]
19. Liu, M.; Wu, Y.; Li, G.; Liu, M.; Hu, R.; Zou, H.; Wang, Z.; Peng, Y. Classification of Cow Behavior Patterns Using Inertial Measurement Units and a Fully Convolutional Network Model. *J. Dairy Sci.* **2023**, *106*, 1351–1359. [CrossRef]
20. Norbu, N.; Alvarez-Hess, P.S.; Leury, B.J.; Wright, M.M.; Douglas, M.L.; Moate, P.J.; Williams, S.R.O.; Marett, L.C.; Garner, J.B.; Wales, W.J.; et al. Assessment of RumiWatch Noseband Sensors for the Quantification of Ingestive Behaviors of Dairy Cows at Grazing or Fed in Stalls. *Anim. Feed Sci. Technol.* **2021**, *280*, 115076. [CrossRef]

21. Peng, Y.; Kondo, N.; Fujiura, T.; Suzuki, T.; Wulandari; Yoshioka, H.; Itoyama, E. Classification of Multiple Cattle Behavior Patterns Using a Recurrent Neural Network with Long Short-Term Memory and Inertial Measurement Units. *Comput. Electron. Agric.* **2019**, *157*, 247–253. [CrossRef]
22. Rial, C.; Laplacette, A.; Caixeta, L.; Florentino, C.; Peña-Mosca, F.; Giordano, J.O. Metabolic-Digestive Clinical Disorders of Lactating Dairy Cows Were Associated with Alterations of Rumination, Physical Activity, and Lying Behavior Monitored by an Ear-Attached Sensor. *J. Dairy Sci.* **2023**, *106*, 9323–9344. [CrossRef]
23. Yin, M.; Ma, R.; Luo, H.; Li, J.; Zhao, Q.; Zhang, M. Non-Contact Sensing Technology Enables Precision Livestock Farming in Smart Farms. *Comput. Electron. Agric.* **2023**, *212*, 108171. [CrossRef]
24. Seneviratne, S.; Hu, Y.; Nguyen, T.; Lan, G.; Khalifa, S.; Thilakarathna, K.; Hassan, M.; Seneviratne, A. A Survey of Wearable Devices and Challenges. *IEEE Commun. Surv. Tutor.* **2017**, *19*, 2573–2620. [CrossRef]
25. Peng, Y.; Wulandari; Kondo, N.; Fujiura, T.; Suzuki, T.; Yoshioka, H.; Itoyama, E. Japanese Black Cattle Call Patterns Classification Using Multiple Acoustic Features and Machine Learning Models. *Comput. Electron. Agric.* **2023**, *204*, 107568. [CrossRef]
26. Higaki, S.; Matsui, Y.; Miwa, M.; Yamamura, T.; Hojo, T.; Yoshioka, K.; Vang, A.; Negreiro, A.; Dórea, J.R.R. Leveraging Computer Vision-Based Pose Estimation Technique in Dairy Cows for Objective Mobility Analysis and Scoring System. *Comput. Electron. Agric.* **2024**, *217*, 108573. [CrossRef]
27. Russello, H.; van der Tol, R.; Kootstra, G. T-LEAP: Occlusion-Robust Pose Estimation of Walking Cows Using Temporal Information. *Comput. Electron. Agric.* **2022**, *192*, 106559. [CrossRef]
28. Fu, H.; Gao, J.; Liu, H. Human Pose Estimation and Action Recognition for Fitness Movements. *Comput. Graph.* **2023**, *116*, 418–426. [CrossRef]
29. Xia, J.; Zhang, H.; Wen, S.; Yang, S.; Xu, M. An Efficient Multitask Neural Network for Face Alignment, Head Pose Estimation and Face Tracking. *Expert Syst. Appl.* **2022**, *205*, 117368. [CrossRef]
30. Yin, M.; He, S.; Soomro, T.A.; Yuan, H. Efficient Skeleton-Based Action Recognition via Multi-Stream Depthwise Separable Convolutional Neural Network. *Expert Syst. Appl.* **2023**, *226*, 120080. [CrossRef]
31. Chen, Z.; Cai, Y.; Wang, Q.; Cao, L. POLO: Pose Estimation with One-Stage Location-Sensitive Coding. *Comput. Electron. Agric.* **2024**, *226*, 109384. [CrossRef]
32. Wang, Z.; Zhou, S.; Yin, P.; Xu, A.; Ye, J. GANPose: Pose Estimation of Grouped Pigs Using a Generative Adversarial Network. *Comput. Electron. Agric.* **2023**, *212*, 108119. [CrossRef]
33. Fan, Q.; Liu, S.; Li, S.; Zhao, C. Bottom-up Cattle Pose Estimation via Concise Multi-Branch Network. *Comput. Electron. Agric.* **2023**, *211*, 107945. [CrossRef]
34. Chen, L.; Zhang, L.; Tang, J.; Tang, C.; An, R.; Han, R.; Zhang, Y. GRMPose: GCN-Based Real-Time Dairy Goat Pose Estimation. *Comput. Electron. Agric.* **2024**, *218*, 108662. [CrossRef]
35. Shorten, P.R.; Welten, B.G. Assessment of a Non-Invasive Accelerometer for Detecting Cattle Urination and Defecation Events. *Smart Agric. Technol.* **2022**, *2*, 100031. [CrossRef]
36. Shorten, P.R.; Welten, B.G. Acoustic Sensor Determination of Repeatable Cow Urinations Traits in Winter and Spring. *Comput. Electron. Agric.* **2022**, *196*, 106846. [CrossRef]
37. Li, Z.; Yan, J.; Zhou, J.; Fan, X.; Tang, J. An Efficient SMD-PCBA Detection Based on YOLOv7 Network Model. *Eng. Appl. Artif. Intell.* **2023**, *124*, 106492. [CrossRef]
38. *NYT/815/2004*; Feeding Standard of Beef Cattle. Ministry of Agriculture and Rural Affairs of the People's Republic of China: Beijing, China, 2011.
39. Mayo, L.M.; Silvia, W.J.; Ray, D.L.; Jones, B.W.; Stone, A.E.; Tsai, I.C.; Clark, J.D.; Bewley, J.M.; Heersche, G. Automated Estrous Detection Using Multiple Commercial Precision Dairy Monitoring Technologies in Synchronized Dairy Cows. *J. Dairy Sci.* **2019**, *102*, 2645–2656. [CrossRef]
40. Senger, P. *Pathways to Pregnancy and Parturition*, 2nd ed.; Current Concepts: Ames, IA, USA, 2005.
41. Taghavi, M.; Russello, H.; Ouweltjes, W.; Kamphuis, C.; Adriaens, I. Cow Key Point Detection in Indoor Housing Conditions with a Deep Learning Model. *J. Dairy Sci.* **2024**, *107*, 2374–2389. [CrossRef]
42. Li, C.; Li, L.; Geng, Y.; Jiang, H.; Cheng, M.; Zhang, B.; Ke, Z.; Xu, X.; Chu, X. YOLOv6 v3.0: A Full-Scale Reloading. *arXiv* **2023**, arXiv:2301.05586.
43. Wang, C.Y.; Mark Liao, H.Y.; Wu, Y.H.; Chen, P.Y.; Hsieh, J.W.; Yeh, I.H. CSPNet: A New Backbone That Can Enhance Learning Capability of CNN. In Proceedings of the IEEE Computer Society Conference on Computer Vision and Pattern Recognition Workshops, Seattle, WA, USA, 14–19 June 2020.
44. Wang, Z.; Li, Y.; Liu, Y.; Meng, F. Improved Object Detection via Large Kernel Attention. *Expert Syst. Appl.* **2024**, *240*, 122507. [CrossRef]
45. Dai, X.; Chen, Y.; Xiao, B.; Chen, D.; Liu, M.; Yuan, L.; Zhang, L. Dynamic Head: Unifying Object Detection Heads with Attentions. In Proceedings of the IEEE Computer Society Conference on Computer Vision and Pattern Recognition, Nashville, TN, USA, 20–25 June 2021.
46. Chen, Y.; Dai, X.; Liu, M.; Chen, D.; Yuan, L.; Liu, Z. *Dynamic ReLU*; Lecture Notes in Computer Science (Including Subseries Lecture Notes in Artificial Intelligence and Lecture Notes in Bioinformatics); Springer: Berlin/Heidelberg, Germany, 2020; Volume 12364 LNCS.

47. Wang, C.-Y.; Bochkovskiy, A.; Liao, H.-Y.M. YOLOv7: Trainable Bag-of-Freebies Sets New State-of-the-Art for Real-Time Object Detectors. *arXiv* **2022**, arXiv:2207.02696.
48. Wang, C.Y.; Liao, H.Y.M.; Yeh, I.H. Designing Network Design Strategies Through Gradient Path Analysis. *arXiv* **2022**, arXiv:2211.04800. [CrossRef]
49. He, K.; Zhang, X.; Ren, S.; Sun, J. Spatial Pyramid Pooling in Deep Convolutional Networks for Visual Recognition. *IEEE Trans. Pattern Anal. Mach. Intell.* **2015**, *37*, 1904–1916. [CrossRef]
50. Liu, S.; Qi, L.; Qin, H.; Shi, J.; Jia, J. Path Aggregation Network for Instance Segmentation. In Proceedings of the Proceedings of the IEEE Computer Society Conference on Computer Vision and Pattern Recognition, Salt Lake City, UT, USA, 18–23 June 2018.
51. da Silva Santos, A.; de Medeiros, V.W.C.; Gonçalves, G.E. Monitoring and Classification of Cattle Behavior: A Survey. *Smart Agric. Technol.* **2023**, *3*, 100091. [CrossRef]

agriculture

MDPI

Article

# Chinese Bayberry Detection in an Orchard Environment Based on an Improved YOLOv7-Tiny Model

**Zhenlei Chen [1], Mengbo Qian [1,*], Xiaobin Zhang [1] and Jianxi Zhu [2]**

[1] College of Optical Mechanical and Electrical Engineering, Zhejiang A&F University, Hangzhou 311300, China; czl888888@stu.zafu.edu.cn (Z.C.); 2022612021032@stu.zafu.edu.cn (X.Z.)
[2] Zhejiang Academic of Agricultural Machinery, Jinhua 321051, China; zhujianxi829@163.com
* Correspondence: qianmengbo@zafu.edu.cn

**Abstract:** The precise detection of Chinese bayberry locations using object detection technology is a crucial step to achieve unmanned harvesting of these berries. Because of the small size and easy occlusion of bayberry fruit, the existing detection algorithms have low recognition accuracy for such objects. In order to realize the fast and accurate recognition of bayberry in fruit trees, and then guide the robotic arm to carry out accurate fruit harvesting, this paper proposes a detection algorithm based on an improved YOLOv7-tiny model. The model introduces partial convolution (PConv), a SimAM attention mechanism and SIoU into YOLOv7-tiny, which enables the model to improve the feature extraction capability of the target without adding extra parameters. Experimental results on a self-built Chinese bayberry dataset demonstrate that the improved algorithm achieved a recall rate of 97.6% and a model size of only 9.0 MB. Meanwhile, the precision of the improved model is 88.1%, which is 26%, 2.7%, 4.7%, 6.5%, and 4.7% higher than that of Faster R-CNN, YOLOv3-tiny, YOLOv5-m, YOLOv6-n, and YOLOv7-tiny, respectively. In addition, the proposed model was tested under natural conditions with the five models mentioned above, and the results showed that the proposed model can more effectively reduce the rates of misdetections and omissions in bayberry recognition. Finally, the improved algorithm was deployed on a mobile harvesting robot for field harvesting experiments, and the practicability of the algorithm was further verified.

**Keywords:** Chinese bayberry detection; YOLOv7-tiny; attention mechanism; partial convolution; target recognition

## 1. Introduction

The Chinese bayberry is a common subtropical fruit with a spherical shape, widely cultivated in the southern regions of China [1]. It is highly favored by the Chinese people for its delicious and juicy taste, as well as its medicinal value [2]. However, Chinese bayberry trees are tall and mainly grown in hilly areas, so manual harvesting poses risks of falling. Additionally, Chinese bayberry fruits have a short ripening period and are sensitive to environmental influences, making the time available for manual harvesting short [3]. These challenges have somewhat constrained the development of the Chinese bayberry industry. In recent years, various devices have been utilized for fruit harvesting [4,5]; however, these devices mainly vibrate the fruit trees for the harvesting of fruit and lack sufficient automation. Therefore, research on automatic bayberry harvesting robots holds great practical significance.

The image detection system is a crucial component of an automatic bayberry harvesting robot, which can only control the actuator to realize automatic picking by accurately recognizing the position of the bayberry [6,7]. Currently, detection algorithms are mainly classified as either traditional [8,9] or deep learning-based [10,11]. Traditional image detection algorithms first select candidate regions in the image, then extract features from these regions, and finally classify them using a trained classifier. These algorithms suffer from window redundancy, long processing times, and poor robustness in complex orchard environments, making them insufficient to meet harvesting requirements.

In recent years, with advancements in computer science and the refinement of deep learning theories, deep learning-based object detection algorithms [12–14] have gradually replaced traditional image detection algorithms. Deep learning object detection algorithms can be categorized into two types: two-stage and one-stage. Two-stage object detection algorithms first generate candidate regions in the image and then locate and regress the boundaries of the target in these regions. Common two-stage detection algorithms include Fast regions with a convolutional neural network (Fast R-CNN) [15], Faster regions with a convolutional neural network (Faster R-CNN) [16], and Mask regions with a convolutional neural network (Mask R-CNN) [17]. Yu et al. (2019) [18] utilized Resnet50 as the backbone network, exploring a strawberry detection algorithm based on Mask R-CNN. Experimental results demonstrated that the model achieved an average precision of 95.78%. Liu et al. (2019) [19] applied Faster R-CNN to persimmon detection, proposing an improved Faster R-CNN detection algorithm. This method used Deterministic Networking (DetNet) as the backbone network and added the Efficient Channel Attention (ECA) attention mechanism to effectively perform persimmon detection in complex environments. Although two-stage object detection algorithms generally have high precision, they are slow. One-stage object detection algorithms do not require the selection of candidate regions; they directly divide the image into multiple small regions and perform feature extraction, target classification, and localization in each small region. You Only Look Once (YOLO) [20] and Single-Shot multiBox Detector (SSD) currently the mainstream one-stage detection algorithms.

Currently, owing to the high detection precision and short recognition time of the YOLO detection algorithms, many researchers have applied them to the field of agricultural harvesting robots. Ji et al. (2021) [21] used an improved YOLOv4 for apple detection in complex orchard environments. This model employed the EfficientNet network as the feature extraction network for apple recognition, reducing the model's size. Cao et al. (2022) [22] proposed a YOLOv4-LightC-CBAM model for real-time mango detection. By adjusting the model's network width and adding attention mechanisms, they improved the network's detection precision. However, this algorithm was tested under dark conditions at night with a single background, and was hardly affected by environmental factors around the orchard. Sun et al. (2022) [23] introduced ShuffleNet and GhostNet structures into the YOLOv5 model and added an attention module, proposing a YOLOv5-PRE model for detecting apples in orchards. Experimental results demonstrated that the improved YOLOv5 model achieved an average precision of 94.03%. Li et al. (2023) [24] developed a recognition algorithm for tomatoes based on the improved YOLOv5s model. However, this algorithm was applied in indoor greenhouse environments and did not reflect real orchard conditions. Zhou et al. (2023) [25] introduced RepGhost and a decoupled head approach, proposing the RDE-YOLOv7 model for dragon fruit recognition. Experimental results showed that the RDE-YOLOv7 precision, recall rate, and average precision improved by 0.4%, 2.5%, and 0.8%, respectively, compared to those of YOLOv7. Chen et al. (2022) [26] introduced small target detection layers, GhostConv, and CBAM for citrus recognition based on the YOLOv7 model. Experimental results demonstrated that the improved YOLOv7 model achieved an average precision of 97.29% and an average detection time of 69.38 ms.

The above studies have contributed to the development of agricultural fruit detection. Fruit harvesting is part of precision agriculture, where image recognition is carried out in real time by the equipped camera sensor, which in turn controls the robotic arm for precise positioning and harvesting. Efficient recognition algorithms can improve the efficiency of mechanized agricultural harvesting and reduce manual intervention. However, Chinese bayberry fruits are small and densely packed, and the YOLO detection algorithm exhibits poor performance in detecting such small targets [27]. Recognition precision is easily affected by factors such as overlapping fruits, lighting conditions, and branches. Therefore, this paper proposes an improved YOLOv7-tiny algorithm for bayberry detection to address the existing recognition issues. In this study, partial convolution, SimAM and SIoU are introduced into the YOLOv7-tiny model to enhance the feature extraction capability of the model; the improved model was then deployed to a harvesting robot to perform mechanical harvesting in the field.

## 2. Materials and Methods

### 2.1. Bayberry Image Collection

The study used a self-built dataset of Chinese bayberry fruits in hilly environments. The images were captured in a renowned Chinese bayberry orchard in the Lanjiang Street, Yuyao, Ningbo city, Zhejiang province in eastern China. The bayberry trees in the orchard have an approximate spacing of 2–3 m and a height of approximately 3 m, meeting the requirements for intelligent robot harvesting. The data collection utilized an iPhone 13 for image capture, with an image resolution of 3024 × 4032 pixels. To ensure optimal dataset quality, the images were captured on 15 June and 16 June 2023, and 1200 images were obtained in total. To ensure the authenticity and diversity of the dataset, the captured images included cases with a single fruit, multiple fruits, backlight and light-facing environments, and sunny and cloudy scenes. Some samples of the dataset are illustrated in Figure 1.

**Figure 1.** Partial dataset samples. (**a**) Single fruit. (**b**) Multiple fruits. (**c**) Backlight environment. (**d**) Light-facing environment.

### 2.2. Data Annotation and Augmentation

To facilitate the training of the neural network model, it is imperative to annotate the acquired dataset. In this experiment, bayberries were annotated as a distinct class, and

LabelImg 1.8.5 version annotation software was employed for manual annotation. An XML file was generated for each annotated image, encompassing the classifications of all labels, positions of the labels, and size of the image, as depicted in Figure 2.

**Figure 2.** Example of Bayberry label fabrication.

A visual analysis of the annotated dataset files is presented in Figure 3. Figure 3a illustrates the distribution of the locations of labels in the dataset. The horizontal coordinate (x) represents the ratio of the horizontal coordinate of the label center to the image width, whereas the vertical coordinate (y) represents the ratio of the horizontal coordinate of the label center to the image height. As shown, there are a large number of bayberries in the self-built dataset and they are mainly distributed in the middle part of the image. Figure 3b illustrates the size distribution of labels in the dataset. The horizontal coordinate (width) represents the ratio of the label width to the image width, whereas the vertical coordinate (height) represents the ratio of the label height to the image height. As shown, the label height and width in the self-built dataset are less than 0.2 times the corresponding image dimensions in most cases; thus, the labeled items belong to the category of small targets, which aligns with real orchard conditions. Ultimately, the dataset of 1200 images was partitioned into training and testing sets in an 8:2 ratio.

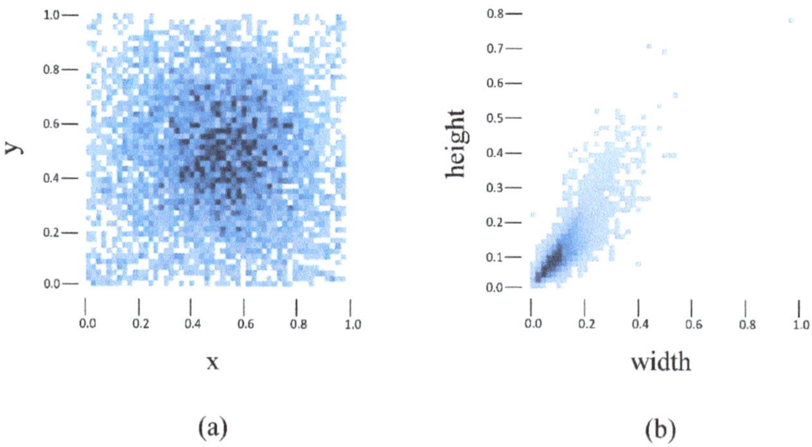


**Figure 3.** Visual analysis of the annotated dataset. (**a**) Distribution of annotated dataset locations. (**b**) Distribution of annotated dataset sizes.

By augmenting the dataset, we can not only effectively simulate the complex environments in real orchards, but also enhance the model's generalization ability, preventing overfitting. This study applied the inherent mosaic data augmentation method of the YOLO model to enhance the training data, including combining four images into one (Mosaic), flipping the image horizontally (Fliplr), overlaying two images (Mixup), and enhancement of the image's HSV in terms of hue, saturation, and value. Figure 4 illustrates the data augmentation process. The data augmentation enhances the diversity of the dataset, thus improving the model's robustness. The hyperparameters for the data augmentation method are presented in Table 1.

**Figure 4.** Flow chart of mosaic data augmentation.

**Table 1.** The hyperparameter settings of data augmentation.

| Hyperparameter | Value |
|---|---|
| Mosaic | 100% probability |
| Fliplr | 50% probability |
| Mixup | 15% probability |
| HSV_Hue | 0.015 fraction |
| HSV_Saturation | 0.7 fraction |
| HSV_Value | 0.4 fraction |

### 2.3. *YOLOv7-Tiny Object Detection Network*

YOLO algorithms are among the most popular object detection algorithms. Their simple network structure and high detection speed make them widely applicable for agricultural harvesting robots. YOLOv7, introduced by Wang et al. [28], is the latest detection algorithm in the YOLO series.

YOLOv7 employs a cascade-based model scaling approach to generate multiple architectures, including YOLOv7, YOLOv7-tiny, YOLOv7x, YOLOv7-w6, YOLOv7-e6, YOLOv7-e6e, and YOLOv7-d6. Regular GPU deployment is used for the former three networks, whereas the latter four networks are intended for high-performance cloud GPU deployment. In this study, a regular GPU was used, and the first three networks were individually trained. We use average precision (mAP), recall and model size as the evaluation metrics of the model. Average precision represents a combined assessment of precision and recall under category and detection thresholds and recall indicates the number of correctly detected targets as a percentage of the total number of targets. The results are presented in Table 2. As shown, YOLOv7-tiny outperformed the other two networks in terms of mean average precision, recall rate, and model size. Therefore, YOLOv7-tiny was selected for the subsequent experiment.

**Table 2.** Comparison of the performance of different YOLOv7 models.

| Model | mAP (%) | Recall (%) | Model Size (MB) |
|---|---|---|---|
| YOLOv7 | 91.9 | 94.5 | 73.1 |
| YOLOv7-tiny | 93.7 | 96.3 | 11.9 |
| YOLOv7x | 92.5 | 94.6 | 138.7 |

YOLOv7-tiny consists of four components: Input, Backbone, Neck, and Head. The Input side preprocesses each input image through procedures such as data augmentation and adaptive anchor box calculation, standardizing the image size to $640 \times 640 \times 3$. The processed images are then fed into the Backbone layer. The Backbone layer serves as the feature extraction layer, comprising multiple CBL convolution modules, Efficient Layer Aggregation Network (ELAN) modules, and MP modules. The CBL convolution module, consisting of a Conv layer, a BN layer, and the Leaky ReLU activation function, extracts feature information from the images. The ELAN module is formed by combining multiple CBL modules through multiple branches to enhance the network's learning ability. The MP module utilizes max-pooling operations for downsampling. The Neck layer acts as the feature fusion layer, incorporating SPPCSP modules, CBL modules, and ELAN modules. The SPPCSP module consists of two parts, SPP and CSP. SPP utilizes the maximum pooling of four different scales to achieve the fusion of information from different feature scales; CSP performs a concat operation to enrich feature information. This layer adopts a path aggregation feature pyramid network (PAFPN) structure, facilitating multi-scale fusion of the network through top-down and bottom-up approaches, which is advantageous for small object detection. The Head layer serves as the output prediction layer, where the three different-scale feature maps of the Neck layer output undergo adjustment of the image channel numbers, using convolution modules to predict the confidence, category, and anchor boxes.

### *2.4. Improvement of the YOLOv7-Tiny Network Model*

#### 2.4.1. Improvement of the Backbone Network

The structural diagram of a conventional convolution is illustrated in Figure 5a. This convolution necessitates feature extraction for each input channel, increasing the model's computational complexity and slowing down network operation. To enhance the speed of neural networks, previous studies have often used the number of floating-point operations (FLOPs) as an evaluation metric for the computational complexity of a model, and achieve the effect of reducing the computational complexity of the model by decreasing the FLOPs, such as Depthwise Convolution (DWConv) [29] and Ghost Convolution (Ghost Conv) [30]. However, a decrease in the number of FLOPs does not necessarily translate to an improvement in network operation speed. This is because the detection speed of the network is related not only to FLOPs but also to floating-point operations per second (FLOPS). To simultaneously reduce computational redundancy and memory access, enabling neural networks to maintain high FLOPS while reducing the number of FLOPs, Chen et al. [31] proposed a novel convolution module, partial convolution (PConv). PConv performs feature extraction with Conv only on a subset of input channels, without affecting the remaining channels, effectively avoiding computational redundancy in feature maps. For continuous or regular memory access, the first or last continuous channel is treated as the representative for the entire feature map to be calculated. The working mechanism of PConv is illustrated in Figure 5b.

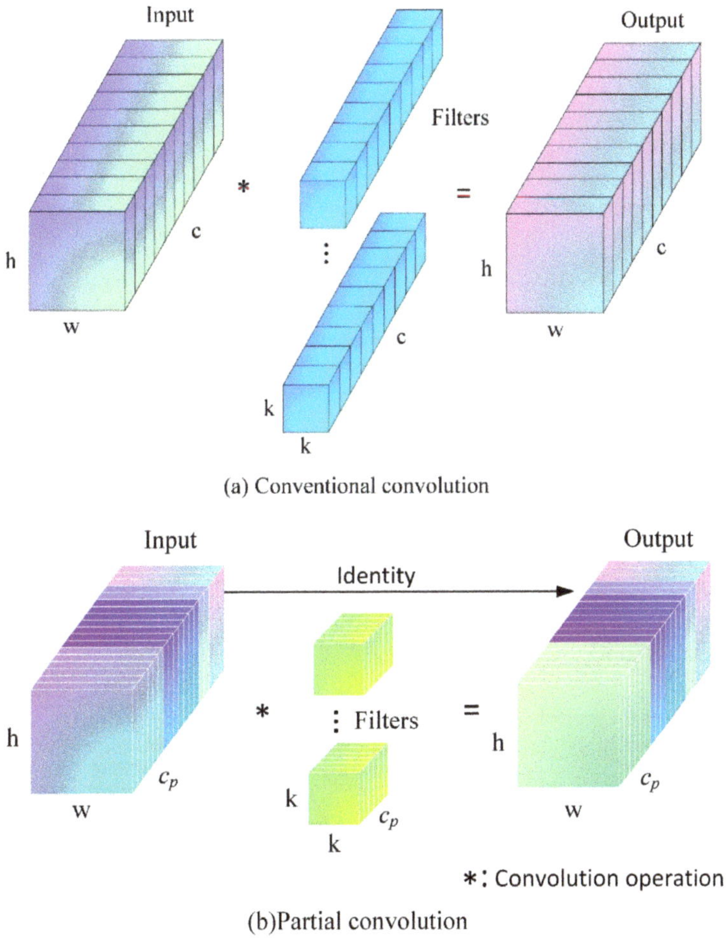


**Figure 5.** Convolution structure diagram. (**a**) Conventional convolution. (**b**) Partial convolution.

In comparison to Conv, PConv only performs feature extraction on a certain number ($c_p$) of input channels, while retaining the original feature information in other channels. The number of FLOPs for Conv and PConv is calculated as in Equations (1) and (2).

$$\text{FLOPs}_{(\text{Conv})} = k^2 \times c^2 \times h \times w = k^2c^2hw \quad (1)$$

$$\text{FLOPs}_{(\text{PConv})} = k^2 \times c_p^2 \times h \times w = k^2c_p^2hw \quad (2)$$

where $h$ is the height of the input feature, $w$ is the width of the input feature, $k$ is the size of the convolution kernel, $c_p$ is the number of PConv channels, and $c$ is the number of Conv channels. The memory access cost (MAC) for Conv and PConv is calculated as in Equations (3) and (4).

$$\text{MAC}_{(\text{Conv})} = k^2 \times c^2 + h \times w \times 2c \approx 2chw \quad (3)$$

$$\text{MAC}_{(\text{PConv})} = k^2 \times c_p^2 + h \times w \times 2c_p \approx 2c_phw \quad (4)$$

Generally, $c_p$ is taken as $1/4$ of $c$; thus, the number of FLOPs and memory access cost for PConv are $1/16$ and $1/4$ of those of Conv, respectively.

In this study, PConv was introduced into the ELAN layer of the backbone, replacing Conv with a kernel size of 3 and a stride of 1, leading to the proposed P-ELAN module.

2.4.2. Neck Improvement

Orchard environments are intricate, and bayberry fruits are likely to be covered by branches and leaves, decreasing detection precision. Introducing attention [32] modules into the model can effectively enhance feature extraction capability. Currently, attention mechanisms typically operate along the channel or spatial dimensions, and are classified into channel and spatial attention modules. Channel attention modules [33] focus solely on channel dimension information, as depicted in Figure 6a, whereas spatial attention modules [34] concentrate on spatial dimension information, as shown in Figure 6b. As both concentrate on one dimension of information while neglecting the other, their contribution to performance improvement is limited. Therefore, Yang et al. [35] proposed the SimAM attention module, which was inspired by neuroscience. SimAM is a 3D attention module that can simultaneously coordinate channel and spatial dimension information, as illustrated in Figure 6c. The SimAM attention mechanism calculates attention weights through an energy function. The definition of the energy function is associated with neural spatial inhibition, i.e., neurons with richer information inhibit surrounding neurons. Thus, the importance of neurons can be expressed through the energy function, which is defined as in Equation (5).

$$e_t(w_t, b_t, y, x_i) = \frac{1}{M-1}\sum_{i=1}^{M-1}(-1-(w_t x_i + b_t))^2 + (1-(w_t t + b_t))^2 + \lambda w_t^2 \tag{5}$$

where $w_t$ represents the weight, $b_t$ is the offset, $M$ is the number of neurons, $t$ denotes the target neurons in the same channel in the input feature map, $x_i$ represents other neurons, and $\lambda$ is a constant. The analytical expressions for the parameters in Equation (5) are as in Equation (6).

$$\begin{cases} w_t = -\frac{2(t-\mu_t)}{(t-\mu_t)^2+2\sigma_t^2+2\lambda} \\ b_t = -\frac{1}{2}(t+\mu_t)w_t \end{cases} \tag{6}$$

where $\mu$ and $\sigma$ represent the mean and variance, respectively. The calculation process is as in Equation (7).

$$\begin{cases} \mu_t = \frac{1}{M-1}\sum_{i=1}^{M-1} x_i \\ \sigma_t^2 = \frac{1}{M-1}\sum_{i=1}^{M-1}(x_i-\mu_t)^2 \end{cases} \tag{7}$$

This process is finally simplified to obtain the minimum energy formula, as shown in Equation (8).

$$e_t^* = \frac{4(\hat{\sigma}^2+\lambda)}{(t-\mu)^2+2\sigma^2+2\lambda} \tag{8}$$

From Equation (8), it is evident that the significance of each neuron can be calculated by $1/e^*$. The greater the importance of neuron t, the lower its energy.

In this study, the SimAM attention module was introduced behind each ELAN layer in the Neck, enabling the model to enhance feature extraction ability without introducing additional parameters.

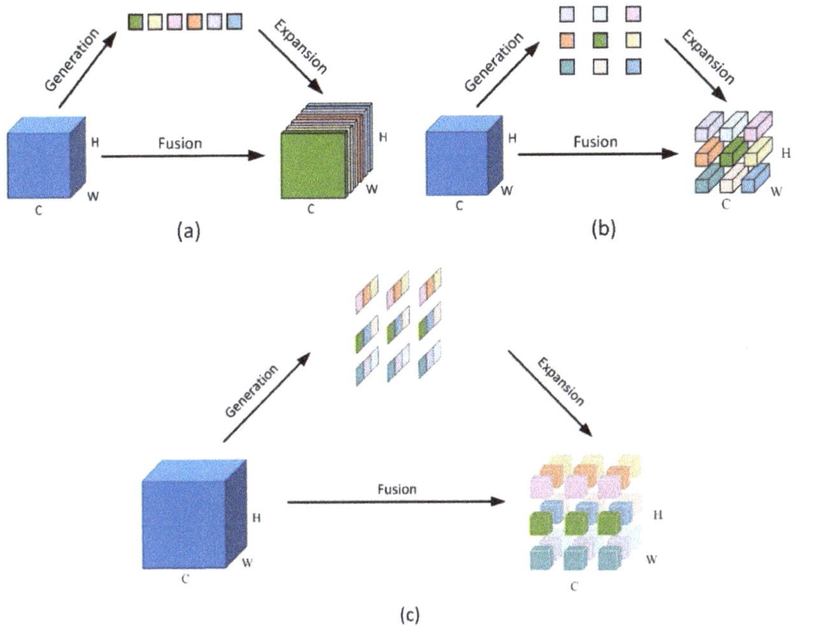


**Figure 6.** Principles of attention mechanism operation. (**a**) Channel attention mechanism. (**b**) Spatial attention mechanism. (**c**) 3D attention mechanism—SimAM.

2.4.3. Loss Function Improvement

YOLOv7-tiny employs CIoU [36] as the bounding box loss function. CIoU, an improvement over DIoU, incorporates three aspects: the overlap area between the prediction box and the ground truth box, the distance between their centers, and the aspect ratio. However, CIoU does not consider the orientation between the ground truth and predicted bounding boxes, leading to potential drifting of the prediction boxes during training and imprecise localization.

To address this, SIoU [37] introduces the vectorial angle between the ground truth and prediction bounding box. By adding an angle penalty term to reduce the number of degrees of freedom, it accelerates the model's convergence. The SIoU loss function comprises four components: angle loss, distance loss, shape loss, and IoU loss.

Angle loss: The angle loss is depicted in Figure 7a, and the formula is as in Equation (9).

$$\Lambda = 1 - 2 \times \sin^2\left(\arcsin\left(\frac{c_h}{\sigma}\right) - \frac{\pi}{4}\right) = \cos\left(2 \times \left(\arcsin\left(\frac{c_h}{\sigma}\right) - \frac{\pi}{4}\right)\right) \quad (9)$$

where $c_h$ is the difference in height between the centers of the ground truth box (A) and prediction box (B), $\sigma$ is the distance between the centers of the ground truth box A and prediction box B, $(b_x^{gt}, b_y^{gt})$ denotes the coordinates of the center of the ground truth box, and $(b_x, b_y)$ denotes the coordinates of the center of the prediction box. The calculation process of $\sigma$ and $c_h$ is as in Equations (10) and (11).

$$\sigma = \sqrt{\left(b_x^{gt} - b_x\right)^2 + \left(b_y^{gt} - b_y\right)^2} \quad (10)$$

$$c_h = \max\left(b_y^{gt}, b_y\right) - \min\left(b_y^{gt}, b_y\right) \quad (11)$$

Distance loss: The distance loss is depicted in Figure 7b, and the formula is as in Equation (12).

$$\Delta = \sum_{t=x,y} \left(1 - e^{-\gamma \rho_t}\right) = 2 - e^{-\gamma \rho_x} - e^{-\gamma \rho_y} \tag{12}$$

where $c_w$ and $c_h$ are respectively the width and height of the minimum bounding rectangle of the ground truth box and prediction box. The calculation process of $\rho_x$, $\rho_y$ and $\gamma$ is as in Equations (13) and (14).

$$\rho_x = \left(\frac{b_x^{gt} - b_x}{c_w}\right)^2, \rho_y = \left(\frac{b_y^{gt} - b_y}{c_h}\right)^2 \tag{13}$$

$$\gamma = 2 - \Lambda \tag{14}$$

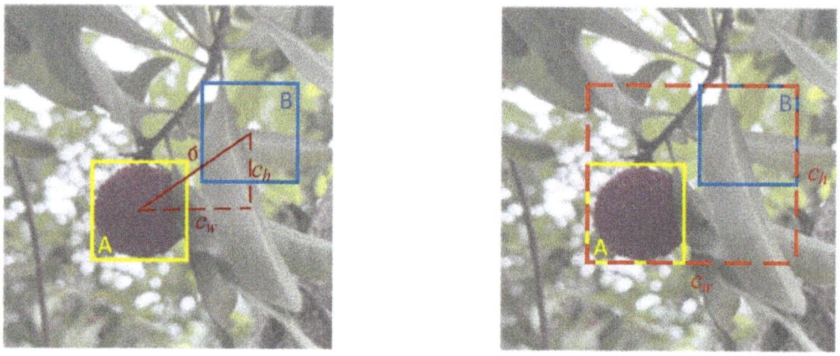


(a) Angle loss (b) Distance loss

**Figure 7.** Angle and distance loss.

Shape loss: The formula is as in Equation (15).

$$\Omega = \sum_{t=w,h} \left(1 - e^{-w_t}\right)^{\theta} = \left(1 - e^{-w_w}\right)^{\theta} + \left(1 - e^{-w_h}\right)^{\theta} \tag{15}$$

where $(w, h)$ and $(w^{gt}, h^{gt})$ are respectively the width and height of the prediction box and ground truth box. The calculation process of $w_w$ and $w_h$ s as in Equation (16).

$$w_w = \frac{|w - w^{gt}|}{\max(w, w^{gt})}, w_h = \frac{|h - h^{gt}|}{\max(h, h^{gt})} \tag{16}$$

IoU loss: The formula is as in Equation (17).

$$IoU = \frac{A}{B} \tag{17}$$

where $A$ is the intersection of the ground truth box and prediction box, and $B$ is their union.

Finally, combining these four loss functions, the formula for SIoU is as in Equation (18).

$$Loss_{SIoU} = 1 - IoU + \frac{\Delta + \Omega}{2} \tag{18}$$

### 2.4.4. Experimental Process and Proposed Algorithm

The experiment was conducted in two phases: training and validation. First, the dataset was manually annotated. Then, the annotated dataset was split into training and validation sets in an 8:2 ratio. Subsequently, the training set was fed into the improved YOLOv7-tiny algorithm for training. Finally, the performance of the trained model was evaluated using the validation set. The experimental process is illustrated in Figure 8.

In summary, this paper proposed an improved YOLOv7-tiny algorithm for Chinese bayberry detection by substituting the Conv modules in the ELAN layer with PConv, introducing the SimAM attention mechanism, and replacing the original model's CIoU loss function with SIoU. The structure of the improved YOLOv7-tiny network is depicted in Figure 9.

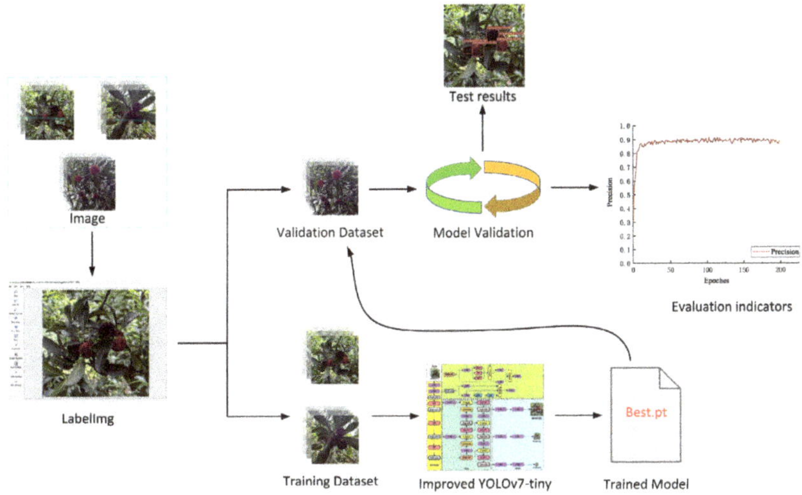


**Figure 8.** Detection process of Chinese bayberry using improved YOLOv7-tiny model.

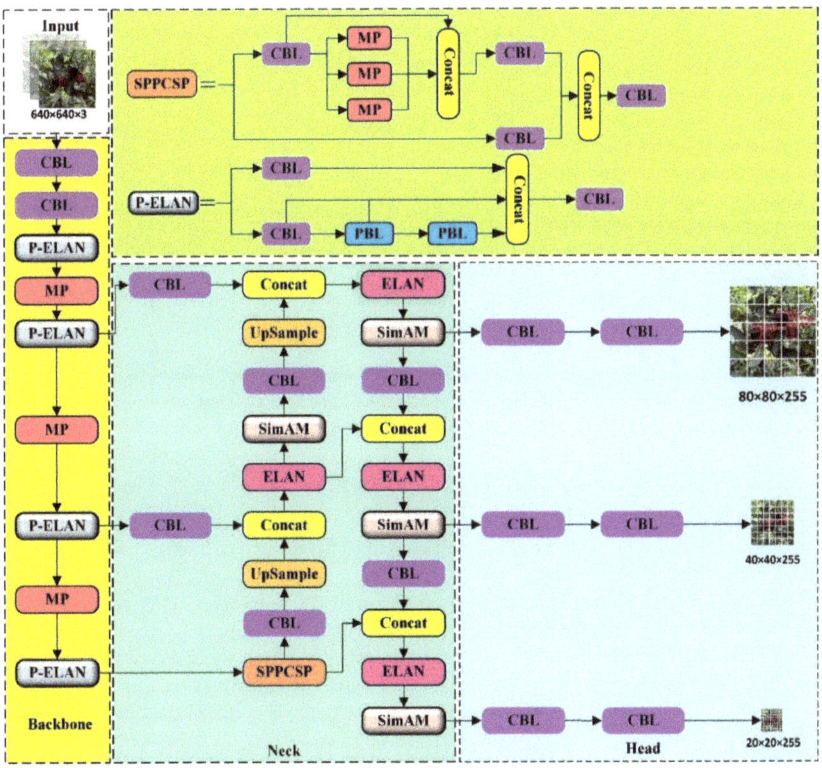


**Figure 9.** Overall structure diagram of improved YOLOv7-tiny model.

### 2.5. *Model Training*

#### 2.5.1. Training Platform and Parameter Settings

All the experiments were performed in the Windows 11 (64 bit) environment (processor: Intel® Core™ i7-12700H, main frequency: 2.3 GHZ; memory: 16 G; graphics card: Nvidia GeForce RTX 4060, video memory: 8 GB). The deep learning framework was PyTorch 2.0. The programming language was Python 3.8.

The training comprised 200 epochs, with an image dimension of 640 × 640 pixels, batch size of 4 and learning rate of 0.01, and utilized pre-trained weights provided by the official source. Table 3 outlines the relevant parameters employed in the training process.

**Table 3.** The hyperparameter settings of the training process.

| Hyperparameter | Value | Hyperparameter | Value |
|---|---|---|---|
| Image Size | 640 × 640 | Momentum | 0.937 |
| Batch Size | 4 | Box loss gain | 0.05 |
| Epochs | 200 | Cls loss gain | 0.3 |
| Learning Rate | 0.01 | Obj loss gain | 0.7 |

#### 2.5.2. Evaluation Metrics

To objectively assess the performance of the model in this experiment, precision (P), recall (R), mean average precision (mAP), and model size were adopted as evaluation metrics.

(1) Precision represents the precision of the model; higher precision indicates better model performance. The formula is as in Equation (19).

$$P = \frac{TP}{TP + FP} \tag{19}$$

where $TP$ denotes the number of correctly identified bayberries and $FP$ represents the number of items incorrectly identified as bayberries.

(2) Recall signifies the percentage of correctly identified bayberries relative to the total number of bayberries. The formula is as in Equation (20).

$$R = \frac{TP}{TP + FN} \tag{20}$$

where $FN$ denotes the number of bayberries that were not correctly identified.

(3) $mAP$ is the area enclosed by the recall rate curve and precision curve. The formula is as in Equation (21).

$$mAP = \frac{1}{n}\sum_{i=1}^{n} AP_i \tag{21}$$

where $n$ is the number of classes in the test samples (in this experiment, $n = 1$) and $AP$ is the average precision for a single class.

(4) Model size denotes the memory occupied by the network model. A smaller size facilitates deployment on mobile devices.

(5) FLOPs is a measure of a model's computational effort, which indicates the number of floating-point operations the model performs while processing a single forward propagation. The smaller the number of FLOPs, the faster the model executes during the inference process.

## 3. Results

### *3.1. Training Results*

In this study, we obtained relevant results by training the model. Figure 10a illustrates the loss curve during the training process. As observed from this graph, the loss value of YOLOv7-tiny rapidly decreased in the initial 70 epochs and stabilized after 120 epochs, indicating the convergence of the loss curve. Figure 10b depicts the precision curve during the training, showing that the precision curve also leveled off around 200 cycles. Figures 10c and 10d respectively describe the recall curve and mAP curve of the model, further verifying the reliability of the improved model. Therefore, the model after 200 epochs was chosen as the bayberry detection model.

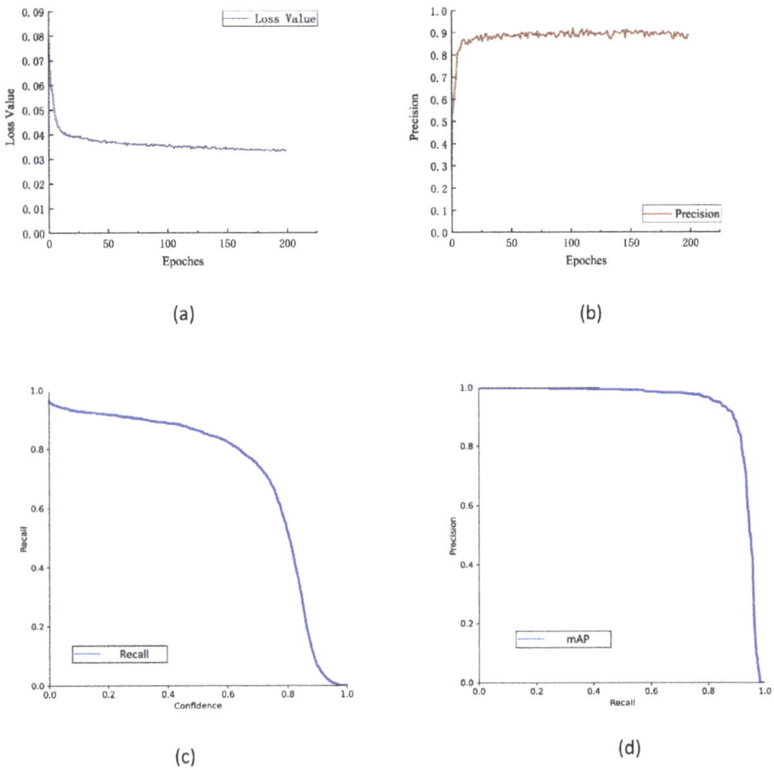


**Figure 10.** Training results. (**a**) Loss curve. (**b**) Precision curve. (**c**) Recall curve. (**d**) mAP curve.

### *3.2. Comparative Study of Attention Mechanisms*

Introducing attention mechanisms to the model can effectively enhance its recognition capability in complex environments. Table 4 presents the performance of the model after separately incorporating four attention mechanisms: CBAM, CA, SA, and SimAM. As shown, all four attention mechanisms improved the model in terms of precision and recall, with SimAM producing the most significant improvement, increasing precision and recall by 1.9% and 0.9%, respectively. Regarding mAP, CBAM exhibited the most notable improvement, although SimAM also showed a 0.1% increase. Figure 11 displays the detection results of the four attention mechanisms. As can be seen from this figure, both CBAM and SA have missed detections. According to the experimental results, SimAM demonstrated the best performance, and thus it was adopted in this study.

**Table 4.** Performance comparison of different attention mechanisms.

| Model | Precision (%) | Recall (%) | mAP (/%) |
|---|---|---|---|
| YOLOv7-tiny | 83.4 | 96.3 | 93.7 |
| +CBAM | 85.2 | 97.0 | 94.1 |
| +CA | 84.1 | 96.5 | 93.7 |
| +SA | 85.0 | 97.0 | 93.9 |
| +SimAM | 85.3 | 97.2 | 93.8 |

**Figure 11.** Detection results of attention mechanisms. (**a**) CBAM Attention. (**b**) CA Attention. (**c**) SA Attention. (**d**) SimAM Attention.

### 3.3. Ablation Experiment

To verify the effectiveness of additional modules, PConv, SimAM, and SIoU modules were sequentially introduced for ablation experiments. The "$\surd$" mark indicates the adoption of the corresponding improvement strategy. Table 5 presents the experimental data results for different combinations.

**Table 5.** Comparison of ablation results.

| SIoU | SimAM | PConv | Precision (%) | Recall (%) | mAP (%) | Model Size (MB) | FLOPs (G) |
|---|---|---|---|---|---|---|---|
| | | | 83.4 | 96.3 | 93.7 | 11.9 | 6.58 |
| √ | | | 84.8 | 96.5 | 93.8 | 11.9 | 6.58 |
| | √ | | 85.3 | 97.2 | 93.8 | 11.9 | 6.58 |
| | | √ | 86.2 | 97.1 | 93.3 | 9.1 | 4.8 |
| √ | √ | | 87.9 | 97.3 | 93.9 | 11.9 | 6.58 |
| √ | √ | √ | 88.1 | 97.6 | 93.4 | 9.0 | 4.8 |

When the SIoU or SimAM modules were added, the model's size or the value of the GLOPs did not increase significantly, but precision, recall, and mAP all improved. Notably, precision showed the most significant improvement, increasing by 1.4% and 1.9%, respectively, demonstrating that SIoU and SimAM can effectively enhance the model's ability to focus on the target. The introduction of the PConv module reduced model size by 22.3% and GLOPs by 27.1%, although mAP decreased by 0.4%, proving that PConv can reduce the computation of the model while maintaining a certain level of precision. The fifth group of experiments indicated that with the combination of SIoU and SimAM, precision, recall, and mAP further increased to 87.9%, 97.3%, and 93.9%, respectively, compared to the previous single-module improvements. Overall, when the SIoU, SimAM, and PConv modules were introduced simultaneously, the model exhibited optimal overall performance, with precision and recall improving by 4.7% and 1.3%, respectively, and the model size decreasing by 22.4%, relative to the model without additional modules. Thus, the ablation experiments validated the effectiveness of adding these modules.

### 3.4. Comparative Study of Different Networks

To demonstrate the bayberry recognition performance of the proposed model, a comparative experiment was conducted under the same experimental conditions with Faster-RCNN, YOLOv3-tiny, YOLOv5-m, YOLOv6-n, and YOLOv7-tiny, using precision, recall, mAP, and model size as performance indices. Table 6 provides the comparison results for the six models.

**Table 6.** Comparison of performance metrics with other network models.

| Model | Precision (%) | Recall (%) | mAP (%) | Model Size (MB) |
|---|---|---|---|---|
| Faster-RCNN | 62.1 | 91.1 | 90.1 | 108.7 |
| YOLOv3-tiny | 85.4 | 94.1 | 87.8 | 16.9 |
| YOLOv5-m | 83.4 | 97.2 | 93.9 | 41.2 |
| YOLOv6-n | 81.6 | 97.1 | 93.2 | 10.2 |
| YOLOv7-tiny | 83.4 | 96.3 | 93.7 | 11.9 |
| Ours | 88.1 | 97.6 | 93.4 | 9.0 |

As shown in Figure 12, the improved YOLOv7-tiny achieved 88.1% precision, 97.6% recall, 93.7% mAP, and 9.0 MB model size, indicating favorable results. Compared to the two-stage object detection algorithm Faster R-CNN, precision, recall, and mAP increased by 26.0%, 6.5%, and 3.3%, respectively, while the model size decreased by 99.7 MB. Compared to YOLOv3-tiny, the improvements were 2.7%, 3.5%, and 5.6%, respectively, with a model size reduction of 7.9 MB. Compared to YOLOv5-m, improvements of 4.7%, 0.4%, and 0.5% were observed, with a model size reduction of 32.2 MB. Compared to YOLOv6-n, improvements of 6.5%, 0.5%, and 0.2% were respectively observed, with a model size reduction of 1.2 MB. Compared to the original YOLOv7-tiny, precision and recall increased by 4.7%

and 1.3%, respectively, mAP decreased by 0.3%, and model size decreased by 2.9 MB. In summary, the improved YOLOv7-tiny was slightly lower in mAP than the original model, but outperformed the other mainstream models in all other indicators. Therefore, the proposed YOLOv7-tiny model exhibited superior performance in bayberry recognition.

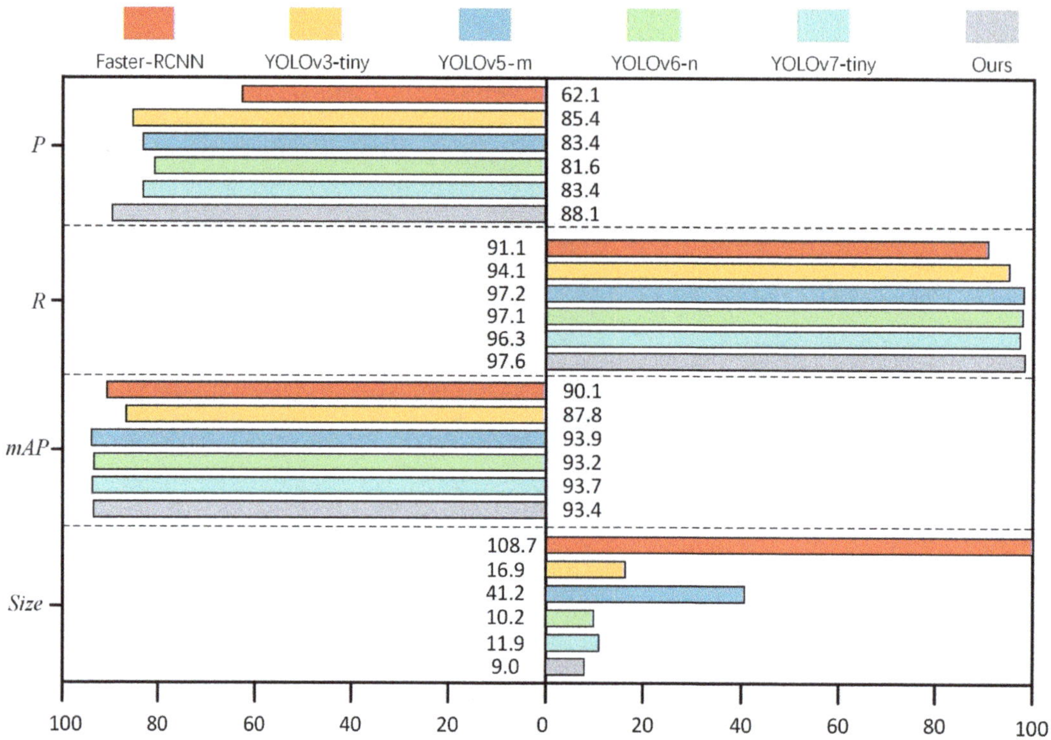


**Figure 12.** Performance comparison of different network models.

For a more intuitive comparison of the improved model's performance, images from four common scenes (single fruit, multiple fruits, and backlight and light-facing) were selected for testing on the six models. The results are shown in Figure 13. As shown in Figure 13a, where there was a single fruit, all six models could accurately identify the bayberry; however, the improved YOLOv7-tiny exhibited higher detection confidence than the other models. Figure 13b illustrates that, as the fruit in the upper right corner was obscured by leaves, Faster R-CNN, YOLO-6-n, and the original YOLOv7-tiny missed it, whereas the improved YOLOv7-tiny detected it. In Figure 13c, the original YOLOv7-tiny misidentified leaves as a fruit, resulting in a false positive, whereas the improved YOLOv7-tiny correctly identified all bayberries. In Figure 13d, the improved YOLOv7-tiny provided more accurate detection boxes than the other five models did. Therefore, the comparative images demonstrate that the improved algorithm achieved more precise recognition, confirming its effectiveness.

**Figure 13.** Detection results in different scenes. (**a**) Single fruit. (**b**) Multiple fruits. (**c**) Backlight environment. (**d**) Light-facing environment.

### 3.5. *Harvesting Robot Model Experiment*

To further validate whether our improved model can meet the practical requirements of harvesting, we tested it using a harvesting robot in a field experiment. The test site of this experiment is the floriculture forest fruit experimental base in Jinhua Academy of Agricultural Sciences, Zhejiang Province, China. The robot consisted of a mobile chassis, robotic arm, and depth camera. The relevant hardware and platform parameters of the harvesting robot are shown in Table 7. Firstly the above improved model is deployed to the platform, secondly the depth camera is turned on using control commands, then the detection input is changed to real time video in the model detection module and finally the robotic arm is controlled to locate and harvest the fruits, as depicted in Figure 14. It is evident that the harvesting robot accurately identified all bayberries within the image range. The final experimental results showed that the improved model has a miss detection rate of 4% and an error rate of 3% for bayberries. In addition, the location and size of the recognized bounding box are accurate, and the spatial localization ability of the model meets the requirements and complies with the harvesting standards.

**Table 7.** Hardware and platform parameters for harvesting robots.

| Parameter | Value | Parameter | Value |
|---|---|---|---|
| Display Card | Nvidia RTX 1050 | System | Ubuntu18.04 |
| Camera | Intel D435 | Framework | PyTorch 1.10 |
| Robot Arm | AUBO-i5 | Python | 3.8 |
| Video Memory | 8 G | Image Size | 640 × 640 |

**Figure 14.** Harvesting robot test.

## 4. Discussion

The traditional method of picking bayberry by hand is no longer suitable for the needs of modern agricultural development due to the increase in labor costs and the low efficiency of picking. Therefore, in recent years, the technology related to picking robots has been continuously developed, which effectively improves the picking efficiency of agricultural fruits, reduces the cost of manual picking, and improves the economic value of agricultural fruits. The visual detection of fruits is a key step for selecting robots to complete the picking task, and the algorithm of fruit detection based on deep learning is better than the traditional detection algorithm in terms of speed and accuracy. Deep learning object detection algorithms can be categorized into two types: two-stage and one-stage. From Table 6, it can be seen that Faster-RCNN, as a two-stage detection algorithm, does not meet the requirements for harvesting recognition either in terms of detection accuracy or model volume. The YOLO algorithm is one of the most popular one-stage detection algorithms. Its simple network structure and high detection speed make it widely applicable for agricultural harvesting robots. YOLOv7 is the latest detection algorithm in the YOLO series, and a lot of improvements have been made to it compared to its predecessor. First, it introduces the Extended Efficient Layer Aggregation Network (E-ELAN). The E-ELAN module strengthens the network's feature extraction ability by controlling the shortest and longest gradient paths. Second, it introduces reparameterized convolution (RepConv), which employs two different structures during training and inference. In the training phase, it utilizes a multi-branch path structure to enhance network performance, whereas in the inference phase, it adopts a single-path structure derived from reparameterization to boost network inference speed. Finally, YOLOv7 employs a cascade-based model scaling method, allowing the selection of an appropriate model size according to specific task requirements to meet detection demands. Table 6 shows that YOLOv7 has the best overall algorithmic performance compared to previous versions. In this study, we added Partial Convolution (PConv), SimAM Attention Mechanism and SIoU to YOLOv7-tiny to improve the accuracy of poplar plum (Chinese bayberry) detection in the complex environment of the orchard. As can be seen from Figure 13, the improved model effectively solves the phenomena of missed detection, wrong detection, and inaccurate localization of the bounding box produced by the original model

However, this study has certain shortcomings. First, the study only dealt with the identification and harvesting of bayberry fruits without distinguishing whether the fruits were ripe and ready for harvesting. Additionally, the dataset only included bayberry varieties from the region, and data were not collected for other varieties of bayberry. Therefore, future work will include collecting data on the ripeness of bayberries, as well as using other varieties of bayberry to enrich the dataset and make the algorithm more accurate.

## 5. Conclusions

To precisely detect bayberry fruits in hilly environments, this study first conducted comparative experiments on three versions of YOLOv7 under the same conditions. The results indicated that YOLOv7-tiny outperformed the other two versions in terms of mAP, recall, and model size. Therefore, YOLOv7-tiny was chosen for further experiments. Second, to address the original YOLO algorithm's suboptimal performance in small object detection, an improved YOLOv7-tiny network model was proposed. The model incorporated PConv, SimAM, and SIoU modules to enhance its efficacy in small object detection. Experimental results demonstrated that the improved model achieved 97.6% recall and 9.0 MB model size. Meanwhile, the precision of the improved model is 88.1%, which is 26%, 2.7%, 4.7%, 6.5%, and 4.7% higher than that of Faster R-CNN, YOLOv3-tiny, YOLOv5-m, YOLOv6-n, and YOLOv7-tiny, respectively. The detection result images revealed that the improved model not only effectively detected highly shaded bayberries but also avoided false positives, thus outperforming the other algorithms. Finally, to validate its feasibility, the improved algorithm was deployed on a harvesting robot for the harvesting experiment. The results demonstrated that the algorithm accurately identified all targets in the bayberry tree images.

In conclusion, this study proposes an algorithm that can efficiently and accurately recognize bayberry. The harvesting robot can detect and localize the location of bayberry fruits in real time by using this algorithm for accurate harvesting, which promotes the development of precision agriculture to a certain extent.

**Author Contributions:** Z.C. contributed to conceptualization, methodology, software, investigation, formal analysis, and writing—original draft. M.Q. contributed to funding acquisition and editing. X.Z. contributed to writing—review and editing. J.Z. contributed to formal analysis and methodology. All authors have read and agreed to the published version of the manuscript.

**Funding:** This study was funded by the Research and Development of Autonomous Operation System and Equipment for the Whole Scene of Orchard Project, Zhejiang Province "Leading Goose" R&D Public Relations Program (No.2023C02049), this research was also supported by Zhejiang Province "Leading Goose" R&D Public Relations Program (No.2022C02057).

**Institutional Review Board Statement:** Not applicable.

**Data Availability Statement:** Data will be made available on reasonable request.

**Acknowledgments:** We extend our gratitude to all authors for their technical assistance in this study.

**Conflicts of Interest:** The authors declare that they have no conflicts of interest or personal relationships related to the work in this paper.

## References

1. Zhu, C.; Feng, C.; Li, X.; Xu, C.; Sun, C.; Chen, K. Analysis of expressed sequence tags from Chinese bayberry fruit (*Myrica rubra* Sieb. and Zucc.) at different ripening stages and their association with fruit quality development. *Int. J. Mol. Sci.* **2013**, *14*, 3110–3123. [CrossRef] [PubMed]
2. Ge, S.; Wang, L.; Ma, J.; Jiang, S.; Peng, W. Biological analysis on extractives of bayberry fresh flesh by GC–MS. *Saudi J. Biol. Sci.* **2018**, *25*, 816–818. [CrossRef]
3. Chen, Y.; Xu, H.; Zhang, X.; Gao, P.; Xu, Z.; Huang, X. An object detection method for bayberry trees based on an improved YOLO algorithm. *Int. J. Digit. Earth* **2023**, *16*, 781–805. [CrossRef]
4. Hoshyarmanesh, H.; Dastgerdi, H.R.; Ghodsi, M.; Khandan, R.; Zareinia, K. Numerical and experimental vibration analysis of olive tree for optimal mechanized harvesting efficiency and productivity. *Comput. Electron. Agric.* **2017**, *132*, 34–48. [CrossRef]
5. Ni, H.; Zhang, J.; Zhao, N.; Wang, C.; Lv, S.; Ren, F.; Wang, X. Design on the winter jujubes harvesting and sorting device. *Appl. Sci.* **2019**, *9*, 5546. [CrossRef]
6. Wang, Y.; Wu, H.; Zhu, Z.; Ye, Y.; Qian, M. Continuous picking of yellow peaches with recognition and collision-free path. *Comput. Electron. Agric.* **2023**, *214*, 108273. [CrossRef]
7. Yang, H.; Chen, L.; Ma, Z.; Chen, M.; Zhong, Y.; Deng, F.; Li, M. Computer vision-based high-quality tea automatic plucking robot using Delta parallel manipulator. *Comput. Electron. Agric.* **2021**, *181*, 105946. [CrossRef]
8. Lin, G.; Tang, Y.; Zou, X.; Li, J.; Xiong, J. In-field citrus detection and localisation based on RGB-D image analysis. *Biosyst. Eng.* **2019**, *186*, 34–44. [CrossRef]
9. Wu, G.; Li, B.; Zhu, Q.; Huang, M.; Guo, Y. Using color and 3D geometry features to segment fruit point cloud and improve fruit recognition accuracy. *Comput. Electron. Agric.* **2020**, *174*, 105475. [CrossRef]
10. Zhang, X.; Zhang, Y.; Gao, T.; Fang, Y.; Chen, T. A novel SSD-based detection algorithm suitable for small object. *IEICE Trans. Inf. Syst.* **2023**, *106*, 625–634. [CrossRef]
11. Zhang, Z.; Shi, R.; Xing, Z.; Guo, Q.; Zeng, C. Improved faster region-based convolutional neural networks (R-CNN) model based on split attention for the detection of safflower filaments in natural environments. *Agronomy* **2023**, *13*, 2596. [CrossRef]
12. Li, J.; Chen, J.; Sheng, B.; Li, P.; Yang, P.; Feng, D.D.; Qi, J. Automatic detection and classification system of domestic waste via multimodel cascaded convolutional neural network. *IEEE Trans. Ind. Inform.* **2022**, *18*, 163–173. [CrossRef]
13. Dai, L.; Sheng, B.; Chen, T.; Wu, Q.; Liu, R.; Cai, C.; Wu, L.; Yang, D.; Hamzah, H.; Liu, Y.; et al. A deep learning system for predicting time to progression of diabetic retinopathy. *Nat. Med.* **2024**, *30*, 584–594. [CrossRef] [PubMed]
14. Dai, L.; Wu, L.; Li, H.; Cai, C.; Wu, Q.; Kong, H.; Liu, R.; Wang, X.; Hou, X.; Liu, Y.; et al. A deep learning system for detecting diabetic retinopathy across the disease spectrum. *Nat. Commun.* **2021**, *12*, 3242. [CrossRef]
15. Girshick, R. Fast R-CNN. Fast R-CNN. In Proceedings of the 2015 IEEE International Conference on Computer Vision (ICCV), Santiago, Chile, 7–13 December 2015; pp. 1440–1448.
16. Ren, S.; He, K.; Girshick, R.; Sun, J. Faster R-CNN: Towards real-time object detection with region proposal networks. *IEEE Trans. Pattern Anal. Mach. Intell.* **2017**, *39*, 1137–1149. [CrossRef]
17. He, K.; Gkioxari, G.; Dollár, P.; Girshick, R. Mask R-CNN. In Proceedings of the IEEE International Conference on Computer Vision (ICCV), Venice, Italy, 22–29 October 2017; pp. 2980–2988.
18. Yu, Y.; Zhang, K.; Yang, L.; Zhang, D. Fruit detection for strawberry harvesting robot in non-structural environment based on Mask-RCNN. *Comput. Electron. Agric.* **2019**, *163*, 104846. [CrossRef]
19. Liu, Y.; Ren, H.; Zhang, Z.; Men, F.; Zhang, P.; Wu, D.; Feng, R. Research on multi-cluster green persimmon detection method based on improved Faster RCNN. *Front. Plant Sci.* **2023**, *14*, 1177114. [CrossRef] [PubMed]
20. Redmon, J.; Divvala, S.; Girshick, R.; Farhadi, A. You only look once: Unified, real-time object detection. *Proc. IEEE Comput. Soc. Conf. Comput. Vis. Pattern Recognit.* **2016**, *2016*, 779–788.

21. Ji, W.; Gao, X.; Xu, B.; Pan, Y.; Zhang, Z.; Zhao, D. Apple target recognition method in complex environment based on improved YOLOv4. *J. Food Process Eng.* **2021**, *44*, e13866. [CrossRef]
22. Cao, Z.; Yuan, R. Real-Time detection of mango based on improved YOLOv4. *Electronics* **2022**, *11*, 3853. [CrossRef]
23. Sun, L.; Hu, G.; Chen, C.; Cai, H.; Li, C.; Zhang, S.; Chen, J. Lightweight apple detection in complex orchards using YOLOV5-PRE. *Horticulturae* **2022**, *8*, 1169. [CrossRef]
24. Li, T.; Sun, M.; He, Q.; Zhang, G.; Shi, G.; Ding, X.; Lin, S. Tomato recognition and location algorithm based on improved YOLOv5. *Comput. Electron. Agric.* **2023**, *208*, 107759. [CrossRef]
25. Zhou, J.; Zhang, Y.; Wang, J. RDE-YOLOv7: An improved model based on yolov7 for better performance in detecting dragon fruits. *Agronomy* **2023**, *13*, 1042. [CrossRef]
26. Chen, J.; Liu, H.; Zhang, Y.; Zhang, D.; Ouyang, H.; Chen, X. A Multiscale lightweight and efficient model based on YOLOv7: Applied to citrus orchard. *Plants* **2022**, *11*, 3260. [CrossRef] [PubMed]
27. Yu, L.; Qian, M.; Chen, Q.; Sun, F.; Pan, J. An improved YOLOv5 model: Application to mixed impurities detection for walnut kernels. *Foods* **2023**, *12*, 624. [CrossRef]
28. Wang, C.-Y.; Bochkovskiy, A.; Liao, H.-Y.M. Yolov7: Trainable bag-of-freebies sets new state-of-the-art for real-time object detectors. In Proceedings of the IEEE/CVF Conference on Computer Vision and Pattern Recognition, Vancouver, BC, Canada, 17–24 June 2023; pp. 7464–7475.
29. Wang, F.; Lv, C.; Dong, L.; Li, X.; Guo, P.; Zhao, B. Development of effective model for non-destructive detection of defective kiwifruit based on graded lines. *Front. Plant Sci.* **2023**, *14*, 1170221. [CrossRef]
30. Huang, P.; Wang, S.; Chen, J.; Li, W.; Peng, X. Lightweight model for pavement defect detection based on improved YOLOv7. *Sensors* **2023**, *23*, 7112. [CrossRef]
31. Chen, J.; Kao, S.H.; He, H.; Zhuo, W.; Wen, S.; Lee, C.H.; Chan, S.H.G. Run, don't walk: Chasing higher FLOPS for faster neural networks. In Proceedings of the IEEE/CVF conference on Computer Vision and Pattern Recognition, Vancouver, BC, Canada, 17–24 June 2023; pp. 12021–12031.
32. Lin, X.; Sun, S.; Huang, W.; Sheng, B.; Li, P.; Feng, D.D. EAPT: Efficient attention pyramid transformer for image processing. *IEEE Trans. Multimedia* **2023**, *25*, 50–61. [CrossRef]
33. Ma, J.; Lu, A.; Chen, C.; Ma, X.; Ma, Q. YOLOv5-lotus: An efficient object detection method for lotus seedpod in a natural environment. *Comput. Electron. Agric.* **2023**, *206*, 107635. [CrossRef]
34. Zhang, Q.; Yang, Y. SA-Net: Shuffle attention for deep convolutional neural networks. In Proceedings of the IEEE International Conference on Acoustics, Speech and Signal Processing (ICASSP), Toronto, ON, Canada, 6–11 June 2021; pp. 2235–2239.
35. Yang, L.; Zhang, R.-Y.; Li, L.; Xie, X. Simam: A simple, parameter-free attention module for convolutional neural networks. In Proceedings of the International Conference on Machine Learning, PMLR, Virtual Event, 18–24 July 2021; pp. 11863–11874.
36. Zheng, Z.; Wang, P.; Liu, W.; Li, J.; Ye, R.; Ren, D. Distance-IoU Loss: Faster and Better Learning for Bounding Box Regression. In Proceedings of the AAAI Conference on Artificial Intelligence, New York, NY, USA, 7–12 February 2020; Volume 34, pp. 12993–13000.
37. Gevorgyan, Z. SIoU Loss: More Powerful Learning for Bounding Box Regression. *arXiv* **2022**, arXiv:2205.12740.

*agriculture*

MDPI

*Article*

# Research on an Identification and Grasping Device for Dead Yellow-Feather Broilers in Flat Houses Based on Deep Learning

**Chengrui Xin [1], Hengtai Li [2], Yuhua Li [2], Meihui Wang [2], Weihan Lin [2], Shuchen Wang [3], Wentian Zhang [4], Maohua Xiao [1] and Xiuguo Zou [2,*]**

1 College of Engineering, Nanjing Agricultural University, Nanjing 210031, China; 2021112042@stu.njau.edu.cn (C.X.); xiaomaohua@njau.edu.cn (M.X.)

2 College of Artificial Intelligence, Nanjing Agricultural University, Nanjing 210031, China; 2022119021@stu.njau.edu.cn (H.L.); lyhresearch@njau.edu.cn (Y.L.); 9213020124@stu.njau.edu.cn (M.W.); 9213021002@stu.njau.edu.cn (W.L.)

3 School of Electrical and Control Engineering, Xuzhou University of Technology, Xuzhou 221018, China; wsc1967@xzit.edu.cn

4 Faculty of Engineering and Information Technology, University of Technology Sydney, Sydney, NSW 2007, Australia; wentian.zhang@alumni.uts.edu.au

* Correspondence: zouxiuguo@njau.edu.cn; Tel.: +86-25-58606585

**Abstract:** The existence of dead broilers in flat broiler houses poses significant challenges to large-scale and welfare-oriented broiler breeding. To ensure the timely identification and removal of dead broilers, a mobile device based on visual technology for grasping them was meticulously designed in this study. Among the multiple recognition models explored, the YOLOv6 model was selected due to its exceptional performance, attaining an impressive 86.1% accuracy in identification. This model, when integrated with a specially designed robotic arm, forms a potent combination for effectively handling the task of grasping dead broilers. Extensive experiments were conducted to validate the efficacy of the device. The results reveal that the device achieved an average grasping rate of dead broilers of 81.3%. These findings indicate that the proposed device holds great potential for practical field deployment, offering a reliable solution for the prompt identification and grasping of dead broilers, thereby enhancing the overall management and welfare of broiler populations.

**Keywords:** dead broiler recognition and grasping; deep learning; visual technology; mechanical arm

## 1. Introduction

As a domesticated poultry type with a long history, broilers (broiler chickens) are highly popular globally. Broilers can provide economical and nutritious eggs. Additionally, broilers have advantages such as high protein content, low fat levels, and low calorie counts that cannot be compared to pork and beef [1]. In recent years, the development of flat farming at the scale for broilers has been driven by the increasing demand for low-fat and high-protein broilers [2]. The flat breeding mode refers to directly raising a flock of broilers on the ground or a floor composed of mesh [3]. Broilers raised in this mode have full mobility, high bone strength, and good meat quality [4].

Identifying and clearing dead broilers is a time-consuming and laborious task. Moreover, a large amount of auditory input and cognitive information processing will compete with the visual search, leading to a decrease in attention when performing a single repetitive task [5]. Secondly, the breeding environment contains a large amount of gasses that are harmful to humans, such as $NH_3$, $H_2S$, and CO [6], as well as a large amount of dust [7]. Both of these factors are detrimental to human health, particularly for individuals who work in these environments. Finally, as the inspections have to be performed between different broiler coops [8], the risks of cross-infection between different broiler coops and disease transmission between poultry and mammals will be increased when identifying and cleaning the dead broilers manually. Therefore, intelligent dead broiler removal devices can be used to mitigate the abovementioned drawbacks.

The early identification of dead broilers mainly relies on traditional methods through features such as broiler voice, body temperature, and posture [9–12]. With the continuous advancement of technology, deep learning and visual technologies have been implemented to identify the conditions of livestock and poultry [13]. Visual technology has the advantages of high efficiency and can identify the conditions of broilers without touching them [14]. Deep learning technology learns through massive amounts of data [15], further improving the accuracy and efficiency of recognition.

A number of studies have proposed methods to recognize dead broilers based on the different characteristics of chickens. Lu et al. monitored the colors of chicken crowns using machine vision technology to identify dead broilers [16]; however, this required three consecutive even shakes of the coop, which could cause stress behaviors in yellow-feathered broilers. Zheng et al. used machine vision to judge the posture of feeding laying hens to realize the classification of sick and dead hens [17]; however, this method is prone to misidentification when the posture of healthy hens is abnormal or the camera capture position is shifted. Qu et al. proposed an algorithm based on LibSVM visual detection of dead broilers through judging the morphological features of the chicken claws [18], but this detection method overly relied on the chicken claws and was prone to misjudgment. Veera et al. identified dead broilers using infrared thermal images and the contours of broilers [19], but the algorithm was not suitable for the classification of day-old (after nine weeks) chickens or when there was a high density of dead chickens. All of these studies have proven valuable for subsequent applications, but their dead chicken detection algorithms generally rely on the environment and a single chicken feature.

In terms of dead chicken removal devices, research and applications have shown great promise. In the 1980s, robots began to be deployed in the field [20] to perceive environmental information and achieve robotic movement. Liu et al. designed a visual technology-based dead broiler removal device [21]. The stainless-steel plates on both sides of the front end of the device "swept" the dead broilers onto the conveyor belt, which transported the dead broilers to the storage warehouse at the back end. However, the slow movement of the device made it difficult to clean up dead chickens from other locations of the flat chicken house. Hu et al. designed a dead broiler picking actuator with three joints and four fingers, based on the underactuated principle, for dead broilers raised in captivity for 3 to 7 weeks [22]. However, the device was more complex in terms of construction and had not been applied in the poultry industry. Zhao et al. conducted in-depth research on the kinematic characteristics of a five-degrees-of-freedom dead broiler-picking robot arm. Simulation analysis was conducted using the MATLAB 2020a software [23], which provided preliminary theoretical support for the design of subsequent control systems. However, this research was only at the theoretical level and did not yet include actual construction and experimental validation of the robotic arm.

Research on dead broilers has primarily concentrated on cage-raised broilers, with scant attention paid to the recognition of dead broilers in free-range scenarios. Furthermore, the ability to extract the features and details of dead chicken images is relatively low. In terms of devices, despite the relatively comprehensive theoretical exploration regarding the structures and devices for picking up dead broilers, the majority of studies have been confined to the theoretical level. In view of this, this study designed a simple, easy-to-use, and cost-effective dead broiler grasping and moving device, with an average success rate of 81.3%. In addition, this study proposed an enhanced deep learning method for recognizing dead broilers based on YOLOv6n (hereinafter referred to as YOLOv6), solving the problems of missing image details and missed detection in dense situations.

## 2. Materials and Methods

### *2.1. Experimental Base*

The experiment was performed in a large-scale free-range yellow-feather broiler farm at Jinniuhu Street, Luhe District, Nanjing City, Jiangsu Province, China (118.52.38° E, 32.26.54° N). Each broiler was about 15 weeks old, weighing about 1.45 kg, and had a chest width of about 12.95 cm. The interior scene of the experimental broiler house is shown in Figure 1. The data were collected from 8 March to 28 March 2023, and the testing experiment was conducted from 8 March to 12 March 2024.

**Figure 1.** The interior scene of the experimental broiler house.

### *2.2. Moving Chassis*

The mobile chassis was a R550 (AKM) PLUS chassis (Wheel Technology Co., LTD., Dongguan, China) with a depth camera and suspended Ackerman structure carrying a laser radar, and it could realize the functions of mapping navigation, obstacle avoidance, sound source location, wireless communication with the mechanical arm, and image acquisition as well as a processing subsystem. Figure 2 shows the structure of the dead broiler identification and grabbing device.

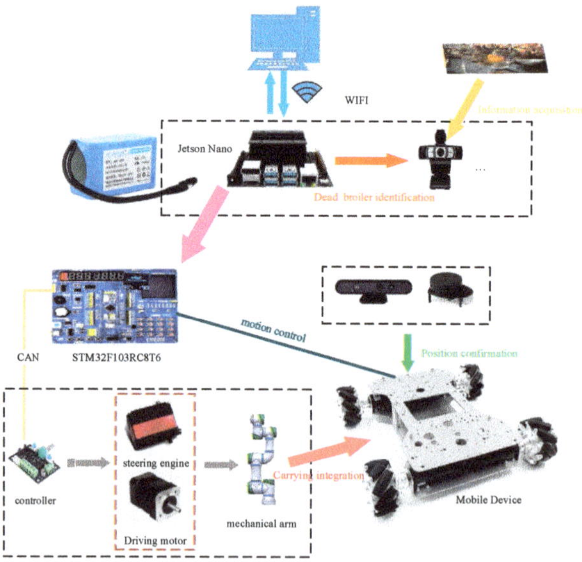


**Figure 2.** The structure of the dead broiler identification and grabbing device.

## *2.3. Image Acquisition and Processing System*

The image acquisition and processing subsystem is composed of a binocular camera and Jetson Nano (NVIDIA, Santa Clara, CA, USA). It identifies dead broilers and their coordinates in scattered broiler farms in real-time via binocular vision positioning. The camera has 5 million pixels with a video resolution of 1920 × 1080. The Jetson Nano is a compact and feature-rich AI computing module developed by NVIDIA, equipped with a 128-core Maxwell GPU [24] and embedded with a trained recognition model for dead broilers. Figure 3 shows the image acquisition and processing system.

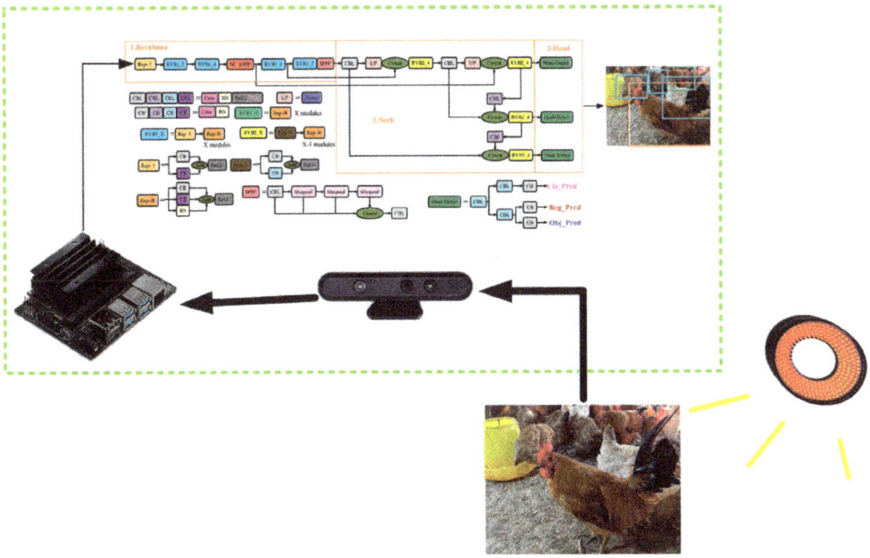

**Figure 3.** The image acquisition and processing system.

Binocular vision positioning utilizes the principle of stereo vision to construct a three-dimensional model of the scene, thereby determining the specific position of the target in space [25]. The imaging model is shown in Figure 4.

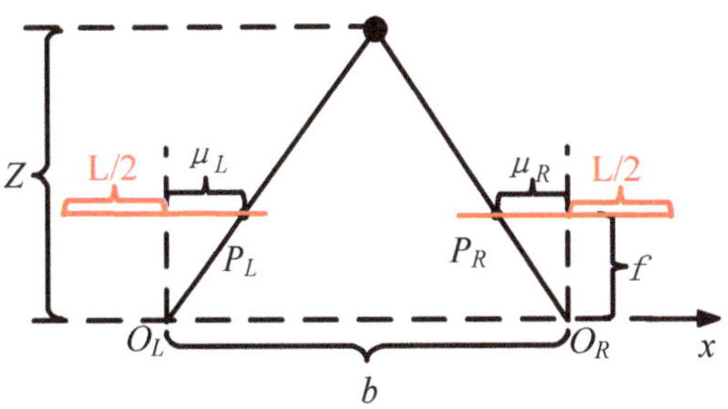


**Figure 4.** Binocular camera imaging model.

Equation (1) was derived from similar triangles.

$$\begin{cases} Z = \frac{f \times b}{d} \\ X = \frac{Z \times \mu L}{f} \end{cases}, \quad (1)$$

where $Z$ is the normal distance from the chicken to the baseline (b) of the camera, $X$ refers to the lateral distance from the chicken to the center of the camera, $f$ refers to the camera focal length, and $d$ is the binocular camera parallax, with a numerical value of $\mu_L + \mu_R$.

## 2.4. Mechanical Arm

The dead broiler grasping manipulator was based on the Dobot manipulator (Dobot Technology Co., LTD., Shenzhen, China), and the model and parameters of the manipulator and motor were improved according to the specifications of broilers. The material was 3D-printed resin, and the control system communicated with Jetson Nano through Bluetooth for the sake of data transmission and instruction interaction. The maximum load was 5 kg, and the repetitive positioning was 0.2 mm. Physical diagrams of the manipulator and the end effector are shown in Figure 5.

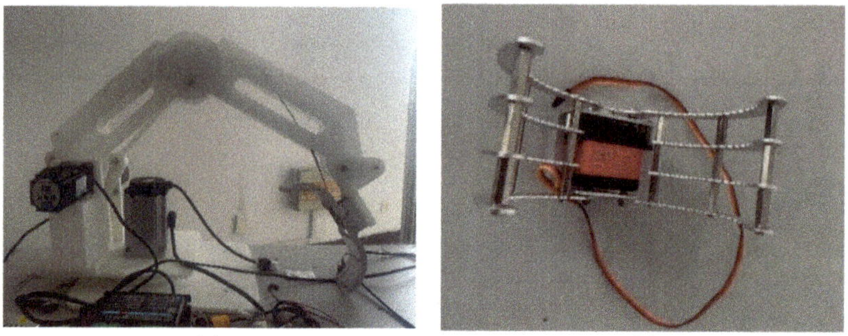

(a) (b)

**Figure 5.** The dead broiler grasping manipulator: (**a**) mechanical arm backbone; (**b**) end effector.

## 2.5. Camera Calibration and Hand–Eye Calibration

In order to realize the transformation from the real-world coordinate system to the manipulator's coordinate system, camera calibration and hand–eye calibration were needed [26]. According to the principle of camera imaging, the camera converts the actual 3D chicken image into 2D information to infer three-dimensional chicken information from two-dimensional chicken image information, it was necessary to calibrate the camera and locate the dead chicken's spatial position. In addition, the mobile chassis and robotic arm control system functioned based on the position of the dead chicken, while the binocular camera was not fixed on the robotic arm, and its position could also change. To ensure that the image information captured by the camera matched the device control system, it was necessary to calibrate the camera coordinate system and the robotic arm base coordinate system uniformly to achieve accurate visual positioning. This process is commonly referred to as hand–eye calibration, and the schematic diagram of the relationship between the three is shown in Figure 6.

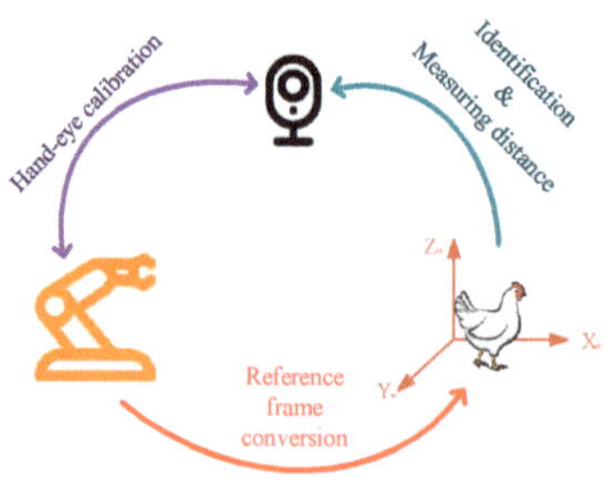


**Figure 6.** Diagram of coordinate system transformation.

In camera calibration and hand–eye calibration operations, calibration plates such as a checkerboard or solid circular arrays are usually used as auxiliary calibration objects. In this experiment, a checkerboard calibration plate was selected, as shown in Figure 7a. The checkerboard calibration board comprised a 10 × 7 checkerboard, where the size of each square was 26 mm × 26 mm. The camera calibration result is shown in Figure 7b.

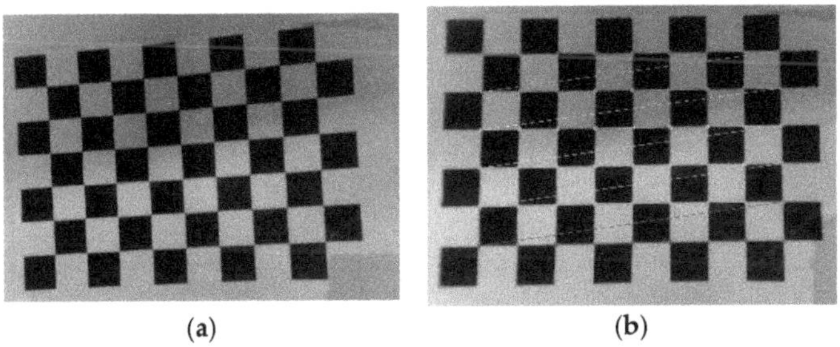

**Figure 7.** The checkerboard calibration board. (**a**) Original image; (**b**) result.

In camera calibration, checkerboard or solid circular array calibration boards are usually used as auxiliary calibration objects [27]. In this experiment, a checkerboard calibration board was selected. The process was as follows: First, a world coordinate system was established on the calibration board, and the corners of the calibration board were touched by the end of the robotic arm, as shown in Figure 8. As the world coordinate system was manually set on the calibration board, the positions of these corners in the world coordinate system were known. We used the position information of the above corners to solve the transformation matrix between the world coordinate system ({World}) and the robotic arm coordinate system ({Base}) [26,28]. The transformation matrix ${}_{B}^{W}\mathbf{T}$ from the robotic arm coordinate system to the world coordinate system is shown in Equation (2).

$$ {}_{B}^{W}\mathbf{T} = \begin{pmatrix} {}_{B}^{W}\mathbf{R} & \mathbf{P} \\ 0 & 1 \end{pmatrix}, \tag{2} $$

where ${}_{B}^{W}\mathbf{R}$ is the rotation matrix and $\mathbf{P}$ is the translation transformation vector.

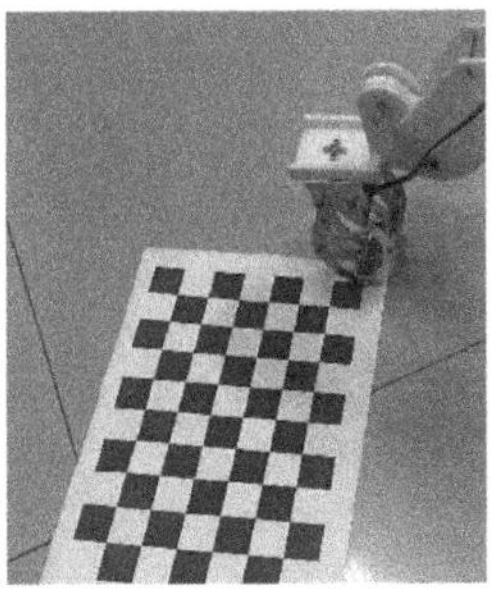

**Figure 8.** Hand–eye calibration operation.

${}_{B}^{W}\mathbf{R}$ and $\mathbf{P}$ can be represented by the position of the origin of the world coordinate system in the coordinate system of the robotic arm. The solution of the rotation matrix ${}_{B}^{W}\mathbf{R}$ can be achieved by introducing a transition matrix, which translates the coordinate system of the robotic dead chicken grasping arm to coincide with the origin of the world coordinate system, as shown in Figure 9 [29].

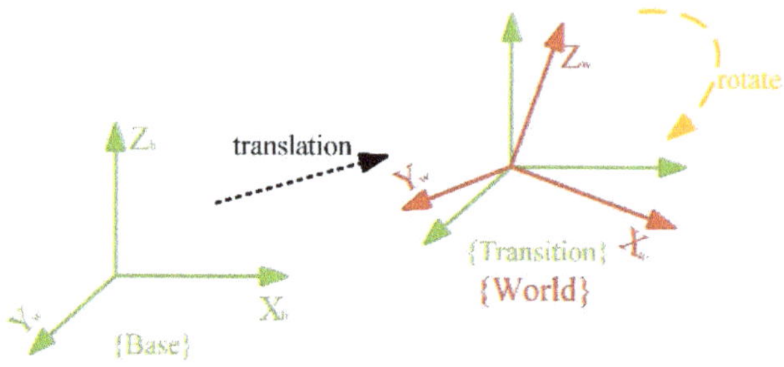


**Figure 9.** The transition coordinate system.

The coordinate system of the robotic arm can obtain the transition matrix ({Transition}) through the translation vector, as shown in Equation (3). Since the origin of the transition coordinate system is consistent with the origin of the world coordinate system, when the end of the robotic arm contacts the origin, the specific position of the world coordinate system in the robotic arm coordinate system can be determined. The translation vector **P** can then be calculated.

$$\begin{pmatrix} xt \\ yt \\ zt \end{pmatrix} = \begin{pmatrix} xb \\ yb \\ zb \end{pmatrix} + \mathbf{P}, \tag{3}$$

where $x_b$, $y_b$, and $z_b$ are the three-dimensional coordinates in the {base} coordinate system, and $x_t$, $y_t$, and $z_t$ are the three-dimensional coordinates in the {tool} coordinate system.

By rotating the **R** vector in the transition coordinate system, the coordinate system of the robotic arm can be obtained, as shown in Equation (4).

$$\begin{pmatrix} xb \\ yb \\ zb \end{pmatrix} = \mathbf{R} \times \begin{pmatrix} xt \\ yt \\ zt \end{pmatrix} = \begin{pmatrix} a11 & a12 & a13 \\ a21 & a22 & a23 \\ a31 & a32 & a33 \end{pmatrix} \tag{4}$$

### 2.6. Data Collection and Image Data Set Production

The quality and quantity of data play a vital role in the effective application of deep learning algorithms. High-quality data can significantly improve the performance of deep learning models [30]. Dead yellow-feather broilers exhibit tightly closed eyes, bodies pressed to the ground, or weak, stiff bodies. The other behavior states of yellow-feather broilers in scattered broiler farms include eating, lying down, and walking [31]. The different behavior states of broilers are shown in Table 1.

**Table 1.** Different behavioral definitions of yellow-feather broilers.

| Behavior Classification | Classification Definition |
| --- | --- |
| Dead | The yellow-feather broiler's eyes close, the body clings to the ground, or the body is weak and stiff. |
| Others | This includes motions such as walking, pecking, inactivity, and resting. |

Among these, "Dead" broilers need to be grabbed. "Others" need not be grabbed. A schematic of the behavior is shown in Figure 10.

After screening, 1565 original images were obtained. The images were labeled using LabelMe 3.16.7 with the label categories shown in Table 2 and converted to PASCAL VOC format for subsequent use. More than one behavior of yellow-feathered chickens could be seen in each image. After image preprocessing (resizing, random flipping, and translation transformation), 2456 images were obtained, which were divided into training, verification, and test sets according to a ratio of 8:1:1.

(a) (b) (c) (d) (e)

**Figure 10.** Different behavior images of yellow-feather broilers: (**a**) walking; (**b**) pecking; (**c**) resting; (**d**) inactive; (**e**) dead.

**Table 2.** The definitions of categories for classification and the number of observations for each category.

| Label Name and Number | Label Definition |
|---|---|
| Walking (20,864) | Actions such as standing, walking, and arranging feathers |
| Pecking (15,648) | Yellow-feathered broilers with their heads touching the ground, troughs or water troughs, or tails cocked up |
| Resting (10,432) | Yellow-feathered broilers lying on the ground |
| Inactive (5216) | Yellow-feathered broilers lying on their backs with their bodies curled up or their tails drooping |
| Dead (32,166) | Yellow-feathered broilers lying flat on the ground with their bodies in a rigid state |

### 2.7. Recognition Model of Dead Yellow-Feather Broiler

Existing target detection algorithms based on deep learning are divided into single-stage or two-stage [32]. To select a suitable network model, three mainstream object detection algorithms—namely, SSD, Faster-RCNN, and YOLOv6—were trained and predicted using the same data set, and the model with the best comprehensive performance was selected. The deep learning methods used and their advantages and disadvantages are detailed in Table 3.

**Table 3.** Comparison of advantages and disadvantages of three algorithms.

| Models | Exact Deep Learning Method | Pros | Cons |
|---|---|---|---|
| SSD | Single deep network for both object classification and localization using default boxes | Fast, simple, and effective for a wide range of object sizes | Less accurate on very small objects |
| Faster-RCNN | Two-stage detector with a region proposal network (RPN) for generating candidate regions | High accuracy, flexible backbone networks, faster-than-previous R-CNN version | Slower than one-stage detectors, more complex to train |
| YOLOv6 | Single-pass detector with efficient backbones and improved training techniques | Very fast, simple architecture, competitive accuracy | Limited information available, may struggle with small objects |

It can be seen from Table 3 that the Faster-RCNN model is more complex to train and is not easy to migrate to the image acquisition and processing system. However, YOLOv6 has the advantages of being lightweight, easy to migrate, and having low requirements for the development environment [33]. Therefore, the network model for detecting dead broilers was improved based on the YOLOv6 model.

The Squeeze-and-Excitation (SE) attention mechanism is a lightweight module designed to enhance the representational power of convolutional neural networks, as shown in Figure 11. It improves the network's characterization ability by explicitly capturing the correlation between convolution channels [34].

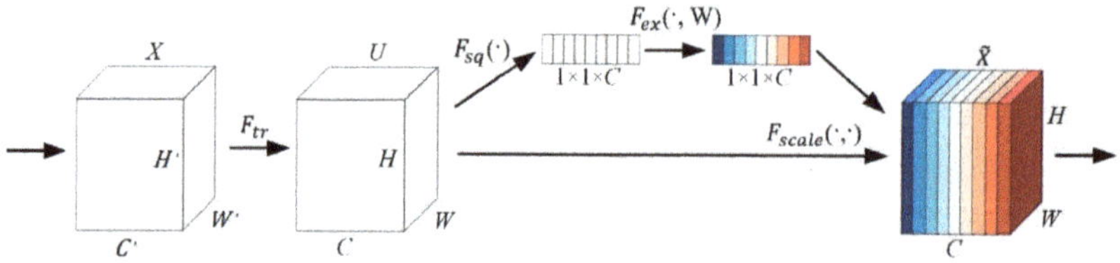


**Figure 11.** The SE module's structure diagram.

The improved YOLOv6 network model is shown in Figure 12. The SE compresses feature maps with a scale of $H \times W \times C$ through global average pooling, retaining only the size of C at the channel scale, which is converted to $1 \times 1 \times C$. After compression, ReLU and restoration are performed and, finally, sigmoid is used for activation. The values of each channel are converted into weight values and multiplied, with the original input features as input features.

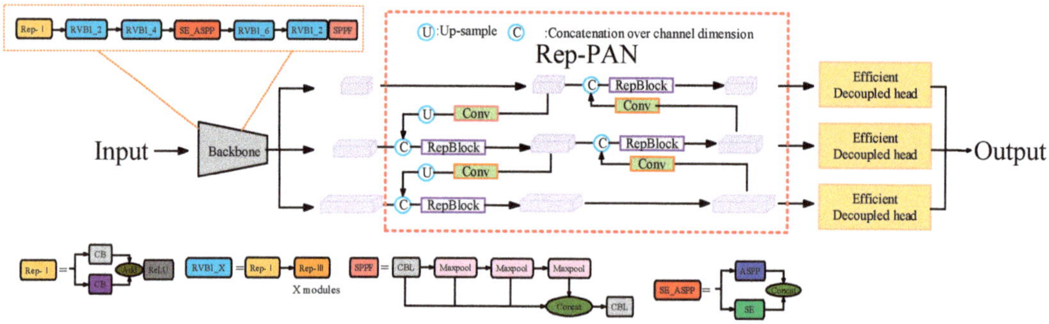


**Figure 12.** Schematic diagram of the improved YOLOv6 network structure.

### 2.7.1. Model Training Parameters

The hardware platform used for this training was configured with a Tesla V100 graphics card with 24 G of memory and an AMD EPYC 9654 CPU. The Python version is 3.8, and the PyTorch framework version is 1.9.0.

### 2.7.2. Model Evaluation Metrics

When evaluating the performance of deep learning algorithms, the following key indicators are used: precision, which is used to measure the proportion of real positive samples in instances where the model predicts positive samples; recall, which reflects the ability of the model to find all positive samples; and mean average precision (maP), which reflects the performance of the comprehensive evaluation model in various categories of multi-category classification tasks. The $F1$ score is the harmonic mean of precision and recall. Together, these indicators constitute a standard system for comprehensively evaluating the performance of deep learning algorithms [35]. Their formulas are shown in Equations (5)–(8):

$$Precision = \frac{TP}{TP + FP}, \quad (5)$$

$$Recall = \frac{TP}{TP + FN}, \quad (6)$$

$$maP = \frac{1}{N}\sum_{i=1}^{N} Api, \quad (7)$$

$$F1 = \frac{2 \times Precision \times Recall}{Precision + Recall} \quad (8)$$

where *TP* is the number of true positive samples, *FP* is the number of false positive samples, *FN* is the number of false negatives, *N* represents the number of target classes detected, *A* is the corresponding accuracy, and $p_i$ is the change value of the recall.

The FPS (frame rate per second of the screen) is also used to measure the processing speed of the model in practical applications [36]. The larger the FPS value, the faster the model detection speed.

### *2.8. The Design of the Dead Broiler Grasping Manipulator*

The Robotics Toolbox module in MATLAB provides powerful functions that can be used to simulate the kinematics and trajectory planning of the manipulator [37]. The manipulator model was established, and the posture of the end effector of the manipulator in the base coordinate system was calculated by the "forward_kinematics" function [38]. The D-H parameters of the manipulator were constructed, and the simulation model of the manipulator is shown in Figure 13.

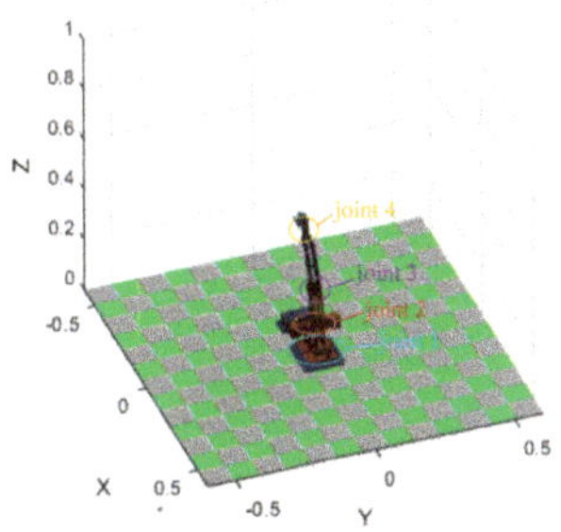


**Figure 13.** Simulation model of manipulator.

Path planning for the manipulator was realized via the quintic polynomial interpolation method. The rationality of the quintic polynomial interpolation method was verified by Robotics Toolbox, and the resulting diagram is shown in Figure 14.

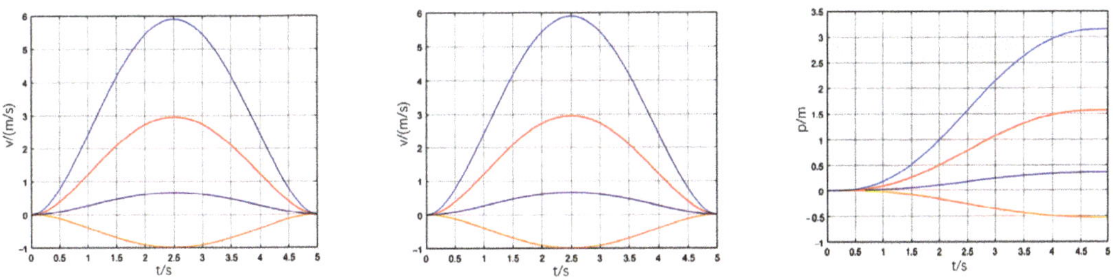


(a) (b) (c)

**Figure 14.** The simulation result graph: (**a**) acceleration simulation results; (**b**) velocity simulation result; (**c**) position simulation result graph. P.s.: The blue line represents the movement of joint 1, the red line represents the movement of joint 2, the purple line represents the movement of joint 3, and the orange line represents the movement of joint 4.

## 3. Results and Analysis

### *3.1. The Different Model's Performance*

After the model was trained and the parameters were adjusted, the final evaluation was conducted using the test set. The overall results for the category of yellow-feathered broilers using the three algorithms SSD, Faster-RCNN, and YOLOv6 are shown in Table 4 and Figure 15. The standard errors of Faster-RCNN and YOLOv6 are shown in Table 5.

**Table 4.** The overall object detection results of the category of yellow-feathered broilers under three models.

| Models | Precision | Recall | F1 Score | maP |
|---|---|---|---|---|
| YOLOv6 | 0.80 | 0.81 | 0.80 | 0.86 |
| SSD | 0.78 | 0.78 | 0.78 | 0.80 |
| Faster-RCNN | 0.81 | 0.81 | 0.81 | 0.87 |

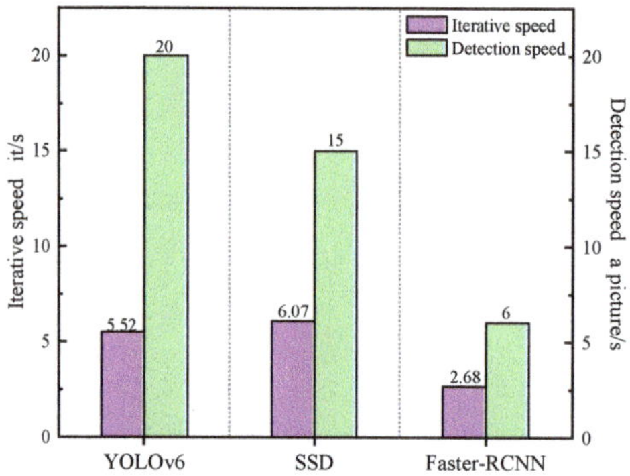


**Figure 15.** Speed comparison of YOLOv6, SSD and Faster-RCNN algorithms detection models.

**Table 5.** The standard errors of Faster-RCNN and YOLOv6.

| Models | Standard Errors of Precision | Standard Errors of Recall | Standard Errors of F1 Score |
|---|---|---|---|
| YOLOv6 | 0.62% | 0.51% | 0.68% |
| Faster-RCNN | 0.81% | 0.77% | 0.73% |

It can be seen from Tables 4 and 5 and Figure 15 that the detection accuracy of YOLOv6 is the same as that of Faster-RCNN. However, the YOLOv6 has lower standard errors of precision and recall, which means that YOLOv6 has a low degree of dispersion. The YOLOv6 and SSD algorithms show obvious advantages in terms of running speed, which are related to their one-stage models [39,40], among which YOLOv6 has the fastest detection speed, as shown in Table 3.

Although YOLOv6 achieves good detection, YOLOv6 cannot identify broilers in the distance if the image background is complex, as shown in Figure 16. It can also be seen from Figure 16 that broiler state behavior in the upper left corner is not ideal, leading to misidentification.

We decided to optimize the YOLOv6 model to improve its performance. The rationality of the SE module was verified through ablation experiments, as shown in Table 6, and Figures 17 and 18.

**Table 6.** Comparison of results of ablation experiments for overall categories.

| Models | Precision | Recall | F1 Score | maP |
|---|---|---|---|---|
| YOLOv6 + SE | 0.84 | 0.88 | 0.83 | 0.90 |
| YOLOv6 + CBAM | 0.82 | 0.86 | 0.84 | 0.90 |

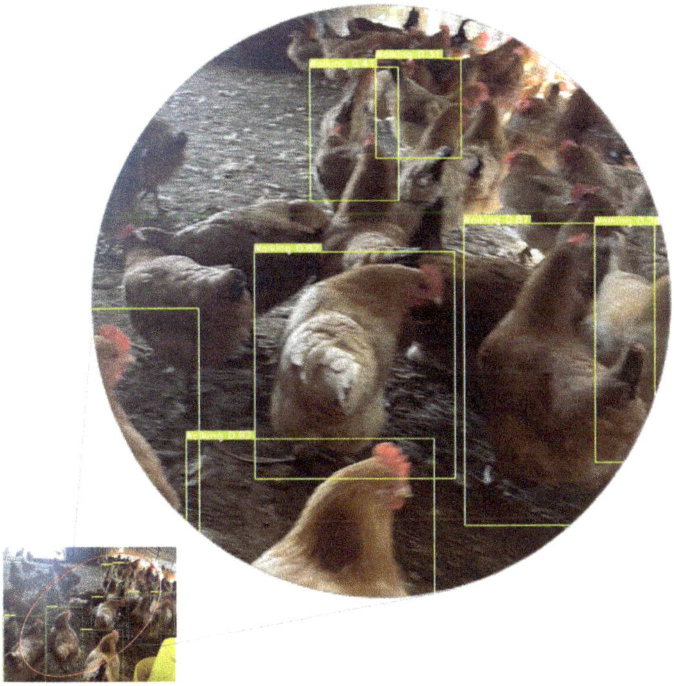

**Figure 16.** Recognition result of the YOLOv6 model.

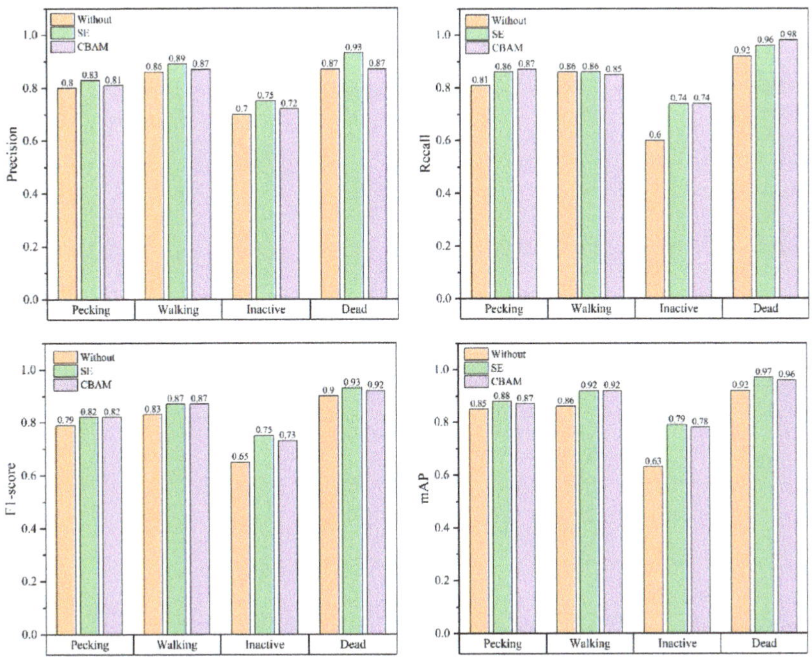


**Figure 17.** Comparison of the results of ablation experiments with different labeling categories.

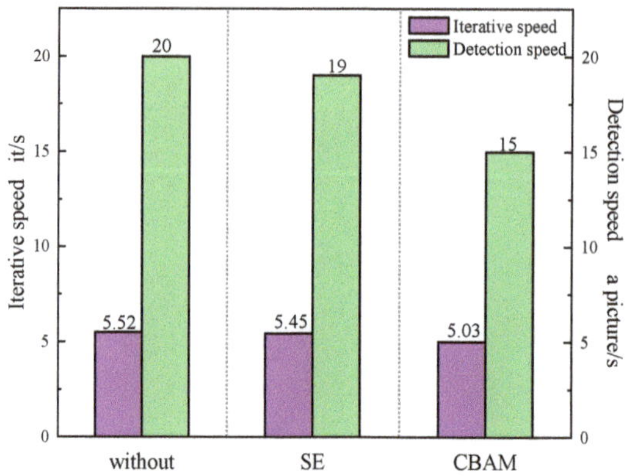


**Figure 18.** Speed comparison of YOLOv6, YOLOv6 + SE and YOLOv6 + CBAM algorithms detection models.

As can be seen from Table 6, Figures 17 and 18, after adding the CBAM and SE attention modules to YOLOv6, different levels of improvement were yielded in maP, F1-score, precision, and recall. The model with the SE attention mechanism incorporated achieved the best results in the recognition of the death label category. Due to the great increase in the number of parameters after the introduction of the CBAM module, although the difference from the SE module in terms of detection accuracy was small, it was significantly less than the difference from the more lightweight SE module in terms of operation speed.

In order to enable the YOLOv6 + SE network model to deal with larger pictures and videos of henhouse scenes, and to improve the global perception, this study improved YOLOv6 + SE based on ASPP (Atrous Spatial Pyramid Pooling). ASPP leverages atrous convolutions with different dilation rates to extract multi-scale features without losing resolution or significantly increasing the computational cost. Through applying parallel Atrous convolutions with varying dilation rates, ASPP can effectively aggregate contextual information at multiple scales. The performance comparison between original and improved modes is shown in Table 7 and Figure 19. The recognition result of the improved YOLOv6 + SE is shown in Figure 20.

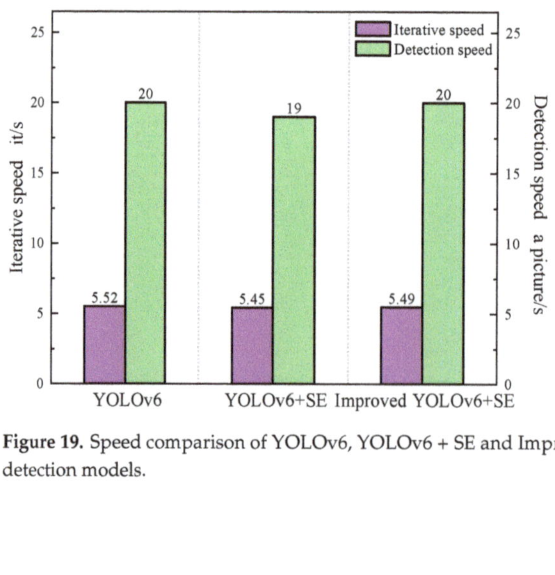


**Figure 19.** Speed comparison of YOLOv6, YOLOv6 + SE and Improved YOLOv6 + SE algorithms detection models.

**Figure 20.** Recognition result of the improved YOLOv6 + SE.

**Table 7.** Comparison of improved and unimproved model results.

| Models | Precision | Recall | F1 Score | maP |
|---|---|---|---|---|
| YOLOv6 + SE | 0.84 | 0.88 | 0.83 | 0.90 |
| Improved YOLOv6 + SE | 0.86 | 0.89 | 0.87 | 0.92 |

Comparing Figure 20 with Figure 16, it can be seen that the improved YOLOv6 significantly increased the number of broilers identified in different behavior states, proving that the ASPP module is able to reduce a certain amount of the leakage in the detection of broiler states in complex backgrounds. At the same time, the introduction of ASPP can also improve the iteration speed and detection speed to a certain extent, which makes up for the problem of the increased number of parameters brought about by the attention mechanism. The curves of loss that the training models suffered during training are shown in Figure 21.

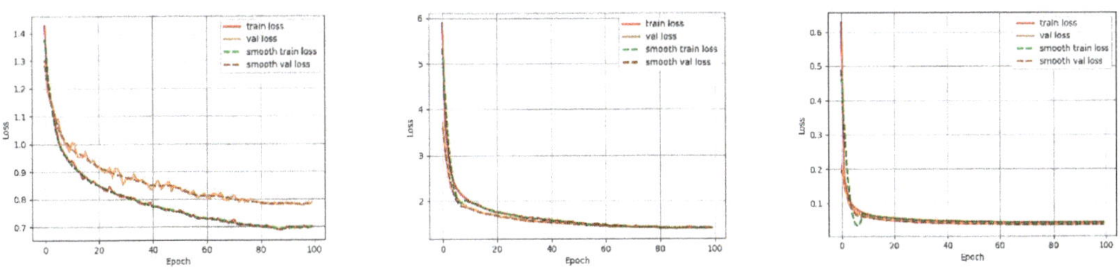


(**a**) (**b**) (**c**)

**Figure 21.** Training loss curves of the improved algorithms: (**a**) SSD; (**b**) Faster-RCNN; (**c**) improved YOLOv6 + SE.

### 3.2. *The Real-Time Detection Effect of the Model*

The improved YOLOv6 network model was migrated to Jetson Nano, and real-time detection was performed by connecting a display screen. The result is shown in Figure 22.

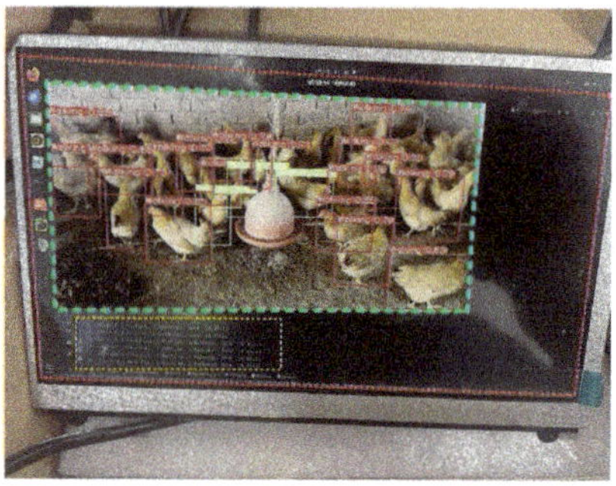

**Figure 22.** The real-time detection results.

The red box displays the results, the green box displays in the form of a video, and the yellow box shows the behavior states of different broilers detected and their detection time. Through this experiment, it can be determined that the improved network model was successfully migrated to Jetson Nano, and the real-time identification effect is good.

### 3.3. *The Design of the Dead Broiler Grasping Experiment*

In this experiment, first, a dead broiler was placed at a point where the mobile device could easily move to and grasp it. Secondly, before being turned on, the device was placed where it could observe the dead broiler. Finally, the device was turned on and operated. The effect of catching a dead broiler achieved by the device is shown in Figure 23.

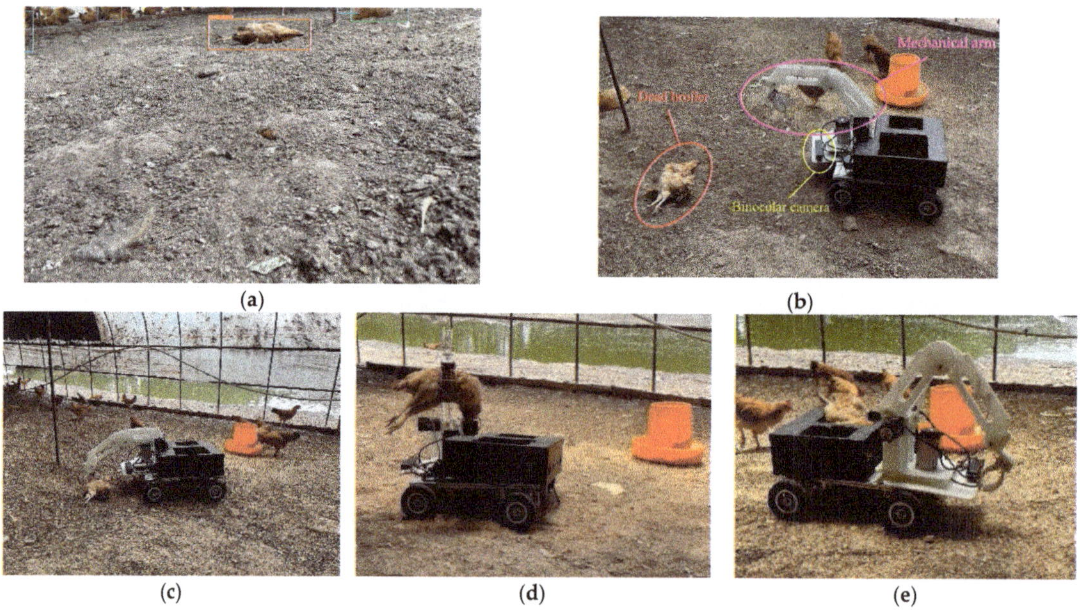


**Figure 23.** (**a**) Identifying the dead broiler. (**b**) The device is moving. (**c**) The device is grabbing the dead broiler. (**d**) The device is receiving the dead broiler. (**e**) The device is transporting the dead broiler.

During the experiment, it was found that there were differences in the parts of dead broilers that were grabbed, as shown in Figure 24. The success rate of the device depends on which body part is grabbed. Three body parts were examined in three groups: (a), (b), and (c). Group (a) showed grabbing by the back, recorded as experiment 1. Group (b) showed grabbing by the back and chest, marked as experiment 2. Group (c) showed grabbing by the chest, marked as experiment 3. After an experiment was finished, the position of the device was kept unchanged; the detailed data on the grabbing results are shown in Table 8.

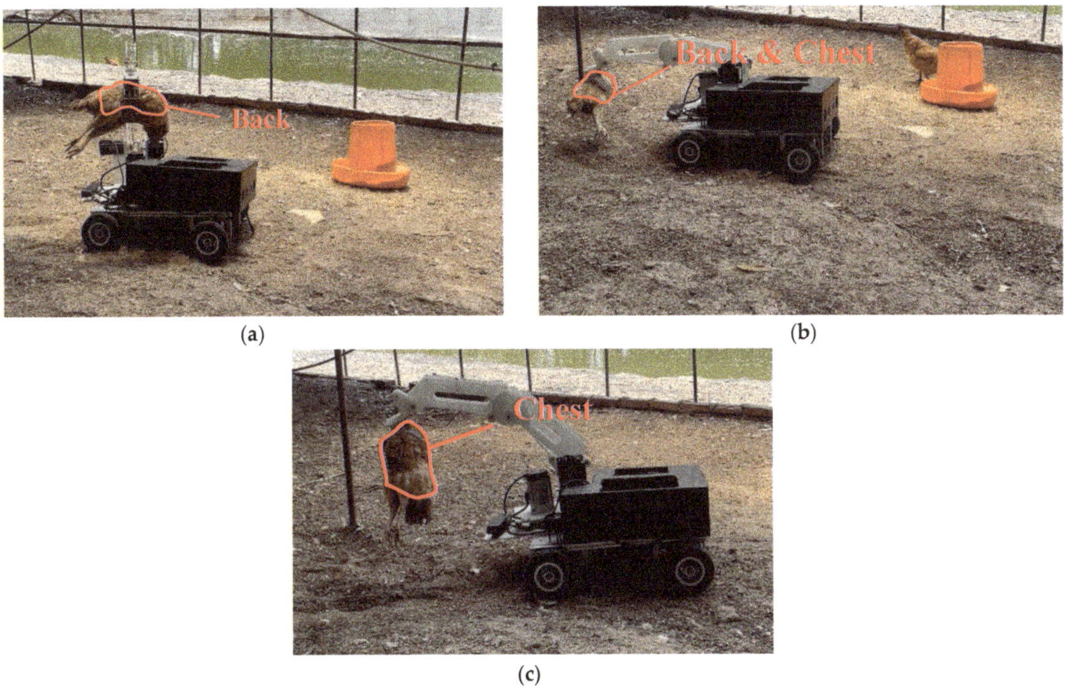


**Figure 24.** The different parts grasped: (**a**) back grab; (**b**) back and chest grab; (**c**) chest grab.

**Table 8.** Comparison of improved model results.

| Experimental Serial Number | Grab Times | Success Rate |
|---|---|---|
| 1 | 30 | 0.87 |
| 2 | 30 | 0.80 |
| 3 | 30 | 0.77 |

The mobile device for identifying and grasping dead broilers completed the task of removing dead broilers with a success rate of over 77% and an average success rate of 81.3%. The success rate when grabbing dead broiler feet and other parts with small contact areas was the lowest. The success rate was reduced in areas where broilers were densely gathered, which may have been due to the failure of the device to collect information on dead broilers; however, the mobile device can disperse the broilers during its operation and capture the information required to complete the grabbing of dead broilers.

## 4. Conclusions

(1) The experimental results demonstrated that the mobile device developed for identifying and grasping dead broilers proved capable of fulfilling the task of removing

dead broilers. It achieved a success rate of over 77%, with an average success rate of 81.3%. However, the success rate was lowest when grasping parts of the broilers such as the neck, feet, or other areas with small contact surfaces. Even though the presence of the mobile device itself exerted a certain influence on the success rate, this aspect will be the focus of future improvements. Additionally, the success rate of grasping deceased broilers decreased in densely populated areas, which can be attributed to the device's limited ability to collect information on dead broilers. Nevertheless, the mobile device could disperse the broilers during movement, capture the necessary information, and complete the grasping task.

(2) This study proposes an enhanced deep learning-based approach for identifying broilers. The YOLOv6 algorithm, with its superior comprehensive performance, was selected as the basic network and underwent in-depth optimization. Specifically, a YOLOv6 network structure based on the SE attention mechanism and ASPP was proposed to address the existing issues found in broiler houses. The experimental outcomes indicated that the recognition accuracy of the improved algorithm model reached 86.1%.

(3) This study designed a mechanical arm for positioning and grasping dead broilers. A model joint simulation of the manipulator was conducted, and the motion trajectory was planned. The experimental results verified that the manipulator model passed the test, the transmission was stable, and the trajectory met the requirements, thereby providing the essential conditions for achieving stable grasping and attaining the design objective.

This study focused on the automatic identification and removal of dead broilers in large-scale flat-breeding yellow-feather broiler farms, aiming to develop a solution that combines vision technology and robotic arm control technology. In order to solve the challenge of dead broiler identification in complex environments, this study proposed a dead broiler detection algorithm with high accuracy, high speed, and easy portability. Compared with traditional machine learning methods, the algorithm achieved significant improvements in accuracy and real-time performance, ensuring that the speed requirements of the dead broiler cleaning process can be met.

In addition, this study independently developed a vision-based mobile device for dead broiler collection that successfully achieved the expected design goals and was capable of efficiently and rapidly identifying and disposing of dead broilers in large-scale free-range broiler farms. The device shows a modularized design, which not only facilitates future function expansion, maintenance, and system upgrades, but also improves the overall flexibility.

Despite these results, there are some limitations of this study, as follows:

(1) Limitations of applicability—The current study was tested and optimized mainly with respect to yellow-feathered broilers. Given the wide variety of broiler breeds available in the market, further verification of the applicability of the device for other breeds is required. If the recognition effect is found to be poor, a breed-specific image database needs to be established as a benchmark for recognition.

(2) Efficiency and energy-saving considerations—When there are multiple dead broilers in the coop at the same time, although the device is able to effectively detect and remove them, further research is needed into how to optimize path planning for a more efficient operation from the point of view of improving efficiency and energy use.

Future work will focus on improvements and refinements in both of these areas, in order to further increase the usefulness and adaptability of the system.

**Author Contributions:** Conceptualization, C.X. and X.Z.; methodology, C.X. and X.Z.; validation, C.X., H.L. and X.Z.; formal analysis, C.X., H.L. and Y.L.; investigation, C.X., H.L., Y.L. and M.W.; data curation, C.X., H.L., Y.L., M.W. and W.L.; writing—original draft preparation, C.X., H.L., Y.L., M.W. and W.L.; writing—review and editing, S.W., W.Z., M.X. and X.Z.; visualization, C.X., H.L. and Y.L.;

supervision, S.W. and X.Z.; project administration, S.W., W.Z., M.X. and X.Z.; funding acquisition, S.W. and X.Z. All authors have read and agreed to the published version of the manuscript.

**Funding:** This research was funded by the International Science and Technology Cooperation Program of Jiangsu Province (Grant No. BZ2023013), Xuzhou Key Research and Development Project (Modern Agriculture) (Grant No. KC21135), and National Entrepreneurship Training Program for University Students (Grant No. 202310307223E).

**Institutional Review Board Statement:** Not applicable.

**Informed Consent Statement:** Not applicable.

**Data Availability Statement:** Data are contained within the article.

**Acknowledgments:** We are thankful to Yungang Bai, Sunyuan Wang, and Zhilong Chen, who contributed to our field data collection and primary data analysis.

**Conflicts of Interest:** The authors declare no conflicts of interest.

## References

1. Zhou, S.; Watcharaanantapong, P.; Yang, X.; Thornton, T.; Gan, H.; Tabler, T.; Zhao, Y. Evaluating broiler welfare and behavior as affected by growth rate and stocking density. *Poult. Sci.* **2024**, *103*, 103459. [CrossRef]
2. Li, G.; Zhao, Y.; Porter, Z.; Purswell, J. Automated measurement of broilers under four stocking densities via faster region-based convolutional neural network. *Animal* **2021**, *15*, 100059. [CrossRef]
3. Chen, G.; Ling, X.; Xie, M.; Xiong, Y.; Li, T.; Shui, C.; Li, C.; Xu, B.; Ma, F. Systematic evaluation of the meat qualities of free-range chicken (Xuan-Zhou) under different ages explored the optimal slaughter age. *Poult. Sci.* **2024**, *103*, 104019. [CrossRef]
4. Park, J.; Kwon, O.; Lee, K.; Heo, Y.; Yoon, C. Ammonia and Hydrogen Sulfide Monitoring in Broiler Barns and Cattle Barns. *J. Environ. Health Sci.* **2015**, *41*, 277–288. [CrossRef]
5. Charman, W. Visual standards for driving. *Ophthal. Physiol. Opt.* **2015**, *5*, 211–220. [CrossRef]
6. Li, X.; Yan, F.; Hu, K.; He, X.; Ma, Z.; Yuan, Q.; Mirzoev, S.I. Design and test of the integrated machine for self-propelled manure collection and bagging in flat chicken coops. *Trans. Chin. Soc. Agric. Eng.* **2024**, *40*, 251–261.
7. Shi, Z.; Xi, L.; Ji, Z.; Cheng, P. LED illuminant improving broilers house environment and growth performance. *Trans. Chin. Soc. Agric. Eng.* **2017**, *33*, 222–227.
8. Hinojosa, C.; Caldwell, D.; Byrd, J.; Ross, M.; Stringfellow, K.; Fowlkes, E.; Lee, J.; Stayer, P.; Farnell, Y.; Farnell, M. Use of a foaming disinfectant and cleaner to reduce aerobic bacteria on poultry transport coops. *J. Appl. Poult. Res.* **2015**, *24*, 364–370. [CrossRef]
9. Okada, H.; Itoh, T.; Suzuki, K.; Tsukamoto, K. Wireless sensor system or detection of avian influenza outbreak farms at an early stage. In Proceedings of the 8th IEEE Conference on Sensors, Christchurch, New Zealand, 25–28 October 2009; pp. 1374–1377.
10. Aydin, A.; Bahr, C.; Viazzi, S.; Exadaktylos, V.; Buyse, J.; Berckmans, D. A novel method to automatically measure the feed intake of broiler chickens by sound technology. *Comput. Electron. Agric.* **2008**, *101*, 17–23. [CrossRef]
11. Cao, Y.; Teng, G.; Yu, L.; Li, Q. Comparison of different de-noising methods in vocalization environment of laying hens including fan noise. *Trans. Chin. Soc. Agric. Eng.* **2014**, *30*, 212–218.
12. Jacob, F.; Baracho, M.; Nääs, I.; Souza, R.; Salgado, D. The use of infrared thermography in the identification of pododermatitis in broilers. *Eng. Agric.* **2016**, *36*, 253–259. [CrossRef]
13. Wang, J.; Wang, N.; Li, L.; Ren, Z. Real-time behavior detection and judgement of egg breeders based on YOLO v3. *Neural Comput. Appl.* **2020**, *32*, 5471–5481. [CrossRef]
14. Zhang, Q. Application and Innovation of New Media Technology in Visual Communication Design. *Agro Food Ind. Hi Tech.* **2017**, *28*, 3170–3173.
15. Shafay, M.; Ahmad, R.; Salah, K.; Yaqoob, I.; Jayaraman, R.; Omar, M. Blockchain for deep learning: Review and open challenges. *Clust. Comput.* **2023**, *26*, 197–221. [CrossRef]
16. Lu, C. Study on Dead Birds Detection System Based on Machine Vision in Modern Chicken Farm. Master's Thesis, JiangSu University, Zhenjiang, China, 2009.
17. Zheng, S.; Wang, L. Development of Monitoring System for Layers Rearing in Multi-Tier Vertical Cages Using Machine Vision. *J. Jilin Agric. Univ.* **2009**, *31*, 476–480.
18. Qu, Z. Study on Detection Method of Dead Chicken in Unmanned Chicken Farm. Master's Thesis, Jilin University, Changchun, China, 2009.
19. Veera, V.; Yang, Z. Automatic Identification of Broiler Mortality Using Image Processing Technology. *ASABE Meet. Present.* **2018**, *18*, 3–5.
20. Sequeira, J.; Ribeiro, M. Human-robot interaction and robot control. *Robot. Motion Control Recent.* **2006**, *35*, 375–390.
21. Liu, H.; Chen, C.; Tsai, Y.; Hsieh, K.; Lin, H. Identifying images of dead chickens with a chicken removal system integrated with adeep learning algorithm. *Sensor* **2021**, *21*, 3579. [CrossRef]

22. Hu, Z. Research on Underactuated End Effector of Dead Chicken Picking Robot. Master's Thesis, HeBei Agricultural University, Baoding, China, 2021.
23. Zhao, W.; Chen, Y.; Cao, T. Design and kinematics analysis of robotic arm used for picking up dead chickens. *J. Chin. Agric. Mech.* **2023**, *44*, 131–136.
24. Li, Y.; Liu, Q.; Li, T.; Wu, Y.; Niu, Z.; Hou, J. Design and experiments of garlic bulbil orientation adjustment device using Jetson Nano processor. *Trans. Chin. Soc. Agric. Eng.* **2021**, *37*, 35–42.
25. Wang, S.; Hu, Y. Binocular visual positioning under inhomogeneous, transforming and fluctuating media. *Trait. Du Signal* **2018**, *35*, 253–276. [CrossRef]
26. Li, X.; Zhang, E.; Fang, X.; Zhai, B. Calibration Method for Industrial Robots Based on the Principle of Perigon Error Close. *IEEE Access* **2022**, *10*, 48569–48576. [CrossRef]
27. Yue, W.; Hua, S.; Ying, D. A Complete Analytical Solution to Hand-Eye Calibration Using Quaternions and Eigenvector-Eigenvalue Identity. *J. Intell. Robot. Syst.* **2023**, *109*, 54.
28. Shang, Y.; Shen, J.; Wei, W.; Zheng, B. Optimization of Ball Mill Cylinder Structure Based on Response Surface Optimization Module and Multi-objective Genetic Algorithm. *J. Mech. Sci. Technol.* **2024**, *38*, 3631–3640. [CrossRef]
29. Feng, X.; Tian, D.; Wu, H. A matrix-solving hand-eye calibration method considering robot kinematic errors. *J. Manuf. Process.* **2023**, *99*, 618–635. [CrossRef]
30. Cui, G.; Qiao, L.; Li, Y.; Chen, Z.; Liang, Z.; Xin, C.; Xiao, M.; Zou, X. Division of Cow Production Groups Based on SOLOv2 and Improved CNN-LSTM. *Agriculture* **2023**, *13*, 1562. [CrossRef]
31. Lao, F.; Teng, G.; Li, J.; Yu, L.; Li, Z. Behavior recognition method for individual laying hen based on computer vision. *Trans Chin. Soc. Agric. Eng.* **2018**, *28*, 157–163.
32. Guo, J.; He, G.; Deng, H.; Fan, W.; Xu, L.; Cao, L.; Feng, D.; Li, J.; Wu, H.; Lv, J.; et al. Pigeon cleaning behavior detection algorithm based on lightweight network. *Comput. Electron. Agric.* **2022**, *199*, 107032. [CrossRef]
33. Bist, B.; Subedi, S.; Yang, X.; Chai, L. A Novel YOLOv6 Object Detector for Monitoring Piling Behavior of Cage-Free Laying Hens. *AgriEngineering* **2023**, *5*, 905–923. [CrossRef]
34. Xie, W.; Kimura, M.; Takaki, K.; Asada, Y.; Iida, T.; Jia, X. Interpretable Framework of Physics-Guided Neural Network with Attention Mechanism: Simulating Paddy Field Water Temperature Variations. *Water Resour. Res.* **2022**, *58*, e2021WR030493. [CrossRef]
35. Yun, Y.; Choi, J.; Chung, H.; Bae, K.; Moon, J. Performance evaluation of an occupant metabolic rate estimation algorithm using activity classification and object detection models. *Build. Environ.* **2024**, *252*, 252111299. [CrossRef]
36. Sun, F.; Wang, Y.; Lan, P.; Zhang, X.; Chen, X.; Wang, Z. Identification of apple fruit diseases using improved YOLOv5s and transfer learning. *Trans Chin. Soc. Agric. Eng.* **2022**, *38*, 171–179.
37. Corke, P. MATLAB toolboxes: Robotics and vision for students and teachers. *IEEE Robot. Autom. Mag.* **2007**, *14*, 16–17. [CrossRef]
38. Vila-Rosado, D.; Domínguez-López, J. A MATLAB toolbox for robotic manipulators. In Proceedings of the Sixth Mexican International Conference on Computer Science, Puebla, Mexico, 26–30 September 2005; pp. 256–263.
39. Liu, Y.; He, Y.; Wu, X.; Wang, W.; Zhang, L.; Lv, H. Potato Sprouting and Surface Damage Detection Method Based on Improved Faster R-CNN. *Trans Chin. Soc. Agric. Mach.* **2024**, *55*, 371–378.
40. Biswas, D.; Su, H.; Wang, C.; Stevanovic, A.; Wang, W. An automatic traffic density estimation using Single Shot Detection (SSD) and Mobile Net-SSD. *Phys. Chem. Earth* **2019**, *110*, 176–184. [CrossRef]

*agriculture*

MDPI

*Article*

# YOLOv8-Pearpollen: Method for the Lightweight Identification of Pollen Germination Vigor in Pear Trees

**Weili Sun [1,†], Cairong Chen [2,†], Tengfei Liu [1], Haoyu Jiang [1], Luxu Tian [2], Xiuqing Fu [2], Mingxu Niu [3], Shihao Huang [2] and Fei Hu [1,*]**

1 College of Artificial Intelligence, Nanjing Agricultural University, Nanjing 210031, China; 2023819031@stu.njau.edu.cn (W.S.); 2023119012@stu.njau.edu.cn (T.L.); 2022819036@stu.njau.edu.cn (H.J.)
2 College of Engineering, Nanjing Agricultural University, Nanjing 210031, China; crchen@njau.edu.cn (C.C.); 2022812055@stu.njau.edu.cn (L.T.); fuxiuqing@njau.edu.cn (X.F.); 9223010725@stu.njau.edu.cn (S.H.)
3 College of Horticulture, Nanjing Agricultural University, Nanjing 210031, China; mingxu@njau.edu.cn
* Correspondence: hufei@njau.edu.cn
† These authors contributed equally to this work.

**Abstract:** Pear trees must be artificially pollinated to ensure yield, and the efficiency of pollination and the quality of pollen germination affect the size, shape, taste, and nutritional value of the fruit. Detecting the pollen germination vigor of pear trees is important to improve the efficiency of artificial pollination and consequently the fruiting rate of pear trees. To overcome the limitations of traditional manual detection methods, such as low efficiency and accuracy and high cost, and to meet the requirements of screening high-quality pollen to promote the yield and production of fruit trees, we proposed a detection method for pear pollen germination vigor named YOLOv8-Pearpollen, an improved version of YOLOv8-n. A pear pollen germination dataset was constructed, and the image was enhanced using Blend Alpha to improve the robustness of the data. A combination of knowledge distillation and model pruning was used to reduce the complexity of the model and the difficulty of deployment in hardware facilities while ensuring that the model achieved or approached the detection effect of a large-volume model that can adapt to the actual requirements of agricultural production. Various ablation tests on knowledge distillation and model pruning were conducted to obtain a high-quality lightweighting method suitable for this model. Test results showed that the mAP of YOLOv8-Pearpollen reached 96.7%. The Params, FLOPs, and weights were only 1.5 M, 4.0 G, and 3.1 MB, respectively, and the detection speed was 147.1 FPS. A high degree of lightweighting and superior detection accuracy were simultaneously achieved.

**Keywords:** pear pollen; sprouting vigor; lightweight; object detection; YOLOv8

## 1. Introduction

Pollen is the male reproductive cell of seed plants and is an important agent of sexual reproduction, which is closely related to the improvement of plant varieties and the selection of high-quality genotypes. Good-quality male pollen can greatly improve pollination success, and pollen germination vigor is one of the indicators of male pollen quality [1]. Therefore, measuring pollen germination vigor is crucial for plant science and agricultural production to achieve a high and excellent yield of plants and facilitate low-cost and efficient pollination. Most existing pollen viability assays employ staining and in vitro culture methods [2]. However, staining cannot accurately indicate pollen viability and is not suitable for studying the effect of a treatment on pollen viability. Meanwhile, in vitro culture methods require cumbersome experimental manipulations and high economic and labor costs. Therefore, developing an efficient and simple method for the determination of pollen viability is important [3].

Pear is a native fruit tree of China. Its original species can be traced back to the tertiary period (66 million years ago to 2.6 million years ago) or earlier in the mountainous areas of western or southwestern China, and it has a long history of more than 3000 years of cultivation in China. Pears are popular because of their sweet flavor and high nutritional value. China's pear production reaches up to 19 million tons (2023), accounting for 3/4 of the world's total pear production [4]. Given that pear is a typical self-fertilizing nonaffinity fruit tree, only the same variety of pear trees can blossom and they cannot pollinate each other. Therefore, the production must be reasonably configured using pollinating trees or artificial pollination to obtain a normal yield. In addition, the pear flowering period is only a short 4–5 days. Inappropriate pollination tree configuration, such as poor flowering, could easily cause yield reduction. Therefore, pears need efficient pollination methods and male pollen with excellent germination vigor. In different varieties of pear pollen there are small differences in the rate of germination potential, but the consistent trend shows that while in pollen tube germination length in the early stage of germination there are individual differences, the germination characteristics are obvious; both pollen grains and pollen tubes are consistent in shape, and easy to observe during study.

Machine learning has been increasingly utilized in the field of palynology to classify and predict pollen concentrations. Punyasena et al., developed a layered machine learning classification system that discriminates variations in pollen shape, size, and texture, demonstrating the capability of machine learning systems to solve challenging palynological classification problems [5]. By combining pollen observations with meteorological and land surface variables, Liu et al., used machine learning to estimate atmospheric ambrosia pollen concentrations in Tulsa, OK [6]. Sobol et al., investigated the use of supervised machine learning for biome classification using pollen datasets, assigning modern pollen samples to biome classes [7]. In addition, Sobol et al., applied machine learning to reconstruct past biome states using modern pollen assemblages in Southern Africa [8]. Huete et al., utilized machine learning approaches to forecast grass pollen evolution by analyzing satellite-based landscape information and phenology [1]. Zewdie et al., presented a method for estimating airborne pollen concentrations using deep neural networks and ensemble machine learning methods, testing the performance of machine learning models on a dataset from 2012 to 2017 [9]. Furthermore, Zewdie et al., applied machine learning methods to NEXRAD (Next Generation Weather Radar) weather radar data to estimate daily Ambrosia pollen concentrations over a region [9]. Cordero et al., used supervised machine learning algorithms to predict daily Olea pollen concentrations in central Spain, showcasing the potential of machine learning in sunderstanding and forecasting pollen risk levels [10]. Yamazaki et al., introduced a simple method for measuring pollen germination rates using machine learning; specifically, the Yolov5 package for transfer learning [11]. This method could detect germinated and non-germinated pollen, allowing for the estimation of pollen germination rates across different plant species.

The above research results show that machine learning and target detection have a wide range of applications and play a good role in the field of sporology, meeting the needs of life and production and conforming to the trend of combining practical agricultural production with machine learning target detection [12]. In practice, a large amount of image data must be provided to enrich the model and cope with the effects of different lighting conditions and changes in pollen size, shape, and attitude on the model performance; however, this process requires substantial time and effort. Data augmentation is particularly important and helps the model to generalize unseen data by introducing additional variations in the data [13]. Therefore, the model can perform well on the training set and maintain high accuracy on new and different data, reducing the risk of overfitting, improving the robustness of the model to small perturbations or noise in the input data, and enriching the application scenarios of the model [14]. To further adapt to the needs of agricultural production fields in environments with limited computational resources (e.g., mobile devices and embedded systems) and to reduce the cost of applications, we constructed a model named YOLOv8-Pearpollen based on the YOLOv8 target detection model for the pollen emergence viability detection of pear pollen. With this model, we designed two deep lightweight improvements, namely, model pruning and knowledge distillation, to reduce unnecessary parameters and connections in the neural network, minimize the Params and FLOPs, optimize the efficiency and performance of the model, maximize the use of limited data resources, and improve the model's pervasiveness in the deployment of real hardware.

## 2. Materials and Methods

### *2.1. Microphenotypic Trait Observation System*

The microphenotypic trait observation system was equipped with a model NSZ818M stereoscopic microscope (NOVEL, Nanjing, China) with a total magnification range of 7.5–135×. A magnification of 100 was chosen in this experiment to help improve the accuracy of the pollen grain classification and identification tasks. Installed on the stereoscopic microscope was a Sony industrial microscope digital camera E3ISPM20000KPA (Hangzhou, China), which was connected to the stereoscopic microscope through the C-mount interface and had an image acquisition resolution of 5440 × 3648 pixels. Clear and detailed images are helpful for machine learning model training and data analysis, and the model comprehensively learned and understood the sample's features [12]. The system was equipped with a four-type light source (coaxial light, ring light, adjustable angled light, and bottom fill light) that enhanced the contrast of the samples and made the fine structure visible by changing the angle and intensity of the illumination. The use of different illumination modes such as transmitted light and reflected light resulted in a special imaging effect, which was conducive to the observation of special microstructures or microscopic phenotypes [15]. Specialized data acquisition software was used to adjust the parameters, such as the exposure and contrast of the captured image data, and to capture and store the images. Finally, the processed dataset was labeled to train the detection model of pollen germination status. Figure 1 illustrates the structure of the microphenotypic trait observation system used in this experiment.

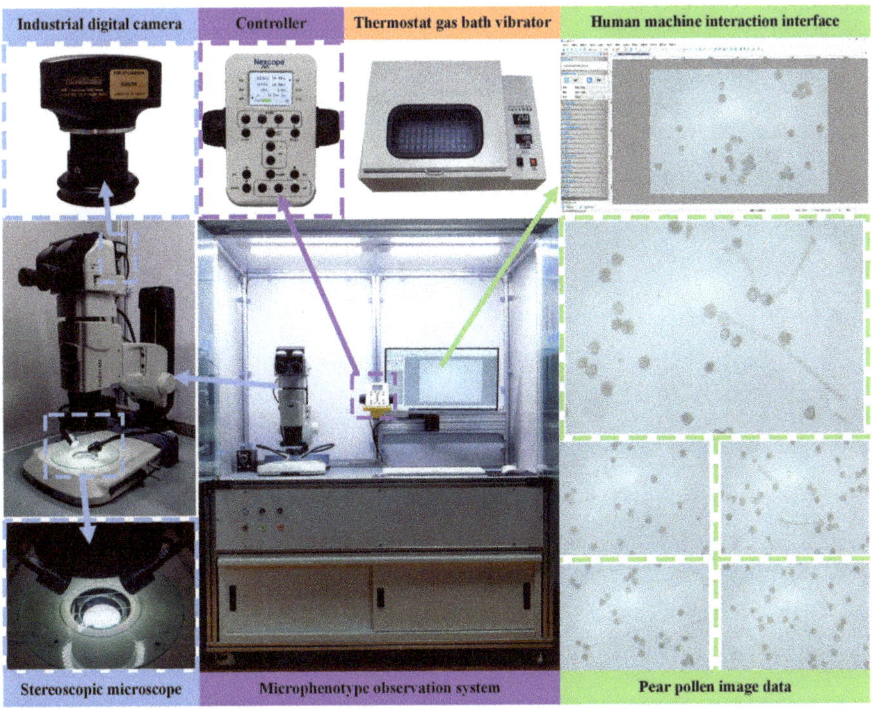


**Figure 1.** Structure of the microphenotypic trait observation system.

### 2.2. Germination Image Acquisition and Dataset Construction

The pear pollen used in this experiment was the variety "yellow flower" obtained from the Baima Teaching and Research Base of Nanjing Agricultural University, Nanjing, Jiangsu Province, China. During the macrobud stage, a large number of flower buds were collected, placed inside a room to dry, and rubbed and sieved to obtain a number of pear anthers. The anthers were then packaged in sulfuric acid paper bags, placed in dry containers with color-changing silica gel desiccant, and then frozen and sealed at a low temperature for preservation. Prior to use, the anthers were first placed in a 4 °C environment to wake up for 8–12 h and resealed after each use as shown in Figure 2a,b. After the pollen culture solution was configured (10% sucrose + 0.01% boric acid + 0.03% calcium gluconate + 0.04% xanthan gum), a pipette gun was used to transfer 1.5 mL of the culture solution to a 2 mL centrifuge tube. A trace amount of pear pollen (approximately 0.2 mL) was then dispersed and suspended in the culture solution [16]. After the centrifuge tube was fixed, it was placed in a ZD-85 (Changzhou, China) thermostatic gas-bath vibrator (light-proof environment, 25 °C, 120 rpm speed, reciprocating mode amplitude of 25 mm) for 2 h so that the pollen was uniformly suspended and distributed in the culture medium and sprouted pollen tubes. The cultured pollen suspension was evenly spread on a glass slide, which was then placed under a microscope to observe and take images of the pollen germination condition. Seven trials were repeated. After screening, 1500 images saved in the JPEG format were collected throughout the experimental period; the experimental process is shown in Figure 2c.

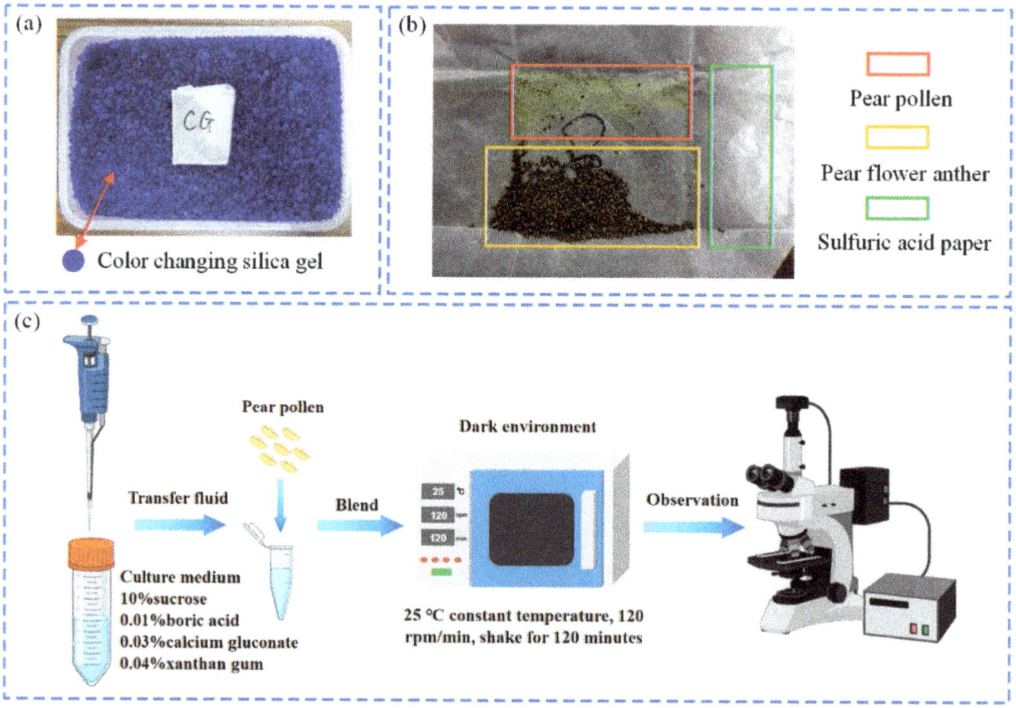


**Figure 2.** Data acquisition: (**a**) Pollen storage conditions. (**b**) "Yellow flower" pear pollen material. (**c**) Experimental procedure.

Labellmg was used to annotate the images of pollen germination status [17]. Whether the length of pollen tube was greater than or equal to the length of pollen grain diameter was used as the criterion to judge if the pear pollen germinated or not, and the results were stored as an .xml file. The labeled xml file was batch converted to a txt coordinate label file to obtain the normalized length, width, and coordinates of the center point of the labeled box and the target categories of "sprout" and "not sprout" [18]. "Sprout" represents pear pollen that meets the conditions for sprouting, and "not sprout" represents pear pollen that has not sprouted or does not meet the conditions for sprouting. The collected images of pear pollen sprouting conditions are shown in Figure 3a–c.

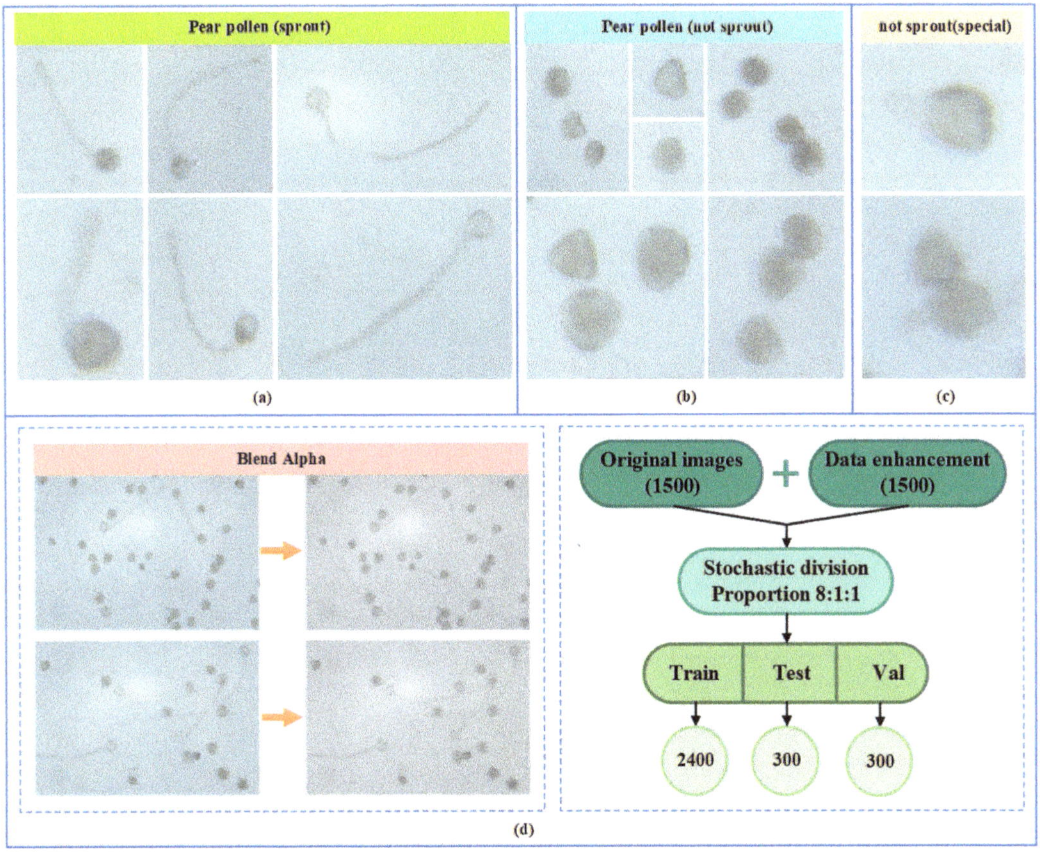


**Figure 3.** Dataset construction: (**a**) Schematic of germinated pollen. (**b**) Schematic of ungerminated pollen. (**c**) Schematic of ungerminated pollen that does not satisfy the judgment conditions for germination. (**d**) Data enhancement and dataset construction.

The main factors affecting the accuracy of pollen germination viability detection in pear trees are pollen color, detection background color, and pollen growth attitude. To avoid unnecessary labor and time costs, increase the robustness of the detection method, and effectively expand the dataset [19], we chose the Blend Alpha method to enhance the obtained image data. The graph contains three color channels, red, green, and blue (RGB), and an Alpha channel for recording the transparency of each pixel in the image, with values ranging from 0 (completely transparent) to 1 (completely opaque). Blend Alpha processes the image through transparency mixing and achieves the synthesis and enhancement of the image by adjusting the image Alpha value. The formula for calculating the synthesized pixel value is shown in Equation (1), where $C_{result}$ is the synthesized pixel color value; $C_{foreground}$ and $C_{background}$ are the pixel color values of the foreground and background, respectively; and $\alpha$ is the transparency value of the foreground pixel. The result from Alpha was blended with 50% of the original image, resulting in the removal of 50% of all colors to create a certain grayscale effect. The image result formula is shown in Equation (2), which is linearly interpolated between the foreground branch (*FG*) and background branch (*BG*) according to the overlay factor (0.5 in this experiment). The obtained final dataset had 3000 images, including 1500 original images and 1500 images enhanced by data, which were randomly divided into training, validation, and test sets according to the ratio of 8:1:1, as shown in Figure 3d.

$$C_{result} = C_{foreground} * \alpha + C_{background} * (1 - \alpha) \quad (1)$$

$$FinalPic = factor * FG + (1 - factor) * BG \quad (2)$$

*2.3. YOLOv8-Pearpollen Design*

To improve the detection accuracy of the model and reduce its computation, complexity, and cost for hardware deployment [11], we designed YOLOv8-Pearpollen based on YOLOv8n. Its detailed structure is shown in Figure 4. Relative to YOLOv8, the following optimizations were made for YOLOv8-Pearpollen: (1) The lightweight design of knowledge distillation was added, offline distillation was chosen for migrating the knowledge of the pretrained teacher model to the student network [20], and the distillation method that combines feature and logical distillation was adopted to distill the feature layer in the model and the probability distribution of the model output so that the model's performance would be close to or reach the performance of a model with small volume, improving its performance with small volumes. The performance of a large-volume model was used as the basis to improve the efficiency of the proposed model. (2) The lightweight design of model pruning was incorporated, and the method of sparse group Lasso was used through the efficient coordinate descent method algorithm to achieve rapid convergence even when dealing with large datasets [21]. The iterative optimization of the coefficients within each cluster ensured model accuracy while maintaining computational efficiency. The features of the model were automatically selected to effectively cut out unimportant feature coefficients, thus pruning the model to improve its generalization performance and enhance its interpretability, which is invaluable for practical applications.

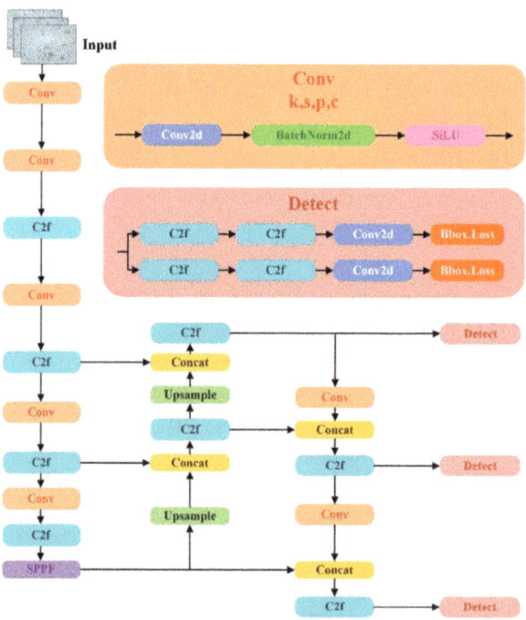


**Figure 4.** Structure of YOLOv8-Pearpollen.

### 2.3.1. Knowledge Distillation

A complex detection structure is usually required for the accurate and precise detection of small targets such as pear tree pollen. This structure must be specially designed to capture the fine features of small targets, leading to an increase in the complexity and volume of the model. To address these difficulties, this study introduced knowledge distillation for improving the small target detection performance of YOLOv8-Pearpollen. Knowledge distillation is a deep learning technique that migrates the knowledge from a large and complex model ("teacher" model) to a simple and small model ("student" model) [22]. We trained the dataset for 400 rounds using the YOLOv8l large-volume model to obtain a voluminous but accurate teacher model. YOLOv8n was then used as the basis for the student model and trained it by knowledge distillation to learn the detection effect of the teacher model. In terms of distillation methods, knowledge distillation is divided into logic or feature distillation. The soft labels of the teacher model and the manually labeled real labels were used to cotrain the student model. The soft labels contain information about the relationships between the different categories, not just the hard labels (the final classification results). Meanwhile, the soft output of the teacher model served as a regularization to help generalize the student model to unseen data and allow it to learn the subtle patterns that the teacher model learned on the training data.

Logic distillation utilizes the outputs prior to the last fully connected layer of the model network as a carrier of knowledge, allowing the student model to learn the nuances of the teacher model's recognition of categories [23]. Rather than just the final probabilistic outputs or categorical labels, the output distributions of both models are also matched. The complexity of the model is controlled, and overfitting is prevented by introducing L1 regularization. The total training loss of the logic distillation process ($Loss_{logical}$) is shown in Equation (3), where $Loss_{distill}$ is the base loss function of the distilled model, and $\lambda$ is a nonnegative regularization hyperparameter. As the value of $\lambda$ increases, the impact of the regularization term on the total loss also increases, thus making the model sparser. Conversely, the smaller the value of $\lambda$, the smaller the regularization term's contribution to the total loss, and the complexity of the model increases accordingly. The weight vector is w. Meanwhile, L1 regularization limits the size of the model parameters by adding the absolute values of the weight vectors to the loss function, thus inducing sparse feature learning in the model [24].

$$Loss_{logical} = Loss_{distill} + \lambda \sum_{i=1}^{n} |w_i| \quad (3)$$

Different from logic distillation that focuses on knowledge transfer in the output layer of the model, feature distillation concentrates on the transfer of feature representations in the internal layers (e.g., convolutional layers). By minimizing the differences between the feature representations of the student and teacher models on these internal layers, the student model can learn the rich representations and the high-level abstraction capabilities of the teacher model. The principle of feature distillation is illustrated in Figure 5. Channel-wise distillation [25] focuses on the channel level of the model's internal features. For the feature maps of the student and teacher models, we used the softmax function to obtain the normalized probability distributions. Equation (4) calculates the distillation loss in the channel direction, where $T$ and $S$ are the teacher and student models, respectively, and $y^T$ and $y^S$ are their corresponding activation maps. Converting the activation values to probability distributions aims to make the feature maps of the student model as close as possible to those of the teacher model in the channel dimension and to realize a fine-grained and efficient knowledge distillation process as shown in Equation (5), where $c = 1, 2, \ldots C$ is the channel, and $i$ is the spatial location of the channel. The temperature parameter $\mathcal{T}$ is introduced to regulate the smoothness of the softmax function and control the "softness" of the output distribution. A high $\mathcal{T}$ leads to a flat distribution, and a low value leads to a sharp distribution [26].

$$L_{CWD} = \varphi\left(\phi\left(y^T\right), \phi\left(y^S\right)\right) = \varphi\left(\phi\left(y_c^T\right), \phi\left(y_c^S\right)\right) \tag{4}$$

$$\phi(y_c) = \frac{exp\left(\frac{y_{c,i}}{\mathcal{T}}\right)}{\sum_{i=1}^{W \cdot H} exp\left(\frac{y_{c,i}}{\mathcal{T}}\right)} \tag{5}$$

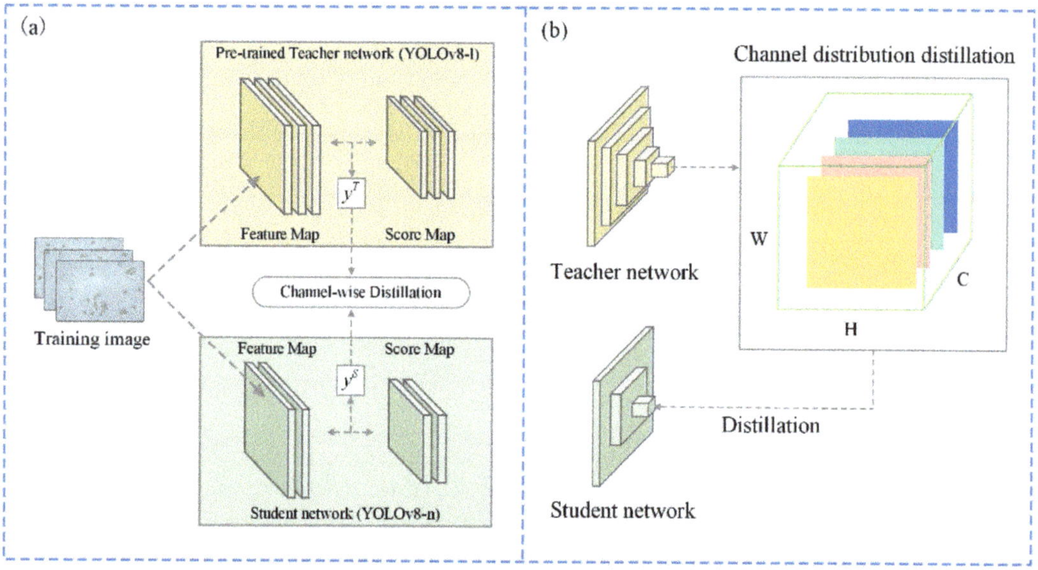


**Figure 5.** (**a**) Structural map of feature distillation. (**b**) Aligning each channel of the student feature map with the channels of the teacher network by minimizing KL divergence.

Knowledge distillation can be applied by determining the Kullback–Leibler (KL) divergence between the feature maps of the teacher and student models [27]. The difference between the outputs of the student and teacher networks was quantified using the KL scatter, which guides the student network to adjust its parameters to mimic the behavior of the teacher network, and was calculated as shown in Equation (6), where $C$ denotes the total number of channels, and $W$ and $H$ denote the width and height of the feature map, respectively. Scatter calculation was performed for each channel and the scatter values from all spatial locations in each channel were summed to ensure that every spatial location within each channel is taken into account. This process helps accurately align the feature responses of the teacher and student networks at the spatial locations and ensures that the student network learns as comprehensively as possible from the teacher network's knowledge [28].

$$\varphi\left(y^T, y^S\right) = \frac{\mathcal{T}^2}{C} \sum_{c=1}^{C} \sum_{i=1}^{W \cdot H} \phi\left(y_{c,i}^T\right) \cdot log\left[\frac{\phi\left(y_{c,i}^T\right)}{\phi\left(y_{c,i}^S\right)}\right] \tag{6}$$

The total training loss of the student model of the knowledge distillation process ($Loss_{Distillation}$) is shown in Equation (7) and consists of two components, $Loss_{feature}$ and

$L_{CWD}$. $Loss_{feature}$ represents the loss of the student model when performing the instance segmentation task and $\alpha$ is an adjustable hyperparameter.

$$Loss_{Distillation} = Loss_{feature} + \alpha L_{CWD} \tag{7}$$

In this experiment, we used a combination of feature and logical distillation to enable the student model to simultaneously acquire knowledge from the internal representation of the teacher's model and the output decision-making level, realizing a comprehensive knowledge transfer. The student model not only learns the final predictions of the teacher model but also understands the abstract feature processing behind arriving at those predictions. Hence, it can further understand the essential features of the input data, improve generalization ability, and achieve or approach the performance of the teacher model while maintaining a small model size. For model deployment under limited resources, such as in mobile devices and embedded systems, the model size and computational requirements can be drastically reduced while maintaining predictive performance, providing great flexibility for a wide range of tasks and model architectures.

### 2.3.2. Model Pruning

The distilled YOLOv8-Pearpollen model was structurally pruned to eliminate the nonessential parts of the model and reduce its complexity while keeping it as accurate as possible.

The sparse group-lasso [29] achieves sparsity at the group and individual feature levels by combining the penalty terms of lasso (L1 regularization) [30] and group-lasso (L2 regularization) [31]. This approach selects important feature clusters and performs feature selection within each selected cluster, further reducing model complexity and improving model interpretability [32]. As shown in Equation (8), $X_\ell$ and $\beta_\ell$ denote the model matrix and coefficient vector of the $l$th cluster, respectively, and $\lambda_1$ and $\lambda_2$ are the regularization parameters that control the strength of the group-lasso and lasso penalties. Equation (9) is the objective function to minimize the sum of squares of the differences between the model-predicted and actual values [33]. Equation (10) is the group-lasso penalty, which applies the L2 paradigm to the coefficient vector $\beta_\ell$ of each group and adjusts the effect of group size by $\sqrt{p_l}$ to achieve group-level sparsity, i.e., the entire group can be selected in or excluded. Equation (11) is the lasso penalty, which applies the L1 paradigm to all coefficients to achieve sparsity at the individual feature level, i.e., only a fraction of the important predictor variables are included in the model. $\lambda_1$ and $\lambda_2$ are the regularization parameters that control the strength of the group-lasso and lasso penalties. A balance between model complexity and sparsity can be achieved by tuning these parameters. The generation of sparsity was changed to the group and individual feature levels by group-lasso to accurately select predictors between and within groups.

$$min_{\beta\in R^p}\left(\| y - \sum_{\ell=1}^{L} X_\ell\beta_\ell \|_2^2 + \lambda_1 \sum_{\ell=1}^{L} \sqrt{p_l}\| \beta_\ell \|_2 + \lambda_2 \| \beta \|_1\right) \tag{8}$$

$$\| y - \sum_{\ell=1}^{L} X_\ell\beta_\ell \|_2^2 \tag{9}$$

$$\lambda_1 \sum_{\ell=1}^{L} \sqrt{p_l}\| \beta_\ell \|_2 \tag{10}$$

$$\lambda_2 \| \beta \|_1 \tag{11}$$

The sparse group-lasso can be calculated from the coordinate descent by choosing an initial coefficient vector $\beta_0$, which is usually a zero vector or any estimate based on a priori knowledge. For each group $l$, the residuals are shown in Equation (12), i.e., the target value minus the current predicted values for all other clusters excluding group $l$. The predictors were denoted as $X_\ell$ = Z = ($Z_1$, $Z_2$, ..., $Z_k$), and the coefficients were denoted as $\beta_\ell$=$\theta$= ($\theta_1$,

$\theta_2, \ldots, \theta_k$). For each feature $j$ within the group, $w_j = (w_1, w_2, \ldots, w_N) = r - \sum_{k \neq j} Z_k \theta_k$ is the updated residual, excluding the contribution of the current feature $j$. If $\left|Z_j^T w_j\right|$ is less than $\lambda_2$, then it will be directly set to 0, otherwise, $\theta_j$ will be set to the minimized value of Equation (13). If the whole coefficient vector is zero is not satisfied, then each coefficient must be optimized and the steps must be repeated until the coefficients of all groups reach the convergence condition [34]. Only one coefficient or one group of coefficients must be updated at a time, and the global optimal solution can be obtained through stepwise approximation. The key advantage of the coordinate descent method is its high efficiency, especially when dealing with high-dimensional data with a group structure. It enables fast and accurate feature selection, thus enhancing the explanatory and predictive performance of the model.

$$r_\ell = y - \sum_{k \neq \ell} X_k \hat{\beta}_k \tag{12}$$

$$\frac{1}{2} \sum_{i=1}^{N} \left(w_i - \sum_{j=1}^{k} Z_{ij} \theta_j\right)^2 + \lambda_1 \parallel \theta \parallel_2 + \lambda_2 \sum_{j=1}^{k} |\theta_j| \tag{13}$$

## 3. Results and Discussion

### *3.1. Training Environment and Hyperparameter Settings*

The hardware configuration for this experiment comprised an Intel(R) Xeon(R) Gold 6248R @ 3.00 GHz processor, an NVIDIA GeForce RTX3090 graphics card, and an operating system of Windows 11. The environment configuration for deep learning consisted of Python 3.11, Pytorch 2.1.0, CUDA 12.2, and Torchvision 0.16.0. The important hyperparameter settings during the training are shown in Table 1. No pretraining weights were used in any part of this experiment to ensure the fairness and effectiveness of the detection model for pear pollen germination vigor.

**Table 1.** Model hyperparameter setting.

| Parameters | Setup |
|---|---|
| Epoch | 150 |
| Batch size | 32 |
| Optimizer | SGD |
| Close_mosaic | 20 |
| Cache | True |
| Image size | 640 × 640 |
| Initial learning rate | $1 \times 10^{-2}$ |
| Final learning rate | $1 \times 10^{-4}$ |
| Momentum | 0.937 |
| Weight-decay | $5 \times 10^{-4}$ |
| Warmup-epochs | 3 |

### *3.2. Ablation Tests*

To test the feasibility and effectiveness of the detection model for pear pollen germination vigor, we designed several sets of ablation experiments, evaluated the degree of contribution of each module to the model performance, and excluded and removed the modules that had a small or negative effect on the final results to make the model concise and efficient.

First, in the design of knowledge distillation, three types of distillation (feature, logical, and hybrid) were set. For feature distillation, the following three types were chosen: channel-wise distillation (CWD), masked generative distillation (MDG), and mimic. For logical distillation, the following three types were selected: L1, L2, and BCKD. The corresponding loss rates $\lambda$ and $\alpha$ were set. The loss rate quantified the difference between the predictions of the student and teacher models. It must be changed to adjust the weight between the lost items. Adjustments must be made to the similarity between the outputs of the intermediate layers of the student and teacher models in terms of shape, direction, or distribution and the difference between the probability distribution of the outputs of the student and teacher models to ensure that the student model can mimic the teacher model to deal with internal mechanism data and thus learn an efficient and sophisticated representation of the data. After the above parameters were changed and the accuracy and efficiency of the model were adjusted, the results of the ablation test for knowledge distillation were compared as shown in Table 2. In terms of feature distillation, CWD and MDG were generally better than the other types. Meanwhile, the results of the mimic method for several data were different from those of the above two methods. Even if the loss rate was adjusted, the results were still not satisfactory. In terms of logical distillation, a general improvement was observed over feature distillation. The $AP_{not\ sprout}$, mAP@0.5, precision, and recall were comparable with those of feature distillation, but the $AP_{sprout}$ was relatively lower. When feature and logical distillation were combined, the data became stable and improved compared with using feature or logical distillation alone. Logical distillation l1 type (loss rate 0.8) and characteristic distillation CWD (loss rate 0.8) stood out among the 17 compared methods, showing the highest values for mAP@0.5, $AP_{sprout}$, and $AP_{not\ sprout}$ at 97.6%, 97.6%, and 97.7%, respectively, relative to the baseline models. YOLOv8-n (without distillation) showed improvement in all data, verifying the excellent effect of knowledge distillation on the small-volume model's performance and accuracy improvement.

**Table 2.** Comparison of test results for knowledge distillation ablation.

| Models | mAP@0.5 (%) | $AP_{sprout}$ (%) | $AP_{not\ sprout}$ (%) | Precision (%) | Recall (%) | FPS |
|---|---|---|---|---|---|---|
| YOLOv8-n | 97.0 | 97.0 | 97.1 | 91.4 | 94.2 | 149.3 |
| feature cwd loss1.0 | 97.2 | 97.6 | 96.8 | 94.0 | 92.6 | 142.9 |
| feature cwd loss0.8 | 97.3 | 97.1 | 97.4 | 93.9 | 93.6 | 138.9 |
| feature mdg loss1.0 | 97.2 | 97.0 | 97.4 | 93.4 | 94.5 | 147.1 |
| feature mimic loss1.0 | 96.8 | 96.7 | 96.9 | 92.7 | 92.7 | 117.6 |
| feature mimic loss0.8 | 96.6 | 96.2 | 97.1 | 93.6 | 92.6 | 128.2 |
| logical l1 loss1.0 | 97.1 | 97.1 | 97.2 | 93.7 | 94.5 | 140.8 |
| logical l1 loss0.8 | 96.8 | 96.3 | 97.2 | 93.1 | 93.5 | 151.1 |
| logical l2 loss1.0 | 97.1 | 96.8 | 97.5 | 93.5 | 93.8 | 151.1 |
| logical l2 loss0.8 | 97.6 | 97.5 | 97.6 | 93.9 | 94.2 | 144.9 |
| logical BCKD loss1.0 | 97.1 | 96.6 | 97.5 | 92.6 | 94.5 | 142.9 |
| logical BCKD loss0.8 | 97.5 | 97.5 | 97.5 | 92.4 | 95.3 | 153.8 |
| all l2 loss1.0 cwd loss1.0 | 97.2 | 97.3 | 97.1 | 93.9 | 94.4 | 137.0 |
| all l2 loss0.8 cwd loss0.8 | 97.4 | 97.5 | 97.4 | 94.3 | 94.5 | 153.8 |
| all l1 loss1.0 cwd loss1.0 | 97.2 | 96.9 | 97.5 | 93.7 | 94.3 | 151.5 |
| all l1 loss0.9 cwd loss0.9 | 97.3 | 97.4 | 97.2 | 94.2 | 94.2 | 125.0 |
| all l1 loss0.8 cwd loss0.8 | 97.6 | 97.6 | 97.7 | 94.0 | 94.7 | 151.5 |

To intuitively reflect the superiority of our choice of knowledge distillation method, we drew a hexagonal radar chart using the normalization method to compare the results of the six parameters as shown in Figure 6 below. Through min–max normalization, the original data were linearly transformed so that the results of the processed data were between [0, 1]. The closer the value to 1, the more effective the model; the closer the value to 0, the worse the effect. The transformation formula is shown in Equation (14), where $X_{max}$ is the maximum value of the data, $X_{min}$ is the minimum value of the data, and $R_A$ is the converted ratio of the difference between the maximum and minimum values of the current indicator. The larger the true value, the higher the normalized score on this item.

$$R_A = \frac{X - X_{min}}{X_{max} - X_{min}} \tag{14}$$

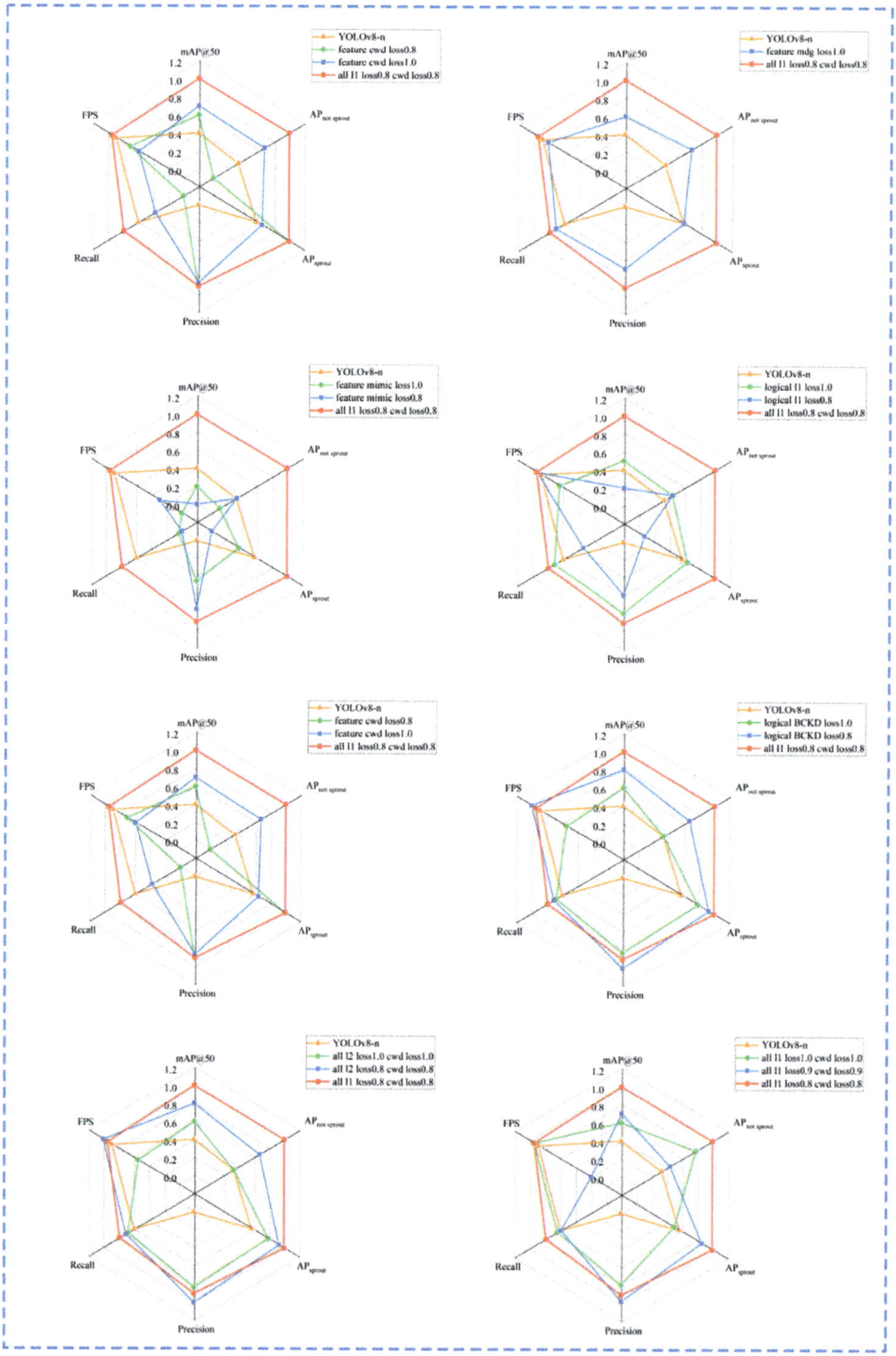


**Figure 6.** Normalized analysis of multiple indicators of knowledge distillation.

To further explore the effect of feature layer selection in feature distillation, we took the model with the best results and designed an ablation test using this model to determine the effect of knowledge distillation by changing the number of distillation layers under the same parameters. The results are shown in Table 3. When using the feature vectors of the 15th, 18th, and 21st layers as pseudo labels for distillation, the mAP@0.5, $AP_{sprout}$, and $AP_{not\ sprout}$ were improved to 97.6%, 97.6%, and 97.7%, respectively, and the precision and recall were improved to 94.0% and 94.7%, respectively, relative to those in the baseline model, YOLOv8-n (which does not use distillation). When the number of distillation layers was increased to six (layers 6, 8, 12, 15, 18, and 21), the $AP_{not\ sprout}$ of the model improved to 97.7% compared with the use of three layers (layers 15, 18, and 21). Meanwhile, the mAP@0.5 and $AP_{sprout}$ decreased to 97.3% and 96.9%, respectively. Although the recall slightly improved to 95.2%, the precision decreased by one percentage point to 92.9%. When the number of distillation layers was further raised to eight, the $AP_{not\ sprout}$ of the model did not change, and the mAP@0.5 and $AP_{sprout}$ continued to drop to 97.0% and 96.3%, respectively. These findings demonstrate that choosing the appropriate number of feature layers is the key to optimizing feature distillation.

**Table 3.** Comparison of ablation test results with the same parameters and changing the number of distillation layers.

| Number of Distillation Layers | mAP@0.5 (%) | APsprout (%) | APnot Sprout (%) | Precision (%) | Recall (%) |
|---|---|---|---|---|---|
| none | 97.0 | 97.0 | 97.1 | 91.4 | 94.2 |
| 15, 18, 21 | 97.6 | 97.6 | 97.7 | 94.0 | 94.7 |
| 6, 8, 12, 15, 18, 21 | 97.3 | 96.9 | 97.7 | 92.9 | 95.2 |
| 2, 4, 6, 8, 12, 15, 18, 21 | 97.0 | 96.3 | 97.7 | 94.1 | 93.9 |

Second, in terms of model pruning design, different pruning methods were adopted. The effect of global pruning and local pruning on precision, Params, and FLOPs was examined, and the results of the ablation test for model pruning were compared as shown in Tables 4 and 5. Four model pruning methods were selected for comparison; namely, lamp [35], groupnorm [36], grouptaylor [37], and groupsilm [38]. Although the lamp method greatly reduced the Params and weights, its mAP@0.5, $AP_{sprout}$, and $AP_{not\ sprout}$ were slightly inferior to those grouptaylor and groupsilm. The mAP@0.5, $AP_{sprout}$, and $AP_{not\ sprout}$ of groupnorm were lower than those of the other methods. Especially in the detection of sprouting pollen, its effect was poor. Grouptaylor had a clear advantage in detection accuracy, showing the highest mAP@0.5, $AP_{sprout}$, and $AP_{not\ sprout}$. However, its frame per second (FPS) was significantly reduced, and the model inference speed was slow. Meanwhile, the Params, FLOPs, and weights of groupsilm showed decreases of 50%, 50%, and 48%, respectively, and its accuracy exhibited minimal reduction. Groupsilm was effective in the detection of sprouted pollen, but its accuracy for detecting unsprouted pollen was low. For global pruning, which calculates the proportion of the model's pruning according to all the parameters, we found that the model's pruning was more effective than that of groupsilm. Global pruning, which uses all parameters to calculate the proportion of model clippings, showed a slight improvement in accuracy over local pruning by calculating clippings according to the parameters of each layer. However, its lightweighting effect was significantly inferior to that of local pruning. With the increase in the pruning rate, the Params and FLOPs of the model decreased and its detection performance decreased significantly. Model pruning can effectively and simultaneously reduce the model size and computational cost, ensure the detection accuracy of the model and effectively optimize its detection efficiency, fully utilize the computational resources, reduce the model complexity in the hardware deployment of the threshold, and enrich the application of the model scene.

Table 4. Analysis of ablation test results of model pruning methods.

| Pruning Method | mAP@0.5 (%) | APsprout (%) | APnot Sprout (%) | Params (M) | FLOPs (G) | Weights (MB) | FPS |
|---|---|---|---|---|---|---|---|
| none | 97.6 | 97.5 | 97.6 | 3.0 | 8.1 | 6.0 | 151.5 |
| lamp | 96.3 | 95.8 | 96.8 | 0.8 | 4.0 | 1.7 | 147.1 |
| groupnorm | 92.4 | 89.2 | 95.6 | 1.2 | 4.0 | 2.6 | 156.3 |
| grouptaylor | 97.3 | 97.7 | 96.8 | 1.6 | 4.0 | 3.2 | 122.0 |
| groupsilm | 96.7 | 96.4 | 97.1 | 1.5 | 4.0 | 3.1 | 147.1 |

Table 5. Analysis of the ablation test results of different model pruning rates with the same methodology.

| Pruning Rate | mAP@0.5 (%) | APsprout (%) | APnot Sprout (%) | Params (M) | FLOPs (G) | Weights (MB) | FPS |
|---|---|---|---|---|---|---|---|
| 1.5 (global) | 97.6 | 97.3 | 97.8 | 2.7 | 5.4 | 5.3 | 144.9 |
| 2.0 (global) | 96.9 | 96.7 | 97.2 | 2.4 | 4.0 | 4.8 | 142.9 |
| 2.0 (nonglobal) | 96.7 | 96.4 | 97.1 | 1.5 | 4.0 | 3.1 | 147.1 |
| 2.5 (global) | 95.1 | 94.5 | 95.7 | 2.1 | 3.2 | 4.3 | 144.9 |
| 3.0 (global) | 93.7 | 91.7 | 95.6 | 1.7 | 2.7 | 3.4 | 128.2 |

The designed detection model for pear pollen germination vigor was compared with the YOLOv8-n benchmark detection model, and the experimental results are demonstrated in Table 6. After the model was subjected to knowledge distillation, its mAP@0.5, APsprout, and APnot sprout improved to 97.6%, 97.5%, and 97.6%, respectively. After the model was pruned, its Params, FLOPs, and weights were further reduced by 50%, 50.6%, and 48.3% respectively, relative to their original values. Meanwhile, its FPS reached 147.1, which was a reduction of 4.4 compared with that of the unpruned one. In terms of accuracy, the APnot sprout only decreased by 0.6%, and the mAP@0.5 and APsprout decreased by only 0.9% and 1.2%, respectively. The ablation test demonstrated the effectiveness of the detection model for pear pollen germination viability in terms of lightweighting and the superiority of its accuracy, further proving the feasibility of this experimental design for pear pollen detection.

Table 6. Feature distillation and structured pruning validity analysis.

| Models | mAP@0.5 (%) | APsprout (%) | APnot Sprout (%) | Params (M) | FLOPs (G) | Weights (MB) | FPS |
|---|---|---|---|---|---|---|---|
| YOLOv8-n | 97.0 | 97.0 | 97.1 | 3.0 | 8.1 | 6.0 | 149.3 |
| +distill | 97.6 | 97.6 | 97.7 | 3.0 | 8.1 | 6.0 | 151.5 |
| +distill +prune | 96.7 | 96.4 | 97.1 | 1.5 | 4.0 | 3.1 | 147.1 |

### 3.3. Comparative Tests

YOLOv8-Pearpollen was compared with multigeneration models from YOLOv3 to YOLOv8 to further prove its superiority for the detection of pear pollen germination vigor. The following outcome indicators were evaluated: mAP50 (which includes the accuracy of the single identification of pollen germination and pollen nongermination results), Params, FLOPs, and weights. The results are shown in Table 7. Compared with those of the other models, the Params, FLOPs, and weights of YOLOv8-Pearpollen were significantly lower at 1.5 M, 4.0 G, and 3.1 MB, respectively. For YOLOv3, which has a high accuracy, its Params, FLOPs, and weights decreased by 98.6%, 98.6%, and 98.4%, respectively, relative to those of the model in agriculture and by 75.0%, 69.2%, and 73.5%, respectively, compared with those of YOLOv7-tiny, a commonly used target detection model in agriculture. In terms of detection accuracy, except for the large-volume YOLOv3, YOLOv8-Pearpollen had the best $AP_{not\ sprout}$, which was 0.3%, 1.0%, and 0.3% higher than that of YOLOv5-n, YOLOV6-n, and YOLOv7-tiny, respectively, and its mAP@0.5 and $AP_{sprout}$ were also higher than those of YOLOv5-n and YOLOv7-tiny. The comparison test proved that the YOLOv8-Pearpollen model designed in this experiment has outstanding performance and can maintain superior detection accuracy while reducing the computational complexity. It has advantages over other models, is suitable and easy to deploy in hardware devices (such as embedded devices and mobile devices) to satisfy the actual needs of agricultural production, and can accurately detect the pollen germination vigor of pear trees on devices in real time.

**Table 7.** Comparison of detection results of multiple models.

| Models | mAP@0.5 (%) | APsprout (%) | APnot Sprout (%) | Params (M) | FLOPs (G) | Weights (MB) | FPS |
|---|---|---|---|---|---|---|---|
| YOLOv3 | 98.3 | 98.0 | 98.6 | 103.7 | 282.2 | 198.1 | 102.0 |
| YOLOv5-n | 96.2 | 95.6 | 96.8 | 2.5 | 7.1 | 5.0 | 142.9 |
| YOLOv6-n | 94.2 | 92.2 | 96.1 | 4.2 | 11.8 | 8.3 | 158.7 |
| YOLOv7-tiny | 95.2 | 93.7 | 96.8 | 6.0 | 13.0 | 11.7 | 90.0 |
| YOLOv8-n | 97.0 | 97.0 | 97.1 | 3.0 | 8.1 | 6.0 | 149.3 |
| YOLOv8-Pearpollen | 96.7 | 96.4 | 97.1 | 1.5 | 4.0 | 3.1 | 147.1 |

To further demonstrate the superiority of YOLOv8-Pearpollen's detection effect over that of the other models, we plotted its multimetric normalized histogram as shown in Figure 7. The conversion formulas for the three metrics of model parameter count, computation, and weight file size are shown in Equation (15). Different from Equation (13), RB is the ratio of the maximum value of the current indicator minus the value of the indicator that needs to be converted to the difference between the maximum and minimum values of the current indicator. The smaller the true value, the higher the normalized score on this item, fitting the correct comparison of the model parametric quantity, computational quantity, and weight file size indicators. YOLOv8-Pearpollen reaches the highest level in four data items, namely, the number of model parameters, amount of computation, size of the weight file, and $AP_{not\ sprout}$. It is also in the forefront in terms of mAP and $AP_{sprout}$, and its overall performance reaches the optimal level, far ahead of the other models.

$$R_B = \frac{X_{max} - X}{X_{max} - X_{min}} \tag{15}$$

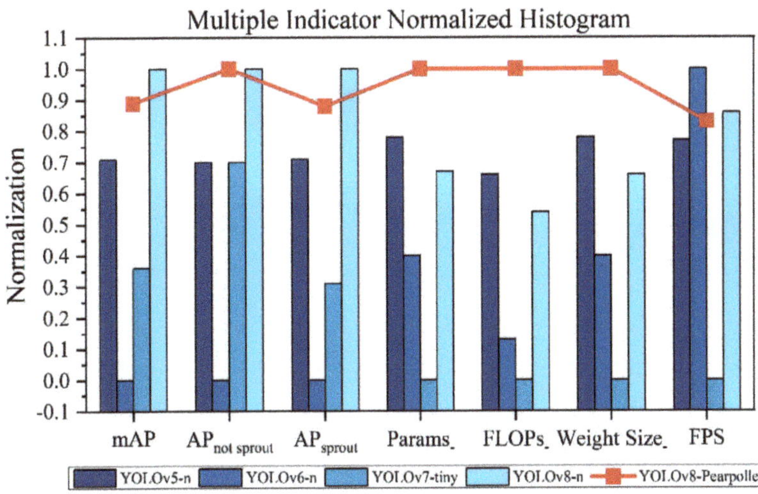


**Figure 7.** Normalized analysis of multiple model detection results.

### 3.4. Evaluation of Model Prediction Performance

We analyzed the recognition effect of YOLOv8-Pearpollen compared with three lightweight models, YOLOv5-n, YOLOv6-n, and YOLOv7-tiny, for pear tree pollen germination. Six images were randomly selected from 1800 images for the comparison test. The detection comparison results are shown in Figure 8 and Table 8. The pink box area in the figure indicates the identified sprouted pear pollen, the red box area indicates the identified not sprouted pear pollen, and the green circle, yellow circle, and blue circle indicate the missed detection, repeated detection, and wrong detection, respectively.

The results show that YOLOv5-n and YOLOv6-n often had repeated detection conditions, especially in pollen-dense or superimposed scenarios. Meanwhile, YOLOv7-tiny showed a certain degree of missed detection, which was rare in the detection results of the other models. Overall, YOLOv8-Pearpollen had almost no repeat detection, which is a huge improvement over the other four models, and accurately recognized pear pollen germination in complex recognition scenarios.

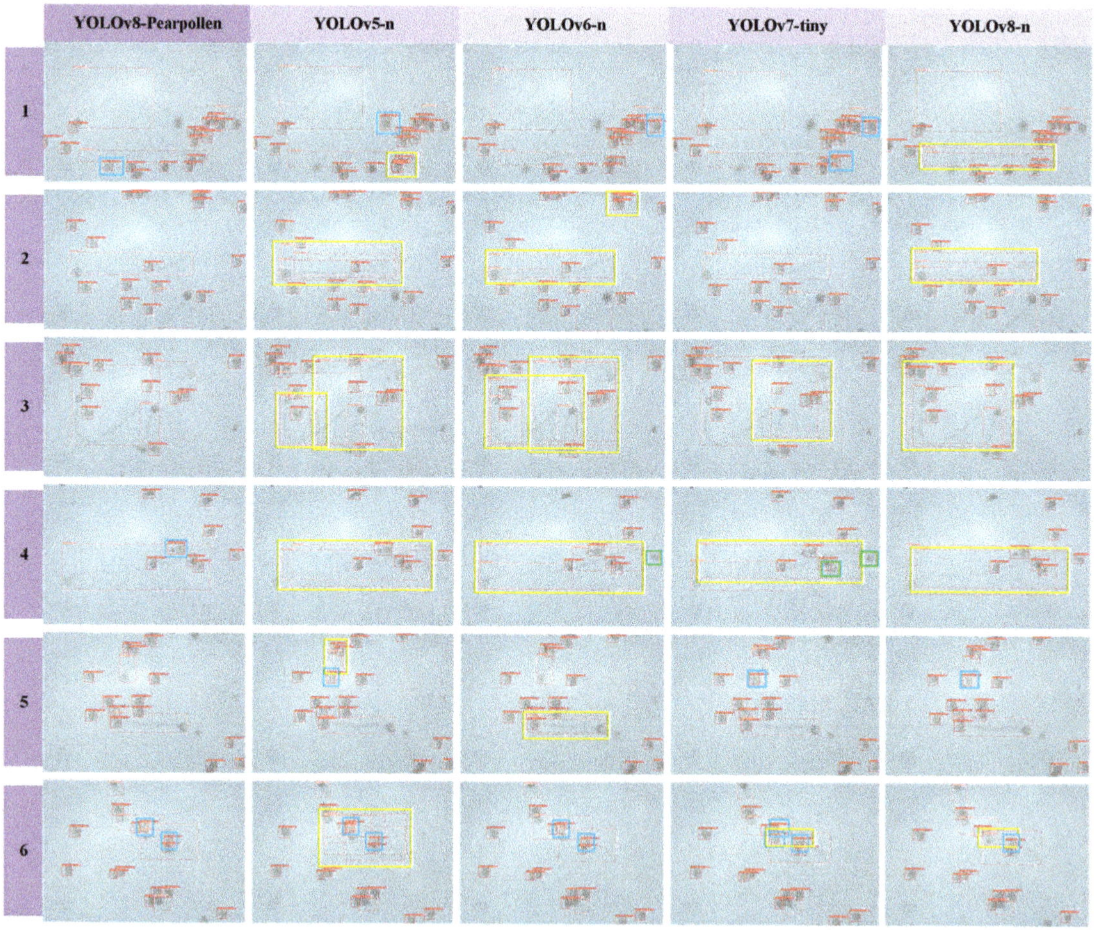


**Figure 8.** Comparison of the detection results of YOLOv8-Pearpollen, YOLOv5-n, and YOLOv6-n, YOLOv7-tiny, and YOLOv8-n.

**Table 8.** Statistical table comparing the detection results of different models.

| Models | Wrong Detection | Repeated Detection | Missed Detection |
|---|---|---|---|
| YOLOv5-n | 4 | 7 | 0 |
| YOLOv6-n | 3 | 6 | 1 |
| YOLOv7-tiny | 6 | 3 | 2 |
| YOLOv8-n | 2 | 5 | 0 |
| YOLOv8-Pearpollen | 4 | 0 | 0 |

## 4. Conclusions

Traditional manual pear tree pollen detection methods are cumbersome and have low efficiency, insufficient accuracy, and high labor costs. Screening high-quality male plant pollen is necessary to improve the success rate of artificial pollination for pear tree pollen. In addition, the cost of deploying the model on hardware must be minimized to further adapt to the requirements of actual agricultural production. To address the above problems, this study proposed a lightweight detection model for pear pollen germination named YOLOv8-Pearpollen. The following are the accomplishments of this work:

(1) A pear tree pollen dataset was collected from several trials, and the original image data were enhanced using Blend Alpha to improve the robustness of the detection method, optimize its detection effect against a complex environmental background, and reduce the labor and time costs required for image collection.

(2) Two designs of knowledge distillation and model pruning were used for the model to improve its lightness of weight, reduce its deployment cost on hardware equipment, and adapt to the actual needs of agricultural production. In terms of knowledge distillation, we chose to use a combination of logical and feature distillation so that the student model could learn from the teacher model and the internal layer and output decision-making to achieve or approach the detection effect of the large-volume teacher model. A number of ablation tests were carried out on different knowledge distillation methods to select the best method, and the detection results were compared under different numbers of distillation layers but the same method to determine effect of the number of distillation layers on the distillation effect. The structured pruning method of sparse-group lasso was used for model pruning to reduce the number of model calculations and maintain the detection performance of the model as much as possible. A variety of model pruning methods were also tested. The detection effects of different pruning rates of the same method were analyzed to obtain the best model pruning effect and ensure that the model has a high degree of lightweighting while showing good detection accuracy.

(3) YOLOv8-Pearpollen was compared with YOLOv3, YOLOv5-n, YOLOV6-n, YOLOv7-tiny, and YOLOv8-n detection models in terms of detection accuracy and model complexity. The results were statistically analyzed using the normalization method so that the strengths and weaknesses of the different models in various indicators are intuitively derived and to prove the effectiveness and superiority of the proposed method.

(4) The test results showed that YOLOv8-Pearpollen achieved a mAP of 96.7%. However, its number of model parameters, amount of computation, and size of the weight file were only 1.5 M, 4.0 G, and 3.1 MB, respectively, which were 50.0%, 50.6%, and 48.3% lower than those of YOLOv8, respectively. It realized a detection speed of 147.1 FPS. Compared with other models, YOLOv8-Pearpollen achieved a high degree of lightweighting while possessing superior detection accuracy, which is suitable for deployment on hardware models and meets the actual needs of agricultural production.

The proposed method still has some limitations. The detection of overlapping or stacked pear pollen is prone to erroneous detection caused by overlapping images that make the model unable to determine which pollen tube belongs to which pollen. Semantic segmentation can be considered to annotate the image data, helping the model to detect the pollen of sprouting pear trees.

We hope that our research on the lightweight identification and detection of pollen germination vigor in pear trees can provide effective methods and valuable references for the screening of high-quality male pollen. The findings will help improve the success rate of artificial pollination and thus increase the fruiting rate of pear trees. The lightweighting degree of the model must be further enhanced so it can be deployed on hardware facilities to promote the practical agriculture production of fruit trees.

**Author Contributions:** Conceptualization, W.S. and C.C.; methodology, W.S. and C.C.; software, W.S. and C.C.; validation, W.S., C.C., T.L. and F.H.; formal analysis, W.S., C.C. and H.J.; investigation, W.S., C.C. and S.H.; resources, M.N.; data curation, W.S. and C.C.; writing—original draft preparation, W.S. and C.C.; writing—review and editing, F.H. and X.F.; visualization, W.S. and C.C.; supervision, H.J. and L.T.; project administration, W.S. and C.C.; funding acquisition, X.F. All authors have read and agreed to the published version of the manuscript.

**Funding:** This work was supported by Jiangsu Province Innovation Support Plan (Rural Industrial Revitalization) (Grant number SZ-SY20231103), Major Science and Technology Projects of Xinjiang Academy of Agricultural and Reclamation Sciences(Grant number NCG202407), the Jiangsu Agriculture Science and Technology Innovation Fund (JASTIF) (Grant number (CX(23)3619), Hainan Seed Industry Laboratory (Grant number B21HJ1005), and Jiangsu Province Seed Industry Revitalization Unveiled Project (Grant number JBGS(2021)007).

**Institutional Review Board Statement:** Not applicable.

**Data Availability Statement:** The data presented in this study are available on request from the corresponding author. The data are not publicly available due to possible further research.

**Conflicts of Interest:** The authors declare no conflicts of interest.

## References

1. Huete, A.; Tran, N.N.; Nguyen, H.; Xie, Q.; Katelaris, C. Forecasting pollen aerobiology with Modis EVI, land cover, and phenology using machine learning tools. In Proceedings of the 2019 IEEE International Geoscience and Remote Sensing Symposium (IGARSS 2019), Yokohama, Japan, 28 July–2 August 2019; pp. 5429–5432.
2. Agudelo, C.G.; Sanati Nezhad, A.; Ghanbari, M.; Naghavi, M.; Packirisamy, M.; Geitmann, A. TipChip: A modular, MEMS-based platform for experimentation and phenotyping of tip-growing cells. *Plant J.* **2013**, *73*, 1057–1068. [CrossRef]
3. Dechkrong, P.; Srima, S.; Nilwaranon, T.; Tongyoo, P.; de Jong, H.; Chunwongse, J. Morphological Characterization of Anther and Pollen Formation in an EMS Induced Tomato Mutant with Blossom Drop Phenotype. *Plant Biol. Crop Res.* **2020**, *1*, 1030.
4. Zhang, M.; Zhao, J.; Hoshino, Y. Deep learning-based high-throughput detection of in vitro germination to assess pollen viability from microscopic images. *J. Exp. Bot.* **2023**, *74*, 6551–6562. [CrossRef] [PubMed]
5. Punyasena, S.W.; Tcheng, D.K.; Wesseln, C.; Mueller, P.G. Classifying black and white spruce pollen using layered machine learning. *New Phytol.* **2012**, *196*, 937–944. [CrossRef] [PubMed]
6. Liu, X.; Wu, D.; Zewdie, G.K.; Wijerante, L.; Timms, C.I.; Riley, A.; Levetin, E.; Lary, D.J. Using machine learning to estimate atmospheric Ambrosia pollen concentrations in Tulsa, OK. *Environ. Health Insights* **2017**, *11*, 1178630217699399. [CrossRef] [PubMed]
7. Sobol, M.K.; Finkelstein, S.A. Predictive pollen-based biome modeling using machine learning. *PLoS ONE* **2018**, *13*, e0202214. [CrossRef] [PubMed]
8. Sobol, M.K.; Scott, L.; Finkelstein, S.A. Reconstructing past biomes states using machine learning and modern pollen assemblages: A case study from Southern Africa. *Quat. Sci. Rev.* **2019**, *212*, 1–17. [CrossRef]
9. Zewdie, G.K.; Lary, D.J.; Levetin, E.; Garuma, G.F. Applying deep neural networks and ensemble machine learning methods to forecast airborne ambrosia pollen. *Int. J. Environ. Res. Public Health* **2019**, *16*, 1992. [CrossRef] [PubMed]
10. Cordero, J.M.; Rojo, J.; Gutiérrez-Bustillo, A.M.; Narros, A.; Borge, R. Predicting the Olea pollen concentration with a machine learning algorithm ensemble. *Int. J. Biometeorol.* **2021**, *65*, 541–554. [CrossRef] [PubMed]
11. Yamazaki, A.; Takezawa, A.; Nagasaka, K.; Motoki, K.; Nishimura, K.; Nakano, R.; Nakazaki, T. A simple method for measuring pollen germination rate using machine learning. *Plant Reprod.* **2023**, *36*, 355–364. [CrossRef] [PubMed]
12. Tian, G.; Li, X.; Wu, Y.; Liu, A.; Zhang, Y.; Ma, Y.; Guo, W.; Sun, X.; Fu, B.; Li, D. Recognition effect of models based on different microscope objectives. In Proceedings of the 3rd International Symposium on Artificial Intelligence for Medicine Sciences, Amsterdam, The Netherlands, 13–15 October 2022; pp. 133–141.
13. Zewdie, G.K.; Lary, D.J.; Liu, X.; Wu, D.; Levetin, E. Estimating the daily pollen concentration in the atmosphere using machine learning and NEXRAD weather radar data. *Environ. Monit. Assess.* **2019**, *191*, 418. [CrossRef] [PubMed]

14. Tan, Z.; Yang, J.; Li, Q.; Su, F.; Yang, T.; Wang, W.; Aierxi, A.; Zhang, X.; Yang, W.; Kong, J.; et al. PollenDetect: An open-source pollen viability status recognition system based on deep learning neural networks. *Int. J. Mol. Sci.* **2022**, *23*, 13469. [CrossRef] [PubMed]
15. Wang, C.; Gui, C.P.; Liu, H.K.; Zhang, D.; Mosig, A. An Image Skeletonization-Based Tool for Pollen Tube Morphology Analysis and Phenotyping. *J. Integr. Plant Biol.* **2013**, *55*, 131–141. [CrossRef] [PubMed]
16. Gallardo, R.; García-Orellana, C.J.; González-Velasco, H.M.; García-Manso, A.; Tormo-Molina, R.; Macías-Macías, M.; Abengózar, E. Automated multifocus pollen detection using deep learning. *Multimed. Tools Appl.* **2024**, *83*, 72097–72112. [CrossRef]
17. Fu, X.; Bai, Y.; Zhou, J.; Zhang, H.; Xian, J. A method for obtaining field wheat freezing injury phenotype based on RGB camera and software control. *Plant Methods* **2021**, *17*, 120. [CrossRef] [PubMed]
18. Iang, H.; Hu, F.; Fu, X.; Chen, C.; Wang, C.; Tian, L.; Shi, Y. YOLOv8-Peas: A lightweight drought tolerance method for peas based on seed germination vigor. *Front. Plant Sci.* **2023**, *14*, 1257947.
19. Zewdie, G.K.; Liu, X.; Wu, D.; Lary, D.J.; Levetin, E. Applying machine learning to forecast daily Ambrosia pollen using environmental and NEXRAD parameters. *Environ. Monit. Assess.* **2019**, *191*, 261. [CrossRef]
20. Shu, C.; Liu, Y.; Gao, J.; Yan, Z.; Shen, C. Channel-wise knowledge distillation for dense prediction. In Proceedings of the IEEE/CVF International Conference on Computer Vision, Montreal, QC, Canada, 10–17 October 2021; pp. 5311–5320.
21. Behnke, M.; Heafield, K. Pruning neural machine translation for speed using group lasso. In Proceedings of the Sixth Conference on Machine Translation, Online, 10–11 November 2021; pp. 1074–1086.
22. Hinton, G.; Vinyals, O.; Dean, J. Distilling the knowledge in a neural network. *arXiv* **2015**, arXiv:1503.02531.
23. Wen, W.; Wu, C.; Wang, Y.; Chen, Y.; Li, H. Learning structured sparsity in deep neural networks. *Adv. Neural Inf. Process. Syst.* **2016**, *29*.
24. Zheng, Z.; Ye, R.; Wang, P.; Ren, D.; Zuo, W.; Hou, Q.; Cheng, M.M. Localization distillation for dense object detection. In Proceedings of the IEEE/CVF Conference on Computer Vision and Pattern Recognition, New Orleans, LA, USA, 18–24 June 2022; pp. 9407–9416.
25. Vidaurre, D.; Bielza, C.; Larranaga, P. A survey of L1 regression. *Int. Stat. Rev.* **2013**, *81*, 361–387. [CrossRef]
26. Cai, T.T.; Zhang, A.R.; Zhou, Y. Sparse group lasso: Optimal sample complexity, convergence rate, and statistical inference. *IEEE Trans. Inf. Theory* **2022**, *68*, 5975–6002. [CrossRef]
27. Fang, G.; Ma, X.; Song, M.; Mi, M.B.; Wang, X. Structure Level Pruning of Efficient Convolutional Neural Networks with Sparse Group LASSO. *Int. J. Mach. Learn. Comput.* **2022**, *12*, 16091–16101.
28. Fang, G.; Ma, X.; Song, M.; Mi, M.B.; Wang, X. Depgraph: Towards any structural pruning. In Proceedings of the IEEE/CVF Conference on Computer Vision and Pattern Recognition, Vancouver, BC, Canada, 17–24 June 2023; pp. 16091–16101.
29. Wu, C.; Pang, W.; Liu, H.; Lu, S. Group pruning with group sparse regularization for deep neural network compression. In Proceedings of the 2019 IEEE 4th International Conference on Signal and Image Processing (ICSIP), Wuxi, China, 19–21 July 2019; pp. 325–329.
30. Li, X.; Chen, L.; Gao, Z.; Zhang, X.; Wang, C.; Chen, H. Lasso regression based channel pruning for efficient object detection model. In Proceedings of the 2019 IEEE International Symposium on Broadband Multimedia Systems and Broadcasting (BMSB), Jeju, Republic of Korea, 5–7 June 2019; pp. 16–29.
31. Oyedotun, O.; Aouada, D.; Ottersten, B. Structured compression of deep neural networks with debiased elastic group lasso. In Proceedings of the IEEE/CVF Winter Conference on Applications of Computer Vision, Snowmass, CO, USA, 1–5 March 2020; pp. 2277–2286.
32. Xie, Z.; Li, P.; Li, F.; Guo, C. Pruning filters base on extending filter group lasso. *IEEE Access* **2020**, *8*, 217867–217876. [CrossRef]
33. Molchanov, P.; Mallya, A.; Tyree, S.; Frosio, I.; Kautz, J. Importance estimation for neural network pruning. In Proceedings of the IEEE/CVF Conference on Computer Vision and Pattern Recognition, Long Beach, CA, USA, 15–20 June 2019; pp. 11264–11272.
34. Lee, J.; Park, S.; Mo, S.; Ahn, S.; Shin, J. Layer-adaptive sparsity for the magnitude-based pruning. *arXiv* **2020**, arXiv:2010.07611.
35. Friedman, J.; Hastie, T.; Tibshirani, R. A note on the group lasso and a sparse group lasso. *arXiv* **2010**, arXiv:1001.0736.
36. Nogueira, P.V.; Coutinho, G.; Pio, R.; da Silva, D.F.; Zambon, C.R. Establishment of growth medium and quantification of pollen grains and germination of pear tree cultivars. *Rev. Ciênc. Agron.* **2016**, *47*, 380–386. [CrossRef]
37. Quinet, M.; Jacquemart, A.L. Troubles in pear pollination: Effects of collection and storage method on pollen viability and fruit production. *Acta Oecol.* **2020**, *105*, 103558. [CrossRef]
38. Yu, L.; Xiong, J.; Fang, X.; Yang, Z.; Chen, Y.; Lin, X.; Chen, S. A litchi fruit recognition method in a natural environment using RGB-D images. *Biosyst. Eng.* **2021**, *204*, 50–63. [CrossRef]

agriculture

Article

# Evaluation of the Habitat Suitability for Zhuji Torreya Based on Machine Learning Algorithms

**Liangjun Wu [1,2], Lihui Yang [3], Yabin Li [4], Jian Shi [5], Xiaochen Zhu [1,*] and Yan Zeng [6]**

1 School of Ecology and Applied Meteorology, Nanjing University of Information Science and Technology, Nanjing 210044, China; 202212340015@nuist.edu.cn
2 Nanning Meteorological Bureau, Nanning 530029, China
3 Fujian Provincial Climate Center, Fuzhou 350025, China; yanglihui08@163.com
4 Heilongjiang Provincial Climate Center, Harbin 150030, China; 18946019717@163.com
5 Zhuji Meteorological Bureau, Zhuji 311800, China; shelleysj@icloud.com
6 Key Laboratory of Transportation Meteorology of China Meteorological Administration, Nanjing Joint Institute for Atmospheric Sciences, Nanjing 210041, China; zengyan@cma.gov.cn
* Correspondence: xiaochen.zhu@nuist.edu.cn

**Abstract:** Torreya, with its dual roles in both food and medicine, has faced multiple challenges in its cultivation in Zhuji city due to frequent global climate disasters in recent years. Therefore, conducting a study on suitable zoning for Torreya habitats based on climatic, topographic, and soil factors is highly important. In this study, we utilized the latitude and longitude coordinates of Torreya distribution points and ecological factor raster data. We thoroughly analyzed the ecological environmental characteristics of the climate, topography, and soil at Torreya distribution points via both physical modeling and machine learning methods. Zhuji city was classified into suitable, moderately suitable, and unsuitable zones to determine regions conducive to Torreya growth. The results indicate that suitable zones for Torreya cultivation in Zhuji city are distributed mainly in mountainous and hilly areas, while unsuitable zones are found predominantly in central basins and northern river plain networks. Moderately suitable zones are located in transitional areas between suitable and unsuitable zones. Compared to climatic factors, soil and topographic factors more significantly restrict Torreya cultivation. Machine learning algorithms can also achieve suitability zoning with a more concise and efficient classification process. In this study, the random forest (RF) algorithm demonstrated greater predictive accuracy than the support vector machine (SVM) and naive Bayes (NB) algorithms, achieving the best classification results.

**Keywords:** Torreya; habitat suitability zoning; machine learning model

## 1. Introduction

In contemporary society, as people's pursuit of health and sustainable lifestyles continues to intensify, more attention is being given to traditional forestry industries, thus creating new development opportunities [1]. Among these, Torreya, known as the "King of Nuts", has multiple purposes, as it is used for food, medicine, oil, timber, decoration, and environmental protection; it is one of China's ancient forestry resources [2,3]. In recent years, Torreya has gained prominence in the market as a supplier of healthy leisure nuts, attracting widespread consumer favor and sustaining market demand [4,5]. The Torreya industry plays a crucial role in poverty alleviation and prosperity for farmers in the Kuaiji Mountain region of Zhejiang Province [6,7]. For instance, by providing an average income of approximately 11,600 yuan per acre, Torreya is considered one of the most economically profitable tree species per unit area. Natural resources such as climate, topography, and soil have a certain influence on the growth of forests or crops [8]. For example, Torreya is a neutral to slightly shady tree species that shows strong adaptability to regions with high humidity and high altitudes characterized by heat [9]. Therefore, in this study, we identified the most suitable areas for Torreya cultivation in Zhuji city by analyzing the topography, climate, soil ecology, and environmental factors of Torreya planting areas in various regions of Zhuji city.

Scholars both in China and internationally have conducted a series of related studies employing two main methodological approaches: physical modeling methods and machine learning methods [10–12]. Traditional physical modeling methods typically rely on linear relationships among ecological factors, potentially overlooking nonlinear and complex interactions [13,14]. Machine learning methods can identify nonlinear relationships among multiple factors, thus more accurately determining suitability ranges [15,16]. In this era of rapid artificial intelligence advancement, compared to traditional physical modeling methods, random forest (RF) and support vector machine (SVM) methods exhibit significant advantages in classification. RF can handle large feature sets and high-dimensional data, performs well with large datasets, and demonstrates high accuracy and robustness. Moreover, SVMs are effective at handling nonlinear data and complex boundaries [17,18]. The advantages of machine learning algorithms in assessing the importance of feature variables and establishing predictive models have been widely applied in various agricultural fields and have included studies on crop climate suitability, yield simulation, and agricultural risk assessment. For instance, Abel Chemura et al. [19] used extreme gradient boosting methods by combining yield data and agronomic variables to model the current climate suitability of maize, sorghum, cassava, and groundnut in Ghana. They used future climate predictions for the 2050s under two greenhouse gas emission scenarios to forecast changes in crop suitability ranges. Guoyong Leng et al. [20] compared the performances of process-based, regression, and machine learning models for simulating US maize yields. Their results showed that machine learning could better reproduce yield variability and extreme conditions, providing more efficient probabilistic risk analysis to assess climate impacts on crop production. Ying Han et al. [21] constructed a frost risk assessment model for tea in Hangzhou by using the RF algorithm. By considering indicators such as the minimum temperature, altitude, tea planting area, and tea yield, they assessed the risk for different cold-resistant tea varieties to guide local agricultural production and provide disaster warning.

However, in most suitability zoning studies, traditional physical modeling methods predominate, with little utilization of machine learning algorithms such as RFs. Therefore, in this study, based on Torreya samples and three categories of raster data, namely, climate, topography, and soil raster data, we fully exploited the feature importance calculation function 'feature_importances' to evaluate the importance of each feature variable, thereby constructing and selecting features. By comparing and analyzing different algorithms, we identified the optimal features and algorithm models capable of rapidly and accurately zoning the suitability of Torreya in Zhuji city.

## 2. Data and Methods

### 2.1. Research Area

As one of China's major Torreya production areas, Zhuji city boasts a long history of Torreya cultivation, spanning over 1300 years. The cultivated area of Torreya reaches 67 $km^2$, with 41,000 Torreya trees aged more than a hundred years within the city. Torreya production in Zhuji city is concentrated mainly in the southeastern region, which includes Zhaojia town, Dongbaihu town, Donghe township, Fengqiao town, Huangshan town, and Chenzhai town. These areas benefit from abundant heat resources conducive to the growth and development of Torreya. Adequate precipitation ensures a sufficient water supply in these planting areas, thereby promoting the healthy growth of Torreya. The soil types are mainly yellow-red soil, mountain red soil, yellow soil, and red soil, which exhibit weak acidity favorable for Torreya root growth and nutrient absorption. Moreover, the existing planting areas are mostly hilly terrain. The specific location and division of Zhuji city are shown in Figure 1.

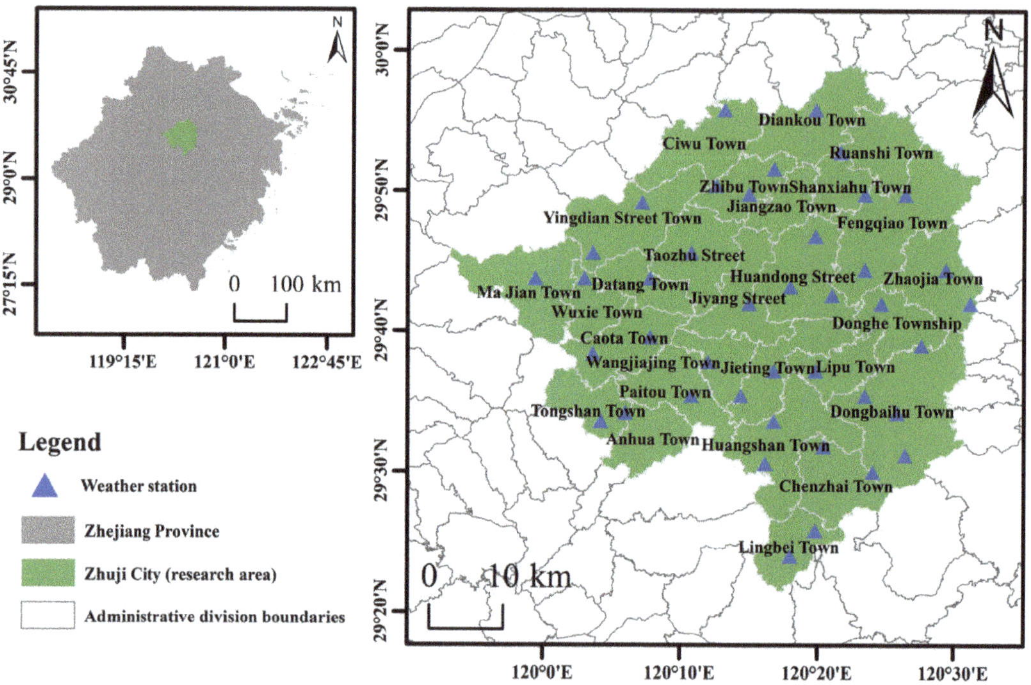


**Figure 1.** Study area map.

### *2.2. Data*

#### 2.2.1. Data Introduction

The data used in this study included climate data, terrain data, soil data, land use data, and Torreya sample data. Climate data were sourced from the Zhuji Meteorological Bureau, comprising meteorological data from 47 regional automatic stations and 1 national basic station spanning from 2006 to 2020. The dataset covered daily average temperature (°C), maximum temperature (°C), precipitation (mm), and average relative humidity (%). The terrain data included elevation (m), slope (°), and aspect and were extracted from a 30 m resolution digital elevation model (DEM) of Zhuji city acquired from the Geographic Spatial Data Cloud Platform. The soil data included soil types, soil layer thickness (m), and soil pH values and were sourced from the Basic Attributes Data Set of the China High-resolution National Soil Information Grid of the National Tibetan Plateau Data Center. The land use data were obtained from the National Earth System Science Data Center. We used GPS instruments to collect latitude and longitude information for the Torreya ancient tree sample points, resulting in approximately 1600 Torreya sample points. Among these, approximately 1200 sample points belonging to the suitable growth category, approximately 400 belonging to the moderately suitable growth category, and approximately 800 belonging to the unsuitable growth category were randomly generated within the urban area of Zhuji.

#### 2.2.2. Data Processing

In our study, we conducted quality checks and comparisons on soil, terrain, and land use data. During this process, various data sources were used and cross-validated to ensure the accuracy and reliability of the data. Specifically, official data provided by local governments and relevant departments were compared with each other and with the data in the existing studies and literature to validate the data consistency and correctness [13,14].

To spatialize the meteorological data, statistical methods combined with geographic information system (GIS) technology were employed. For climate factors such as annual average temperature, annual accumulated temperature above 10 °C, annual effective accumulated temperature above 10 °C, and annual precipitation, excluding potential evapotranspiration, a mixed interpolation method was used. Specifically, this method combined a multivariate regression model with the spatial interpolation of residuals to establish a spatial estimation model for meteorological data. In our study, we used the longitude, latitude, and elevation of meteorological stations to construct multivariate regression models for meteorological elements via SPSS software. This method of inexact interpolation treats the difference between station-estimated values and observed values as residuals; these values are typically influenced by small-scale terrain variations. Subsequently, residual processing was conducted, and spatial secondary interpolation was performed using the inverse distance weighting method in GIS. For the spatialization of potential evapotranspiration, the inverse distance weighting interpolation method was directly applied. We used climate data from 48 meteorological stations in Zhuji city from 2006 to 2020, with 85% of the meteorological station data used for spatial interpolation and the remaining 15% used to validate the accuracy of the spatial interpolation results.

### 2.3. Methods

#### 2.3.1. Growth Climate Indicators

In this study, based on the literature review, several growth climate indices were introduced. These comprehensive indices included Kira's Warmth Index ($WI$) [22], Holdridge's bioclimatic temperature ($BT$) and potential evapotranspiration ($PER$) [23], as well as Xu Wenduo's Humidity Index ($HI$) [24].

(1) Kira's Warmth Index ($WI$):

$$WI = \sum_{i=1}^{12}(t_i - 5) \tag{1}$$

where $WI$ represents the warmth index and $t_i$ is the monthly average temperature greater than 5 °C.

(2) Holdridge's biological temperature ($BT$) and possible evapotranspiration rate ($PER$):

$$BT = \sum T/12 \tag{2}$$

$$PER = 58.93BT/P \tag{3}$$

where $BT$ represents the optimal temperature range required for sustaining life activities within the organism and $T$ denotes the monthly average temperature ranging from 0 °C to 30 °C. A monthly average temperature below 0 °C is regarded as 0; if it exceeds 30 °C, it is considered 30.

(3) Xu Wenduo's Humidity Index ($HI$):

$$HI = P/WI \tag{4}$$

where $P$ represents the annual precipitation (mm) and $WI$ represents the warmth index (°C).

#### 2.3.2. Physical Model Calculation Method for Climate and Habitat Zoning in Torreya

The physical modeling study is based on Torreya latitude and longitude coordinate data and ecological factor grid data. Spatial analysis and mathematical statistics were used to comprehensively analyze the ecological environmental characteristics, such as climate, soil, and terrain, of the Torreya DPs. This analysis identified significant climate variables, ecological factor indicators, and their respective weights affecting Torreya growth. Subsequently, the suitability of Torreya habitats was quantitatively investigated and evaluated, leading to the delineation of planting zoning results.

The single-factor degree is calculated using the following formula:

$$NF_i = \begin{cases} \min\left(\frac{x_i}{L_{imin}}, 1, \frac{L_{imax}}{x_i}\right) & S_{imax} \geq x_i \geq S_{imin} \\ 0 & \text{other} \end{cases} \tag{5}$$

where $NF_i$ represents the suitability value of the i-th crop factor; $L_{imin}$, $L_{imax}$, and $X_i$ denote the minimum optimal value, maximum optimal value, and measured value of the i-th ecological factor, respectively; and $S_{imin}$ and $S_{imax}$ represent the lower and upper limits of the ecological factor's suitable range, respectively. A tiered assignment method is adopted to determine the suitability values of soil types.

According to Equation (5), the suitability index of each evaluation unit is calculated separately for each single factor, with values ranging from 0 to 1. The comprehensive suitability calculation employs the exponential weighted sum method, as shown in the following formula:

$$S_j = \sum_{i=1}^{n} w_{ij} NF_i \tag{6}$$

In the equation, $i$ and $j$ represent the ecological factors and evaluation units, respectively, where the weight value $w_{ij}$ is obtained through the analytic hierarchy process (AHP), with $S_j$ indicating the comprehensive suitability value of each evaluation unit. Based on the evaluation model results, the suitability of each single factor was analyzed cumulatively, resulting in a distribution map of Torreya habitat suitability and subsequent planting zoning.

### 2.3.3. Machine Learning Methods for Climate and Habitat Zoning in Torreya

**Algorithm selection**

To facilitate the comparison and validation of the classification performance of machine learning models in suitability zoning, we used not only the RF model algorithm but also the SVM model algorithm for mutual comparison and validation.

**Sample selection partition**

To determine the distribution of the main Torreya samples in Zhuji city, we fully used the reference sample points provided by the Zhuji Meteorological Bureau and the forestry resource system of the Zhuji Natural Resources and Planning Bureau. To meet the research requirements, the suitable types of Torreya planted in Zhuji city were classified into three major categories: suitable, moderately suitable, and unsuitable. Approximately 800 unsuitable sample points were randomly generated in the urban area of Zhuji. Considering that the ecological conditions differ between moderately suitable and suitable areas for Torreya, to avoid simply dividing Zhuji city into suitable and unsuitable areas, we divided the Torreya sample points into suitable and moderately suitable categories based on the optimal range and suitable range of indicator factors. Overall, there were approximately 1200 suitable sample points and approximately 400 moderately suitable sample points. A total of approximately 2400 sample points were obtained across the entire study area. Before training the model samples, 100 sample points were randomly extracted from each of the three categories for results validation. To ensure the accuracy of model training and testing, a random sampling method was employed, with 75% of the total samples used as the training set and the remaining 25% used as the testing set to establish the RF and SVM models. The specific sample distribution is shown in Figure 2.

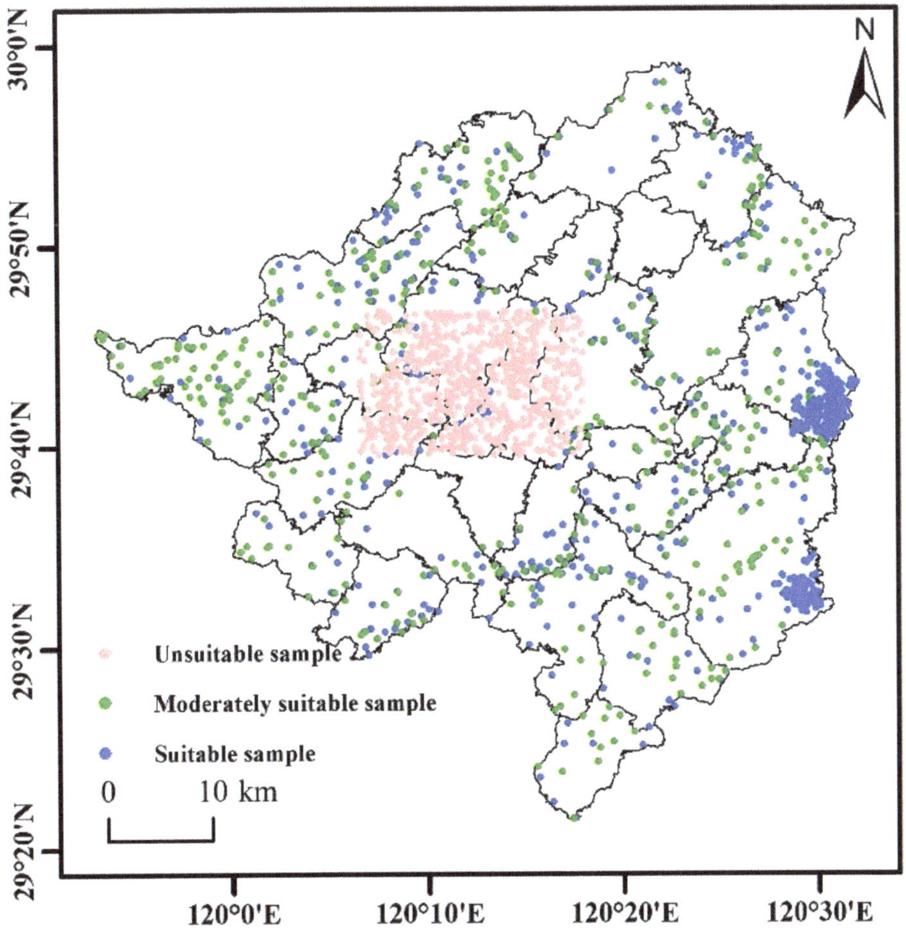


**Figure 2.** Torreya sample distribution map.

**Classification feature evaluation**

To enhance the effectiveness of Torreya suitability zoning, we comprehensively considered the habitat characteristics of the study area and selected multiple feature variables. However, not all preselected feature variables positively influenced the classification of the Torreya suitability zones. Those feature variables that failed to enhance the classification ability of the Torreya suitability zones may result in prolonged algorithm runtime and decreased classification accuracy. Therefore, before classification, we conducted a feature importance assessment to calculate the importance of each feature variable. By ranking the importance of feature variables, we assessed the accuracy of the test set to obtain a subset of classification features that achieved the highest test set accuracy, thereby improving the model's precision. In this study, we used the feature importance calculation function "feature_importances" in RF to evaluate the importance of each feature variable, sorting them in descending order based on their importance scores. Subsequently, we employed a sequential forward feature selection method, starting from zero features and iteratively adding feature variables to the model to maximize cross-validation scores for feature combinations. The specific formula is as follows:

$$P_k = \frac{\sum_{i=1}^{n}\sum_{j=1}^{t} D_{Gkij}}{\sum_{k=1}^{m}\sum_{i=1}^{n}\sum_{j=1}^{t} D_{Gkij}} \times 100\% \quad (7)$$

where $m$, $n$, and $t$ represent the total number of indicators, the number of classification trees, and the number of nodes in a single tree, respectively. $D_{Gkij}$ denotes the Gini exponential reduction value of the $k$-th indicator in the $j$-th node of the tree. $P_k$ represents the importance of the $k$-th indicator.

**Model parameter optimization**

To enhance the performance of the RF model, in our RF model study, we randomly split the dataset into training and testing sets at a ratio of 1:3 and used the sklearn library for model training. Regarding parameter optimization, we employed a combination of random search cross-validation and grid search cross-validation methods. Both the framework parameters and decision tree parameters of the RF model were optimized. Specifically, we set the number of trees in the RF to 100, the maximum number of features to "auto", the minimum number of samples required for node splitting to 2, and the minimum number of samples required at leaf nodes to 1. These parameter optimizations aimed to improve the performance and generalization ability of the RF model, enabling better adaptation to real-world datasets and enhancing prediction accuracy.

Additionally, to ensure an adequate amount of data for training and evaluating the model, in the SVM model study, the ratio of the testing set to the training set was set to 1:3. Regarding the parameter settings for the SVM model, we opted not to specify the maximum depth. For the choice of kernel function, we employed a linear kernel. Finally, we set the regularization parameter C to 1.

2.3.4. Machine Learning Model Accuracy Evaluation Index

Overall accuracy (OA) is an indicator that measures the proportion of correctly classified pixels among all pixels considered for classification, providing an assessment of overall classification accuracy. User accuracy (UA) evaluates the model's classification performance for each class. The kappa coefficient is used to measure the agreement between classification results and actual data, considering the possibility of random errors in classification, thereby making it a more comprehensive evaluation metric. Another important metric is the F1-score (F1 value), which is the harmonic mean of Precision and Recall. The F1-score provides a holistic evaluation of the model's performance in handling imbalanced positive and negative samples. It ranges from 0 to 1, where values closer to 1 indicate a better balance between Precision and Recall, thus superior overall performance.

*2.4. Technical Process*

The input data include 10 climate indicators: annual mean temperature, ≥10 °C annual active temperature sum, ≥10 °C annual effective temperature sum, number of days with daily average temperature ≥ 35 °C, annual precipitation, annual average relative humidity, warmth index, biological temperature, potential evapotranspiration rate, and humid index. They also include 6 soil and topographic indicators: soil pH value, soil depth, soil type, slope, aspect, and elevation. The Technical process chart is shown in Figure 3.

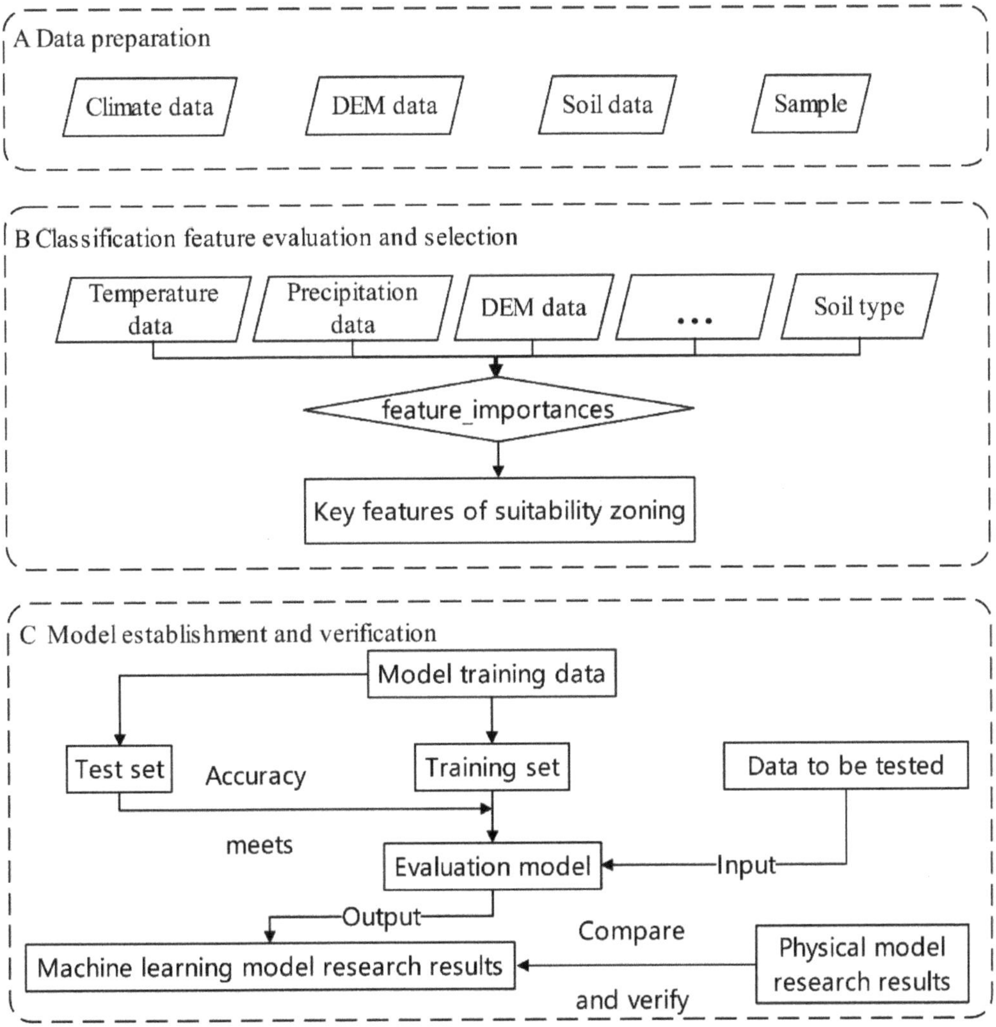


**Figure 3.** Technical process chart.

## 3. Results

### *3.1. Climate Suitability Zoning and Habitat Suitability Zoning Analysis Based on Physical Models*

#### 3.1.1. Spatial Distribution of Climate Suitability Zoning

We used the latitude and longitude coordinate data of Torreya, along with the raster data of climatic factors, to calculate the climate suitability zoning results through GIS, as depicted in Figure 4.

According to the results shown in Figure 4, 85.7% of the surveyed area falls within the climate-suitable zone, 5.1% falls within the moderately suitable zone, and 9.2% is classified as unsuitable due to the climate conditions. Unsuitable areas are primarily distributed in the southern part of Diankou town, Jiangzao town, the northwestern part of Zhibu town, the central part of Taozhu Street, and the northern part of Huandong Street. Overall, the climate resources in Zhuji city generally meet the planting requirements for Torreya. However, differences in spatial distribution may render some areas unsuitable for Torreya cultivation.

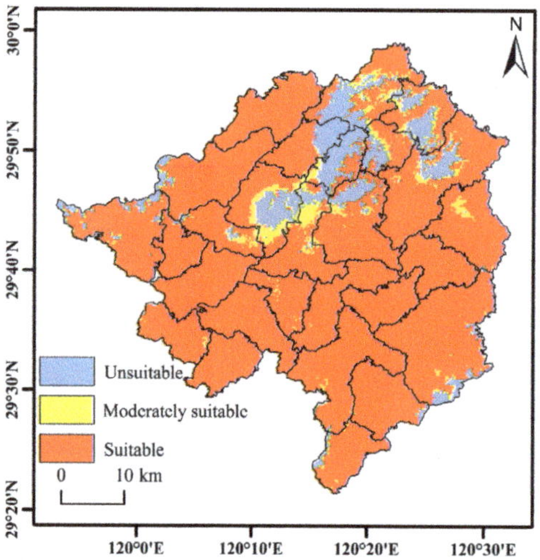


**Figure 4.** Climate suitability zoning map.

3.1.2. Spatial Distribution of Habitat Suitability Zones

Based on the raster data for the climatic factors, we integrated the raster data for the soil and terrain factors and excluded unsuitable areas from the land use data to derive the results of the habitat suitability zoning, as depicted in Figure 5.

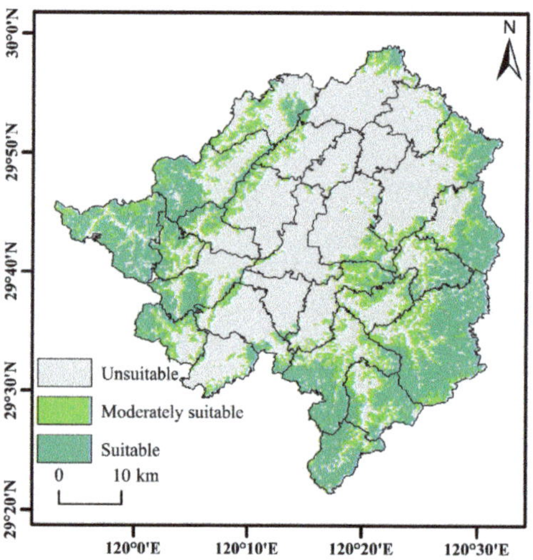


**Figure 5.** Habitat suitability zoning map.

According to the results shown in Figure 5, 26.9% of the surveyed area falls within the suitable habitat zone, and 17.9% falls within the moderately suitable habitat zone; 55.2% is classified as unsuitable due to climate conditions. The western, southern, and eastern regions of Zhuji city are mainly suitable or moderately suitable for Torreya cultivation. The primary townships included Dongbaihu town, Donghe township, Zhaojia town, Huangshan town, Lingbei town, Chenzhai town, Yingdian Street town, and Majian town. Additionally, these habitat zones are distributed in Caota town, Ciwu town, Fengqiao town, Wuxie town, and Tongshan town. The central and northern regions of Zhuji city are mainly unsuitable for Torreya cultivation. The major distributed townships and streets include Shanxiahu town, Jiangzao town, Diankou town, Ruanshi town, Jiyang Street, Huandong Street, Zhibu town, Taozhu Street, Datang town, Wangjiajing town, Anhua town, Paitou town, Jieting town, and Gupu town.

### *3.2. Climate Suitability Zoning and Habitat Suitability Zoning Analysis Based on Machine Learning Methods*

#### 3.2.1. Classification Feature Evaluation Results

The feature importance calculation function yielded the importance ranking of 16 feature variables (Figure 6). The ranking results indicate that compared to climatic features, terrain and soil features are more crucial. Based on this finding, we employed the sequential forward feature selection method and used the test set to train and evaluate the 16 preselected feature variables, as illustrated in Figure 7.

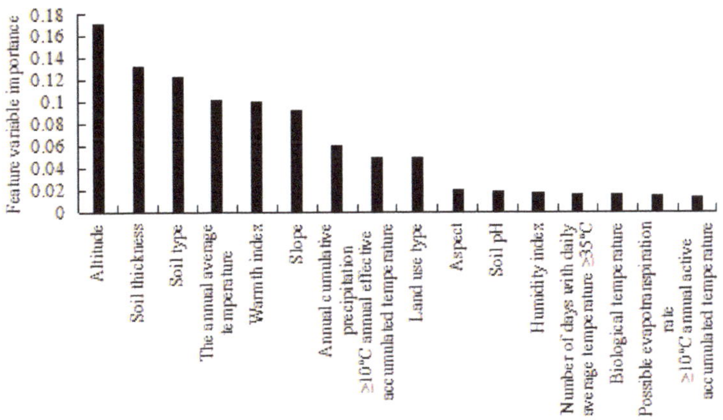


**Figure 6.** Feature importance ranking map.

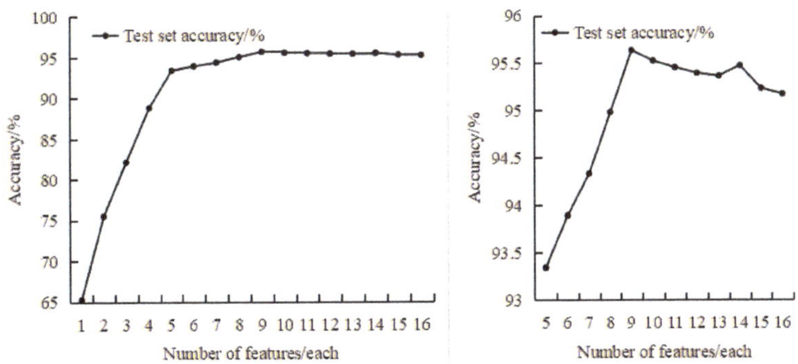


**Figure 7.** Test set accuracy line chart (**left**: test set accuracy with feature numbers 1–16; **right**: partial amplification of test set accuracy).

According to the relationship between the test set accuracy and the number of features illustrated in Figure 7, as the number of feature increases from 1 to 5, the test set accuracy increases from 65.23% to 93.34%. With the increase in the number of features, there is a significant improvement in model accuracy. When the number of features increases from 5 to 9, the accuracy shows a generally stable and slight upward trend, with the highest test set accuracy reaching 95.63%. However, as the number increases from 9 to 17, the overall test set accuracy decreases due to data redundancy. Therefore, for habitat suitability zoning model training, it is advisable to select the top 9 feature variables based on the feature importance rankings shown in Figure 6. For climate suitability zoning model training, the top nine feature variables should be selected from among the climate-related feature variables within the same set.

3.2.2. Spatial Classification Results for Climate Suitability Zones

To compare the classification accuracy of three classifiers, namely, RF, SVM, and naive Bayes (NB), in the process of climate suitability zoning and to select the optimal classification algorithm suitable for suitability zoning in topographically complex areas, we used the confusion matrix method to validate the classification results. The evaluation metrics included the overall accuracy (OA), user accuracy (UA), F1-score, kappa coefficient, and $R^2$. See Table 1 for specific accuracy.

**Table 1.** Climate suitability zone classification model accuracy table.

| Classification | Suitable Type | F1-Score | User Accuracy/% | Overall Accuracy/% | Kappa Coefficient | $R^2$ |
|---|---|---|---|---|---|---|
| RF | Suitable | 0.99 | 99.32% | 98.08% | 0.9937 | 0.9957 |
| | Moderately suitable | 0.98 | 98.46% | | | |
| | Unsuitable | 0.97 | 97.46% | | | |
| SVM | Suitable | 0.89 | 89.03% | 83.36% | 0.7141 | 0.5554 |
| | Moderately suitable | 0.69 | 69.23% | | | |
| | Unsuitable | 0.79 | 78.80% | | | |
| NB | Suitable | 0.76 | 61.29% | 48.30% | 0.2856 | 0.3019 |
| | Moderately suitable | 0.32 | 98.46% | | | |
| | Unsuitable | 0.15 | 8.70% | | | |

According to various model accuracy metrics, the RF algorithm consistently outperforms the SVM and NB algorithms in terms of model classification effectiveness. The classification performance of the SVM algorithm surpasses that of NB. The overall classification accuracy of the RF algorithm reaches as high as 99.08%. The specific classification diagram is shown in Figure 8.

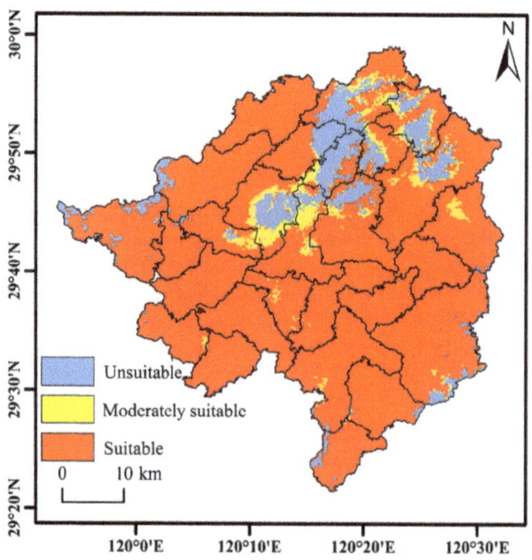


**Figure 8.** Climate suitability zone (RF) classification diagram.

Given that the accuracy and $R^2$ values of the SVM and NB algorithms are significantly lower than those of the RF algorithm, we opted to use the classification results generated by the RF algorithm as the final outcome for climate suitability zoning.

### 3.2.3. Spatial Classification Results for Habitat Suitability Zones

Since NB performed poorly in climate zoning classification and was found through experimentation to be unsuitable for habitat zoning classification, we compared only the classification accuracy of RF and SVM classifiers in habitat suitability zoning. The evaluation metrics were consistent with those used for spatial climate suitability classification. See Table 2 for specific accuracy.

**Table 2.** Habitat suitability zone classification model accuracy table.

| Classification | Suitable Type | F1-Score | User Accuracy/% | Overall Accuracy/% | Kappa Coefficient | $R^2$ |
|---|---|---|---|---|---|---|
| SVM | Suitable | 0.92 | 90.37% | 91.95% | 0.8743 | 0.8933 |
| | Moderately suitable | 0.83 | 88.54% | | | |
| | Unsuitable | 0.96 | 94.61% | | | |
| RF | Suitable | 0.92 | 96.08% | 95.17% | 0.9243 | 0.9268 |
| | Moderately suitable | 0.90 | 91.67% | | | |
| | Unsuitable | 0.97 | 96.30% | | | |

From the perspective of various model accuracy metrics, the RF algorithm consistently outperforms the traditional SVM model in terms of classification effectiveness. Specifically, the RF algorithm exhibits superior performance in terms of OA, UA for each class, the kappa coefficient, and other aspects. The specific classification diagram is shown in Figure 9.

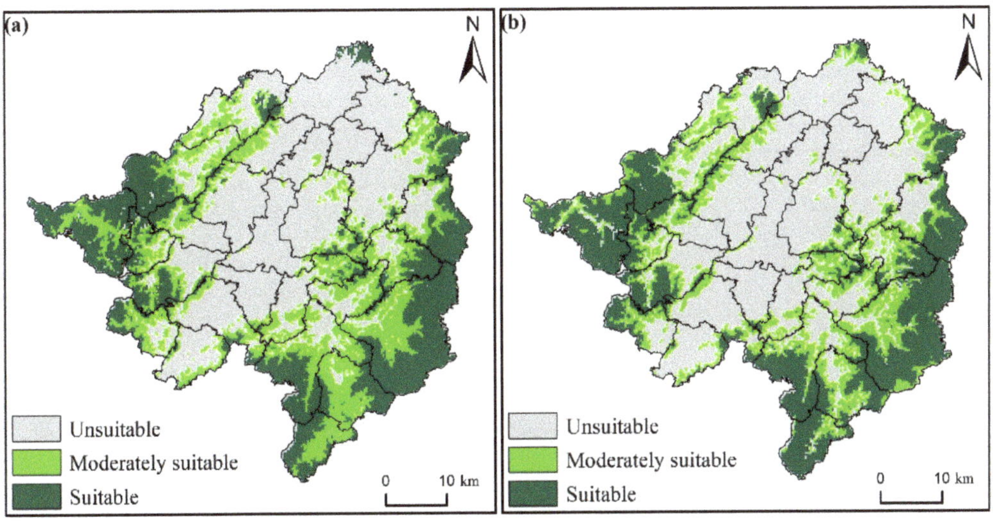


**Figure 9.** Habitat suitability zone classification diagram ((**a**): SVM; (**b**): RF).

The distribution of unsuitable areas is extensive, occupying most of the area of Zhuji city. A consideration of the topographical features of Zhuji city indicated that unsuitable areas are concentrated mainly in the PuYang River valley basin in the central part of the city and the river network plain in the northern part. In contrast, suitable areas are distributed primarily in the low hills of Mount Kuaiji in the east, the low hills of Longmen Mountain in the west, and some towns in the southern part of Zhuji. The distribution of moderately suitable areas is relatively limited; these areas are located mainly in transitional zones between unsuitable and suitable areas. Specifically, regions with lower elevations and relatively flat terrain are less suitable for Torreya cultivation; conversely, areas with low hills and ridges are more suitable for Torreya growth. The classification results of these two machine learning models are relatively consistent with the conclusions drawn from the physical models.

### 3.3. Comparison of Zoning Results

The habitat suitability zoning results for Torreya in Zhuji were resampled to a resolution of 250 m and spatially compared with the classification results of habitat suitability zoning at a 250 m resolution via the SVM and RF methods. The pixel statistics table is shown in Table 3.

**Table 3.** Pixel statistics table.

| Number of Pixels and Overlap Rate | Suitable Category | Moderately Suitable Category | Unsuitable Category | Zhuji |
|---|---|---|---|---|
| Number of physical model simulation pixels/each | 9157 | 6106 | 18,775 | 34,038 |
| SVM model predicts the number of pixels/each | 8935 | 8323 | 16,780 | 34,038 |
| The number of pixels in the overlap between the SVM model and the physical model/each | 7337 | 4362 | 15,866 | 27,565 |
| SVM model pixel overlap rate/% | 80.12% | 71.44% | 84.51% | 80.98% |
| RF model predicts the number of pixels/each | 9304 | 6494 | 18,240 | 34,038 |
| The number of pixels in the overlap between the RF model and the physical model/each | 7846 | 4360 | 17,195 | 29,401 |
| RF model pixel overlap rate/% | 85.68% | 71.41% | 91.58% | 86.38% |

From the perspective of pixel quantity analysis, both the suitable, moderately suitable, and unsuitable categories demonstrated trends closer to the simulated habitat suitability index model results within the prediction outcomes of the RF model. By comparing the predictive classification results of the SVM and RF models with the zoning results of the habitat suitability index model on a per-pixel basis, the overlap rates of the three categories of unsuitable, moderately suitable, and suitable pixels were assessed.

For the SVM model, the overlap rate between the predicted suitable spatial distribution of Torreya and the results of the habitat suitability index model reached 80.12%, with an overlap rate of 71.44% for the moderately suitable spatial distribution and 84.51% for the unsuitable spatial distribution, resulting in an overall comprehensive overlap rate of 80.98%. In contrast, the RF model exhibited an overlap rate of 85.68% for the suitable spatial distribution of Torreya compared to the habitat suitability index model results; an overlap rate of 71.41% for the moderately suitable spatial distribution; and a notably greater overlap rate of 91.58% for the unsuitable spatial distribution. The overall comprehensive overlap rate reached 86.38%. This comparison demonstrates the RF model's superior accuracy and consistency relative to the SVM model in predicting all categories, especially the unsuitable category, for which it achieved a substantially higher overlap rate, providing strong support for the superiority of the model's performance.

In the feature importance ranking chart (Figure 6), elevation and slope are ranked first and sixth, respectively, indicating the significant influences of these two topographic factors on the habitat suitability of Torreya. Figure 10 shows that the suitable areas for Torreya in Zhuji city are distributed mainly in the eastern, southern, and western parts, which is consistent with regions with high elevations and steep slopes, while the predominantly unsuitable areas are distributed in the central and northern parts, which are characterized by lower elevations and gentler slopes. This finding validated the relationship between elevation and slope within a certain range and habitat suitability.

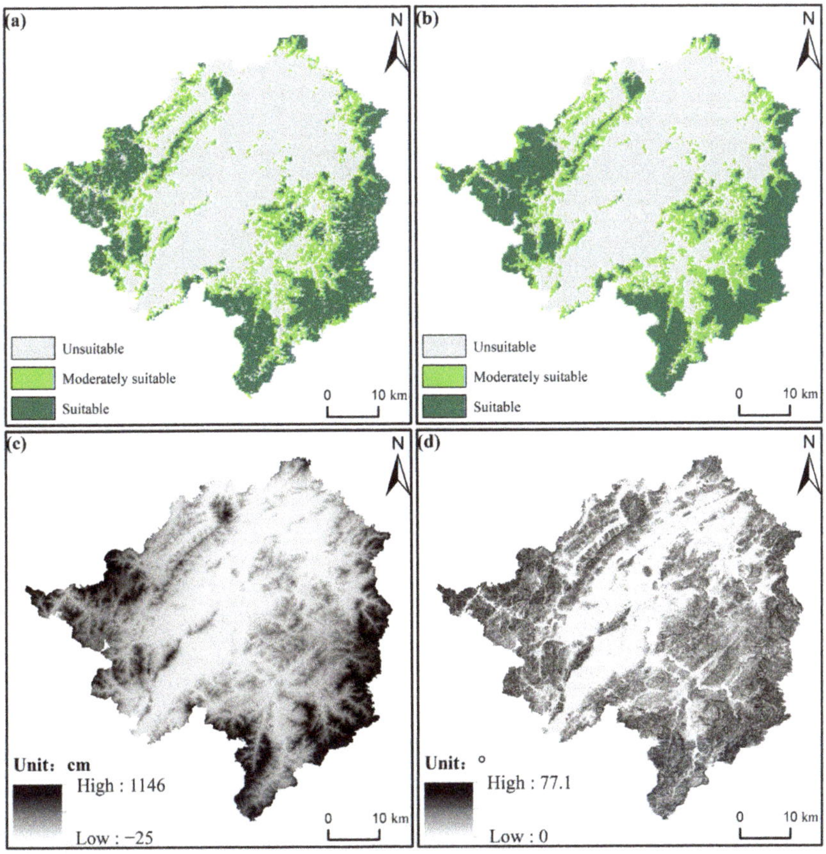


**Figure 10.** Verification analysis diagram ((**a**): physical model habitat suitability zoning; (**b**): RF model habitat suitability zoning; (**c**): altitude; (**d**): slope).

## 4. Discussion

The RF classification method based on the best features achieved an overall classification accuracy of over 95% for both the climate and habitat suitability categories. In contrast, the SVM algorithm still has considerable room for improvement in classification accuracy, especially for the moderately suitable class. Scholars have pointed out two main reasons for this error. Firstly, SVMs are limited in some cases due to data dimensionality and kernel function selection, resulting in relatively limited classification performance on complex problems [25,26]. Secondly, the relatively small number of samples in the moderately suitable class of Torreya leads to a lower classification accuracy [27].

Despite significant progress in solving classification problems, machine learning algorithms still face challenges such as insufficient model interpretability, sensitivity to data quality and features, and poor generalization ability [28,29]. With the increasing complexity of algorithms, understanding the decision-making process of models becomes more difficult, which may affect trust and interpretability. Moreover, the problem of good performance on training data but poor generalization ability on new data needs to be addressed, as uncertainties still exist in the practical application of machine learning algorithms [30,31].

Research indicates that classification algorithms such as RF and SVM algorithms are gradually gaining attention in the agricultural field [32–34]. RFs are widely used in crop classification, pest detection, and soil analysis due to their automatic feature selection and ability to handle multiple samples [35,36]. In contrast, SVMs perform well in handling high-dimensional and complex data and are used for tasks such as crop growth prediction, soil type identification, and disease prediction [37,38]. However, these algorithms still face challenges in practical applications, such as the complexity of parameter tuning, sensitivity to sample imbalance, and lack of model interpretability. Therefore, in agricultural applications, researchers need to choose appropriate classification algorithms based on the characteristics of specific problems and data and integrate domain expertise for comprehensive analysis to improve the accuracy and interpretability of classification models [39,40].

## 5. Conclusions

In this study, based on Torreya samples and various ecological factor grid data, an evaluation index system was constructed for Torreya habitat suitability. Physical model and machine learning model methods were used to simulate climate and habitat suitability zoning, respectively. The main research conclusions are as follows:

(1) The suitable Torreya planting habitats in Zhuji city are distributed mainly in mountainous and hilly areas, while unsuitable areas are located mainly in central basins and northern river plains. Moderately suitable areas are found primarily in transitional zones between suitable and unsuitable areas. The main townships suitable for Torreya cultivation include Dongbaihu town, Zhaojia town, Donghe township, Majian town, Huangshan town, Yingdian Street town, Lingbei town, Chenzhai town, Fengqiao town, and Caota town.

(2) In terms of the relative habitat suitability area, the distribution of climate-suitable areas is more extensive, indicating that climate factors in Zhuji city are not the primary limiting factors for Torreya cultivation. Soil and terrain factors play greater roles in limiting Torreya cultivation.

(3) Machine learning classification algorithms can also achieve suitability zoning for cultivation, and their classification process is more concise and efficient. Compared to the SVM and NB algorithms, the RF algorithm demonstrated higher prediction accuracy in this study, achieving the best classification results.

In research, the selection of indicators is subjective and limited, which may result in biased evaluation outcomes. Simultaneously, the ambiguity and subjectivity of evaluation criteria may diminish interpretability and credibility. Therefore, in future studies, it is imperative to enhance the scientific rigor and comprehensiveness of indicator selection while reinforcing the standardization and operability of evaluation criteria. This process will increase both the scientific validity and practical utility of habitat assessment methods.

**Author Contributions:** L.W.: visualization and writing. L.Y.: conceptualization and supervision. Y.L.: data collection and writing. J.S.: software and formal analysis. X.Z. and Y.Z.: essay writing guidance. All authors have read and agreed to the published version of the manuscript.

**Funding:** This research was funded by the Foundation of Meteorological Technology Innovation Platform, China Meteorological Service Association (CMSA2023MC022).

**Institutional Review Board Statement:** Not applicable.

**Data Availability Statement:** The model prediction results presented in this study are available on request from the corresponding author. The original observations are not publicly available due to the privacy policy.

**Acknowledgments:** Funding from the National Key Research and Development Program of Foundation of Meteorological Technology Innovation Platform, China Meteorological Service Association (CMSA2023MC022) is gratefully acknowledged. We also thank the editors and reviewers for their comments to improve our manuscript.

**Conflicts of Interest:** The authors declare no conflicts of interest.

## References

1. Sansavini, S. The role of research and technology in shaping a sustainable fruit industry: European advances and prospects. *Rev. Bras. Frutic.* **2006**, *28*, 550–558. [CrossRef]
2. Shi, L.K.; Mao, J.H.; Zheng, L.; Zhao, C.W.; Jin, Q.Z.; Wang, X.G. Chemical characterization and free radical scavenging capacity of oils obtained from *Torreya grandis* Fort. ex. Lindl. and *Torreya grandis* Fort. var. Merrillii: A comparative study using chemometrics. *Ind. Crops Prod.* **2018**, *115*, 250–260. [CrossRef]
3. Quan, W.; Zhang, C.; Wang, Z.; Zeng, M.; Qin, F.; He, Z.; Chen, J. Assessment antioxidant properties of *Torreya grandis* protein enzymatic hydrolysates: Utilization of industrial by-products. *Food Biosci.* **2021**, *43*, 101325. [CrossRef]
4. Wu, L.; Wu, L.; Ni, R.; Yan, F. Economic benefits of *Torreya grandis* 'Merrillii' plantings. *J. Zhejiang AF Univ.* **2013**, *30*, 299–303.
5. Laghari, A.; Kandhro, A.; Memon, A. Cold pressed *Torreya grandis* kernel oil. In *Cold Pressed Oils*; Elsevier: Amsterdam, The Netherlands, 2020; pp. 31–38.
6. Chen, X.; Jin, H. Review of cultivation and development of Chinese torreya in China. *For. Trees Livelihoods* **2019**, *28*, 68–78. [CrossRef]
7. Chen, X.; Jin, H. A case study of enhancing sustainable intensification of Chinese Torreya forest in Zhuji of China. *Environ. Nat. Resour. Res.* **2019**, *9*, 53–60. [CrossRef]
8. Mohamed, A.; Reich, R.M.; Khosla, R.; Aguirre-Bravo, C.; Briseño, M.M. Influence of climatic conditions, topography and soil attributes on the spatial distribution of site productivity index of the species rich forests of Jalisco, Mexico. *J. For. Res.* **2014**, *25*, 87–95. [CrossRef]
9. Tang, H.; Hu, Y.-Y.; Yu, W.-W.; Song, L.-L.; Wu, J.-S. Growth, photosynthetic and physiological responses of *Torreya grandis* seedlings to varied light environments. *Trees* **2015**, *29*, 1011–1022. [CrossRef]
10. Al-Ruzouq, R.; Shanableh, A.; Yilmaz, A.G.; Idris, A.; Mukherjee, S.; Khalil, M.A.; Gibril, M.B.A. Dam site suitability mapping and analysis using an integrated GIS and machine learning approach. *Water* **2019**, *11*, 1880. [CrossRef]
11. Radočaj, D.; Jurišić, M. GIS-based cropland suitability prediction using machine learning: A novel approach to sustainable agricultural production. *Agronomy* **2022**, *12*, 2210. [CrossRef]
12. Taghizadeh-Mehrjardi, R.; Nabiollahi, K.; Rasoli, L.; Kerry, R.; Scholten, T. Land suitability assessment and agricultural production sustainability using machine learning models. *Agronomy* **2020**, *10*, 573. [CrossRef]
13. El Baroudy, A. Mapping and evaluating land suitability using a GIS-based model. *Catena* **2016**, *140*, 96–104. [CrossRef]
14. Dengiz, O.; Sezer, İ.; Özdemir, N.; Göl, C.; Yakupoğlu, T.; Öztürk, E.; Sırat, A.; Şahin, M. Application of GIS model in physical land evaluation suitability for rice cultivation. *Anadolu Tarım Bilim. Derg.* **2010**, *25*, 184–191.
15. Chen, Y.; Wu, B.; Chen, D.; Qi, Y. Using machine learning to assess site suitability for afforestation with particular species. *Forests* **2019**, *10*, 739. [CrossRef]
16. Xing, W.; Zhou, C.; Li, J.; Wang, W.; He, J.; Tu, Y.; Cao, X.; Zhang, Y. Suitability evaluation of tea cultivation using machine learning technique at town and village scales. *Agronomy* **2022**, *12*, 2010. [CrossRef]
17. Morais, R. On the suitability, requisites, and challenges of machine learning. *J. Opt. Commun. Netw.* **2021**, *13*, A1–A12. [CrossRef]
18. Cao, H.; Li, H.; Sun, W.; Xie, Y.; Huang, B. A boundary identification approach for the feasible space of structural optimization using a virtual sampling technique-based support vector machine. *Comput. Struct.* **2023**, *287*, 107118. [CrossRef]
19. Chemura, A.; Schauberger, B.; Gornott, C.J.P.O. Impacts of climate change on agro-climatic suitability of major food crops in Ghana. *PLoS ONE* **2020**, *15*, e0229881. [CrossRef] [PubMed]
20. Leng, G.; Hall, J.W. Predicting spatial and temporal variability in crop yields: An inter-comparison of machine learning, regression and process-based models. *Environ. Res. Lett.* **2020**, *15*, 044027. [CrossRef]
21. Han, Y.; He, Y.; Liang, Z.; Shi, G.; Zhu, X.; Qiu, X. Risk Assessment and Application of Tea Frost Hazard in Hangzhou City Based on the Random Forest Algorithm. *Agriculture* **2023**, *13*, 327. [CrossRef]
22. Kira, T. On the altitudinal arrangement of climatic zones in Japan. *Kanchi-Nogaku* **1948**, *2*, 143–173.
23. Holdridge, L.R. *Life Zone Ecology*; CABI: Wallingford, UK, 1967.
24. Xu, W. Ji Liang's caloric index and its application in Chinese vegetation. *J. Ecol.* **1985**, *3*, 35–39.
25. Tharwat, A. Parameter investigation of support vector machine classifier with kernel functions. *Knowl. Inf. Syst.* **2019**, *61*, 1269–1302. [CrossRef]
26. Ghaddar, B.; Naoum-Sawaya, J. High dimensional data classification and feature selection using support vector machines. *Eur. J. Oper. Res.* **2018**, *265*, 993–1004. [CrossRef]
27. Vabalas, A.; Gowen, E.; Poliakoff, E.; Casson, A.J. Machine learning algorithm validation with a limited sample size. *PLoS ONE* **2019**, *14*, e0224365. [CrossRef] [PubMed]
28. Cervantes, J.; Garcia-Lamont, F.; Rodríguez-Mazahua, L.; Lopez, A. A comprehensive survey on support vector machine classification: Applications, challenges and trends. *Neurocomputing* **2020**, *408*, 189–215. [CrossRef]
29. Nayak, J.; Naik, B.; Behera, H. A comprehensive survey on support vector machine in data mining tasks: Applications & challenges. *Int. J. Database Theory Appl.* **2015**, *8*, 169–186.
30. Löw, F.; Michel, U.; Dech, S.; Conrad, C. Impact of feature selection on the accuracy and spatial uncertainty of per-field crop classification using support vector machines. *ISPRS J. Photogramm. Remote Sens.* **2013**, *85*, 102–119. [CrossRef]
31. He, J.; Mattis, S.A.; Butler, T.D.; Dawson, C.N. Data-driven uncertainty quantification for predictive flow and transport modeling using support vector machines. *Comput. Geosci.* **2019**, *23*, 631–645. [CrossRef]

32. Dehghanisanij, H.; Emami, H.; Emami, S.; Rezaverdinejad, V. A hybrid machine learning approach for estimating the water-use efficiency and yield in agriculture. *Sci. Rep.* **2022**, *12*, 6728. [CrossRef]
33. Kok, Z.H.; Shariff, A.R.M.; Alfatni, M.S.M.; Khairunniza-Bejo, S. Support vector machine in precision agriculture: A review. *Comput. Electron. Agric.* **2021**, *191*, 106546. [CrossRef]
34. Son, N.T.; Chen, C.F.; Chen, C.R.; Minh, V.Q. Assessment of Sentinel-1A data for rice crop classification using random forests and support vector machines. *Geocarto Int.* **2018**, *33*, 587–601. [CrossRef]
35. Tatsumi, K.; Yamashiki, Y.; Torres, M.A.C.; Taipe, C.L.R. Crop classification of upland fields using Random forest of time-series Landsat 7 ETM+ data. *Comput. Electron. Agric.* **2015**, *115*, 171–179. [CrossRef]
36. Castro-Franco, M.; Costa, J.L.; Peralta, N.; Aparicio, V. Prediction of soil properties at farm scale using a model-based soil sampling scheme and random forest. *Soil Sci.* **2015**, *180*, 74–85. [CrossRef]
37. Dang, C.; Liu, Y.; Yue, H.; Qian, J.; Zhu, R. Autumn crop yield prediction using data-driven approaches:-support vector machines, random forest, and deep neural network method. *Can. J. Remote Sens.* **2021**, *47*, 162–181. [CrossRef]
38. Kovačević, M.; Bajat, B.; Gajić, B. Soil type classification and estimation of soil properties using support vector machines. *Geoderma* **2010**, *154*, 340–347. [CrossRef]
39. Dziugaite, G.; Ben-David, S.; Roy, D. Enforcing interpretability and its statistical impacts: Trade-offs between accuracy and interpretability. *arXiv* **2020**, arXiv:2010.13764.
40. ElShawi, R.; Sherif, Y.; Al-Mallah, M.; Sakr, S. Interpretability in healthcare: A comparative study of local machine learning interpretability techniques. *Comput. Intell.* **2021**, *37*, 1633–1650. [CrossRef]

*agriculture*

MDPI

*Article*

# Walnut Recognition Method for UAV Remote Sensing Images

**Mingjie Wu [1,2], Lijun Yun [1,2,*], Chen Xue [1,2], Zaiqing Chen [1,2] and Yuelong Xia [1,2]**

1 School of Information, Yunnan Normal University, Kunming 650500, China; 2224100006@ynnu.edu.cn (M.W.); 2223410013@ynnu.edu.cn (C.X.); zaiqingchen@gmail.com (Z.C.); xyl@ynnu.edu.cn (Y.X.)

2 Engineering Research Center of Computer Vision and Intelligent Control Technology, Department of Education of Yunnan Province, Kunming 650500, China

* Correspondence: yunlijun@ynnu.edu.cn

**Abstract:** During the process of walnut identification and counting using UAVs in hilly areas, the complex lighting conditions on the surface of walnuts somewhat affect the detection effectiveness of deep learning models. To address this issue, we proposed a lightweight walnut small object recognition method called w-YOLO. We reconstructed the feature extraction network and feature fusion network of the model to reduce the volume and complexity of the model. Additionally, to improve the recognition accuracy of walnut objects under complex lighting conditions, we adopted an attention mechanism detection layer and redesigned a set of detection heads more suitable for walnut small objects. A series of experiments showed that when identifying walnut objects in UAV remote sensing images, w-YOLO outperforms other mainstream object detection models, achieving a mean Average Precision (mAP0.5) of 97% and an F1-score of 92%, with parameters reduced by 52.3% compared to the YOLOv8s model. Effectively addressed the identification of walnut targets in Yunnan, China, under the influence of complex lighting conditions.

**Keywords:** low-altitude remote sensing; walnut; small object detection; YOLOv8s

## 1. Introduction

The walnut, scientifically known as Juglans regia, is a type of nut tree belonging to the Juglandaceae. Walnuts are rich in protein, unsaturated fatty acids, vitamins, and other minerals beneficial to human health [1]. Currently, there are 21 species of walnuts distributed across the West Indies, Southern Europe, Asia, Central America, North America, and western South America [2]. Among them, China is the world's largest walnut producer, accounting for over half of the global walnut production. The main walnut-producing regions in China include Yunnan Province (880,000 tons), Xinjiang Uygur Autonomous Region (440,000 tons), Sichuan Province (300,000 tons), Shaanxi Province (200,000 tons), and others [3]. Taking Yunnan Province, the largest walnut-producing region, as an example, the primary variety is the Deep-ridged walnut, a unique variety in southwestern China. When Deep-ridged walnuts mature in autumn, their shells harden, change color, and the kernels complete their development. Farmers harvest them during the walnut's ripe period to ensure optimal taste and flavor. However, the ripe period of walnuts is short, and ripe walnuts are prone to oxidation and spoilage. Additionally, overripe walnut kernels tend to stick tightly to the shell, which increases the difficulty of processing after harvesting. Currently, there are two main challenges in walnut production management: firstly, the primary method of walnut harvesting is manual, resulting in very low efficiency and causing many walnuts to overripe and rot on the trees. Secondly, most walnut trees in Yunnan are planted in hilly areas with complex terrain and uneven distribution, making manual counting of walnut fruits extremely difficult.

In recent years, agricultural digitization has continuously improved, promoting the rational utilization of modern production technologies and traditional agricultural production elements, which plays a crucial role in adjusting agricultural production methods and achieving precision agriculture [4,5]. Agricultural digitization refers to the use of advanced technologies such as big data [6], machine learning [7,8], the Internet of Things [9], and deep learning [10–12] in the agricultural production process. Shantam Shorewala et al. [13] proposed a semi-supervised decision method to identify the density and distribution of weeds from color images to locate weeds in fields. Validation results demonstrate that the method generalizes well to different plant species, achieving a maximum recall of 0.99 and a maximum accuracy of 82.13%. Cheng et al. used a deep residual network to detect pests in fields with complex backgrounds. Experimental results showed that the accuracy of this method was higher than support vector machines and backpropagation neural networks and higher than the recognition accuracy of traditional convolutional neural networks (CNNs). However, the network structure complexity of ResNet is relatively high, requiring more computation [14]. Behroozi-Khazaei et al. combined artificial neural networks (ANN) with genetic algorithms (GA) to segment grape clusters similar in color to the background and leaves. Although the improved algorithm can automatically detect grape clusters in images and effectively predict yields, it remains challenging to successfully detect when there is little color difference between grape clusters and leaves [15]. Juan Ignacio Arribas et al. segmented RGB images to separate sunflower leaves from the background and then used a Generalized Sensory Perceptron (GSP) neural network architecture combined with a Posterior Probability Model Selection (PPMS) algorithm to classify sunflower leaves and weeds. However, classification accuracy may be affected when lighting conditions are complex [16]. In summary, algorithms still face challenges such as high computational complexity. Additionally, when background and object features are too similar, models may struggle to meet expectations for crop detection.

With the development of deep learning, object detection algorithms have been widely applied in various fields including remote sensing [17,18], urban data analysis [19], agricultural production [20], embedded development [21] and multispectral image detection [22]. Object detection mainly includes two-stage and one-stage algorithms. Two-stage algorithms include R-CNN [23], Fast-RCNN [24], Faster-RCNN [25], Mask R-CNN [26], etc. These algorithms classify objects based on pre-generated candidate regions, and their detection accuracy is usually higher. However, due to the multi-stage processing required by two-stage algorithms, their complexity is relatively high, and real-time performance is poor, requiring higher hardware requirements. In order to optimize the cumbersome detection process of two-stage algorithms, one-stage detection algorithms have been proposed. Joseph Redmon et al. proposed a one-stage object detection algorithm called YOLO (You Only Look Once), which promoted the development of real-time object detection [27]. Subsequently, many researchers proposed improved one-stage detection algorithms, such as SSD [28], CenterNet [29], YOLOv3 [30], YOLOv7 [31], etc. Chen et al. proposed an improved YOLOv4 model for detecting and counting bayberry trees in images captured by UAVs. Experimental results show that the improved model achieves higher recall while ensuring accuracy [32]. Hao et al. improved the YOLOv3 algorithm for detecting green walnuts. This algorithm utilizes Mixup data augmentation and introduces the lightweight convolutional network MobileNet-v3. In the experiment for detecting green walnuts, the model size is 88.6 MB, and the accuracy reaches 93.3% [33]. Zhong et al. conducted research on walnut recognition in natural environments. They improved the YOLOX algorithm using the Swin Transformer multi-feature fusion module. The improved model achieved an AP50 of 96.72% in natural environments, with a model parameter of 20.55 M [34]. In Li et al.'s study, by improving the feature fusion structure of the YOLOX model, the model's ability to interact with local information in UAV remote sensing images is enhanced, achieving stronger small object detection capabilities [35].

Considering the significant challenge of manually counting walnut fruits in hilly areas and recognizing the superiority of the YOLOv8s algorithm in object detection, we proposed the w-YOLO algorithm to address walnut fruit object detection in hilly terrain and under complex lighting conditions.

The contributions of our work can be summarized in the following points:

1. We utilized UAVs to collect remote-sensing images of walnut trees and established a representative dataset of small walnut targets. The dataset consists of 2490 images with a resolution of 640 × 640, containing a total of 12,138 walnut targets. This work fills the gap in walnut datasets and provides valuable data for walnut target detection and recognition under complex lighting conditions.
2. We made improvements to the YOLOv8s model and designed a w-YOLO model, which includes a lightweight feature extraction network, a better-performing feature fusion network, and a new detection layer. These improvements aim to reduce the model's parameter count, and decrease the size of the model's weight files for deployment on edge computing devices. At the same time, it enhances the model's ability to capture walnut object features, making the model more suitable for walnut detection and recognition under different lighting conditions.
3. The w-YOLO model we designed achieved the recognition of small walnut targets under complex lighting conditions. It significantly improves walnut detection accuracy, with a mAP0.5 of 97% and an F1-score of 92%. The parameter count decreased by 52.3%, and the model's weight file size reduced by 50.7%. Its detection performance surpasses the baseline YOLOv8s and other mainstream object detection models, providing valuable references for walnut detection and management under complex lighting conditions.

The remaining structure of the paper is as follows. In Section 2, we provide an overview of our dataset and introduce the design details of the w-YOLO model. In Section 3, we conduct a series of experiments and analyze the results. Section 4 delves into a detailed discussion of some factors influencing the w-YOLO model. Finally, in the Section 5, we present our conclusions.

## 2. Materials and Methods

In this study, we collect a large amount of walnut image data and preprocess the remote sensing image data, first. Then, we improve YOLOv8s and continuously optimize the training parameters of the model to obtain the optimal walnut detection model, w-YOLO. The basic process of the detection model is illustrated in Figure 1. The input image undergoes data augmentation to increase data diversity, and then undergoes feature extraction through the backbone network. w-YOLO also incorporates BiFPN [36] as the feature fusion network, followed by prediction on the feature outputs. Using the w-YOLO model to compare with other mainstream models, we evaluated the walnut recognition performance under facing light, side light, and backlight. The w-YOLO model achieves both reliable accuracy and the best recall rate. Therefore, the effectiveness of this model can provide effective guidance and high-quality technical support for the management of walnut orchards.

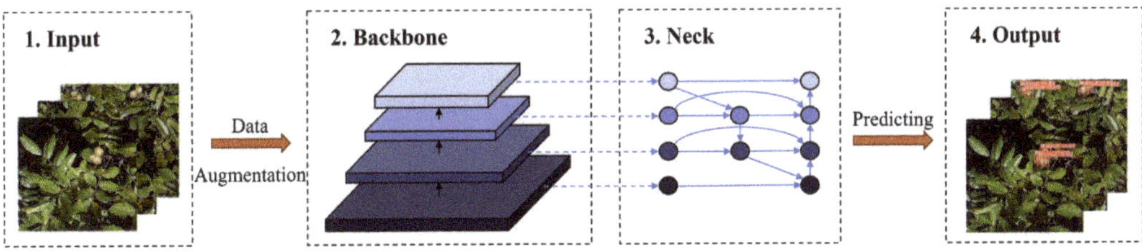


**Figure 1.** The basic flowchart of the w-YOLO model. Unlike YOLOv8, this model utilizes FasterNet for feature extraction and BiFPN for feature fusion. The blue dashed lines represent the transmission of feature information from the backbone network to the neck network, while the blue solid lines indicate the direction of feature information transmission within the neck network.

### 2.1. Research Process

The process of our research work is illustrated in Figure 2. Firstly, we used a DJI Matrice-300-RTK (DJI, Shenzhen, China) equipped with a Zenmuse P1 lens and conducted data collection of walnut tree images using the "following terrain" flight mode. Then, we processed and cropped these images to create a dataset suitable for deep learning models. The dataset was divided into training (64%), validation (16%), and testing (20%) sets, and we trained a preliminary walnut detection model. Subsequently, we selected better parameter combinations to improve the detection performance of the model, resulting in the w-YOLO model. Finally, we evaluated the performance of the w-YOLO model through qualitative and quantitative analysis.

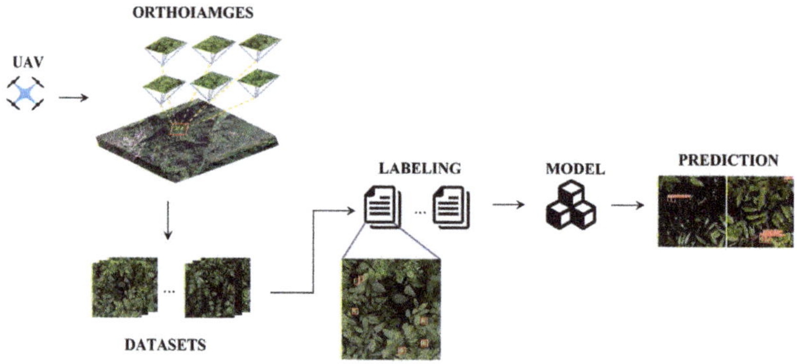


**Figure 2.** The workflow of the research in this paper. Walnut tree images are captured using UAVs in the nadir view, and after preprocessing, they form the dataset and labels. These images are then used to train the walnut detection model, and the best parameter combination is utilized to achieve optimal detection performance.

### 2.2. Study Area

The research site is located in Changning County (WGS 84: 25.024486° N, 99.773675° E), Baoshan City, Yunnan Province, China (Figure 3). Walnuts are a specialty of Changning County and are a Geographical Indication product of China. The area has an average altitude of 1875 m and belongs to a subtropical monsoon climate zone with abundant rainfall (annual precipitation ranging from 700 to 2100 mm), mild temperatures (annual average temperature of 14.8 to 21.3 °C), and long sunshine hours (annual average sunshine of 2335). The favorable climate in this area is very suitable for the growth of walnut trees.

The average height of the walnut sample trees is 10.83 m, and the average base diameter is 39.81 cm. Both the training and testing visible light images in this study were captured using a DJI Matrice-300-RTK equipped with a Zenmuse P1 camera sensor. Both of these devices are manufactured by DJI. The images were taken between 17 August and 20 August 2022. During the flight mission, the preset flight altitude was 100 m, covering an area of 900 square meters. The Ground Sample Distance (GSD) in the horizontal and vertical directions was 0.00464 mm/pixel and 0.00521 mm/pixel, respectively. A total of 180 aerial images were captured, with an image size of 5472 × 3648 pixels (Figure 4), saved in .jpg format.

### 2.3. Dataset Production

We first used a sliding window to crop the 180 aerial images into sizes of 640 × 640 pixels. Then, we discarded some images that did not contain walnut trees, resulting in a dataset consisting of 2490 walnut images. Subsequently, we used the Labelimg to annotate the walnut fruit objects. The detailed information on the walnut dataset we established is shown in Table 1.

**Figure 3.** The study area and its location in this paper.

**Figure 4.** Walnut tree images from the perspective of a UAV.

**Table 1.** The detailed information of the walnut dataset.

| Label Name | Images Number (pcs) | Target Number (pcs) |
|---|---|---|
| Walnut | 2490 | 12,138 |

Note: pcs is the abbreviation for pieces.

## 2.4. A Detection Algorithm for Small Walnut Objects—w-YOLO

In this study, our aim was to design a lightweight walnut detection model without sacrificing detection accuracy, as measured by metrics such as P, R, F1-score, and mAP0.5. Building upon the advantages of YOLOv8s, we made improvements to it, resulting in the w-YOLO model. It is more suitable for detecting small walnut objects in images and can provide technical support for future real-time walnut fruit detection tasks.

2.4.1. YOLOv8s Model

The YOLOv8s model is the latest version of YOLO open-sourced by Ultralytics [37]. YOLOv8s mainly consists of three parts: Backbone, Neck, and Head. The Backbone serves as the feature extraction network, similar to YOLOv5's Backbone, both belonging to CSPDarknet. Here, the input image undergoes initial feature extraction to form an image feature set. The Neck is a feature fusion network, which utilizes a combination of Feature Pyramid Network (FPN) [38] and Path Aggregation Network (PAN) [39] structures to fuse feature maps from different layers of the Backbone, enhancing detection accuracy and robustness. In the Head, a decoupled head structure is used to separate the classification and detection heads, while employing an Anchor-Free approach different from YOLOv5.

2.4.2. w-YOLO

To obtain the w-YOLO model (Figure 5), we first replaced the feature extraction network and the C2f structure in the Neck of YOLOv8 to make it more lightweight. Secondly, we adopted a Weighted Bi-directional Feature Pyramid Network (BiFPN) in the feature fusion part to enhance the feature fusion capability of YOLOv8s. Thirdly, to address the challenge of capturing important feature information, which can be challenging due to multiple downsampling of feature maps, we introduced a self-attention dynamic detection head—DyHead [40]. Finally, considering the characteristics of small walnut objects, we added a detection head of size 160 × 160 and removed the detection head of size 20 × 20.

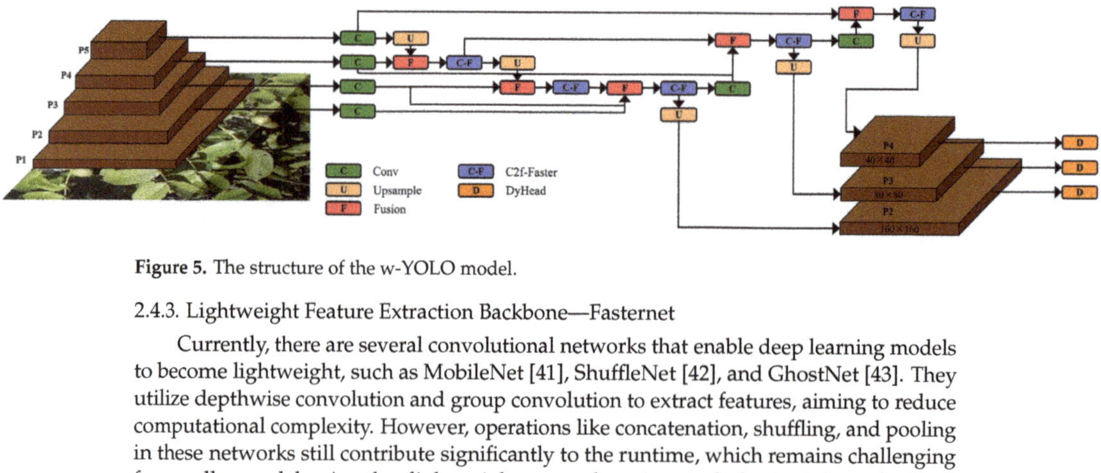


**Figure 5.** The structure of the w-YOLO model.

2.4.3. Lightweight Feature Extraction Backbone—Fasternet

Currently, there are several convolutional networks that enable deep learning models to become lightweight, such as MobileNet [41], ShuffleNet [42], and GhostNet [43]. They utilize depthwise convolution and group convolution to extract features, aiming to reduce computational complexity. However, operations like concatenation, shuffling, and pooling in these networks still contribute significantly to the runtime, which remains challenging for smaller models. Another lightweight network variant includes MobileViT [44] and MobileFormer [45], which combine depthwise convolution (DWConv) with attention mechanisms to reduce computational complexity. Nevertheless, DWConv remains a challenge for further lightweighting in such networks.

FasterNet (Figure 6) achieves lightweight by reducing memory access and computational redundancy in convolutions [46]. FasterNet consists mainly of Embedding layers, Merging layers, and FasterNet Blocks. The FasterNet Block relies on PConv (Partial Convolution) and PWConv (Point-Wise Convolution). PConv is an improvement over DWConv. While DWConv utilizes multiple filters $w \in R^{k \times k}$ to compute the output $O \in R^{c \times h \times w}$, its computational complexity is as shown in Equation (1). In contrast, PConv performs Conv only on a subset of input channels $c_p$ while keeping the rest unchanged. The computational complexity of PConv can be expressed as shown in Equation (2).

$$O_{\text{DWConv}} = h \times w \times k^2 \times c \quad (1)$$

$$O_{\text{PConv}} = h \times w \times k^2 \times c_p^2 \quad (2)$$

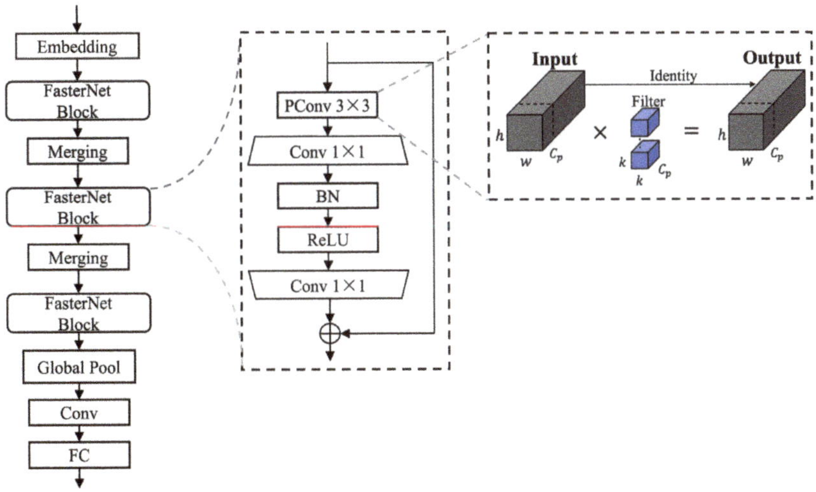


**Figure 6.** The structure of FasterNet and Partial Convolution.

When the typical ratio $\frac{c_p}{c} = \frac{1}{4}$, the computational complexity of PConv is only $\frac{1}{16}$ of DWConv.

To fully utilize information from all channels, PWConv is attached to PConv in a separable manner, as depicted in Figure 7. Its computational complexity can be expressed as

$$O_{\text{PWConv}} = h \times w \times \left(k^2 \times c_p^2 + c \times c_p\right) \tag{3}$$

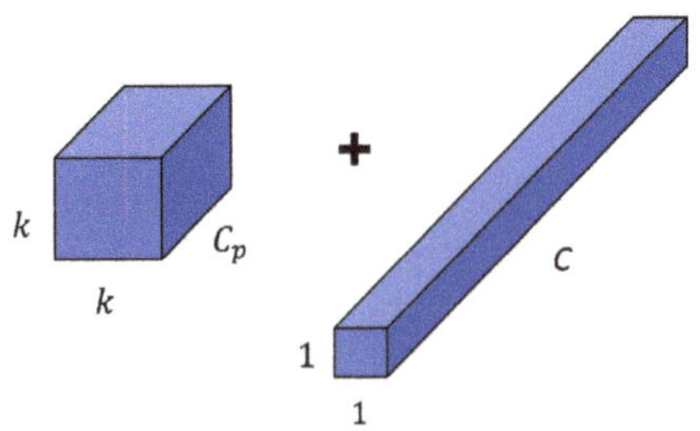

**Figure 7.** The combination approach of Partial Convolution and Point-Wise Convolution.

2.4.4. Multi-Scale Feature Fusion

In the feature fusion network of YOLOv8s, a combination of FPN and PAN is used, aiming to add a bottom-up aggregation pathway to the top-down basis of FPN (Figure 8a). However, this structure introduces a significant number of parameters and computations. We will replace it with the BiFPN structure (Figure 8b) as the feature fusion network for YOLOv8s. This structure learns the importance of different input features and adaptively fuses them. Additionally, the skip connections in BiFPN at the same scale can fuse more features without adding too much computational overhead. The expression of BiFPN can be represented by Equations (4) and (5).

$$P_l^{mid} = \text{Conv}\left(\frac{w_1 \cdot P_l^{in} + w_2 \cdot Resize\left(P_{l+1}^{mid}\right)}{w_1 + w_2 + \epsilon}\right) \tag{4}$$

$$P_l^{out} = \text{Conv}\left(\frac{w_1' \cdot P_l^{in} + w_2' \cdot P_l^{mid} + w_3' \cdot Resize\left(P_{l-1}^{out}\right)}{w_1' + w_2' + w_3' + \epsilon}\right) \tag{5}$$

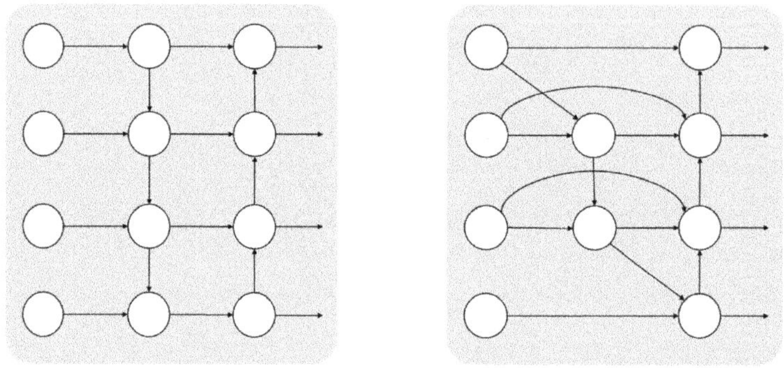

(**a**) (**b**)

**Figure 8.** The structure of the model's feature fusion network. (**a**) The PAN + FPN structure. (**b**) The BiFPN structure.

We improved the Neck network's C2f using the FasterNet Block from FasterNet, making it more lightweight. We refer to the improved C2f as C2f-Faster, and its structure is depicted in Figure 9.

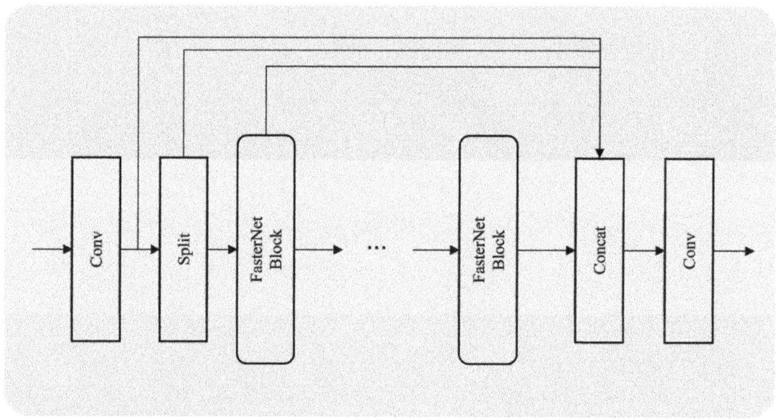


**Figure 9.** The structure of C2f-Faster.

### 2.4.5. Improved Detection Head

Due to the small number of pixels occupied by walnuts in the images, there is limited available feature information, making it difficult to locate small objects on the 80 × 80 feature map of YOLOv8s. Having larger feature maps would help the model capture detailed information about small objects. Therefore, we added a small object detection head with a size of 160 × 160. We also found that the 20 × 20 detection head in YOLOv8s did not perform well in detecting walnut objects, so we removed it to simplify the model.

After multiple downsampling operations, there may be information loss in the feature maps, making it difficult for the detection head to distinguish walnut objects from the background. Therefore, we replaced the detection head of YOLOv8s with the DyHead, which incorporates a self-attention mechanism. After the feature information is input from the Neck network to DyHead, it undergoes a three-dimensional feature tensor $F \in R^{L \times S \times C}$, consisting of scale-aware attention ($\pi_L$), spatial-aware attention ($\pi_S$), and task-aware attention ($\pi_C$). DyHead integrates these three types of attention together, and its expression can be represented as:

$$W(F) = \pi_C(\pi_S(\pi_L(F) \cdot F) \cdot F) \cdot F \tag{6}$$

## 3. Experimental Results

### 3.1. Experimental Setup

All experiments were conducted using the PyTorch 2.0.1 framework and a CUDA 11.7 server on a Quadro RTX 6000 GPU. The dataset images were all 640 × 640 pixels in size, with the training and validation sets divided in an 8:2 ratio, comprising 1594 and 398 images, respectively. During training, we utilized the stochastic gradient descent (SGD) optimizer to update the model parameters, with lr0 set to 0.01, lrf set to 0.07, weight decay set to 0.0005, momentum set to 0.917, batch size set to 32, and epochs set to 300. These settings were consistent across all experiments.

### 3.2. Evaluation Indicators

In this experiment, we use the metrics P (Precision), R (Recall), mAP (mean Average Precision), Parameters, and GFLOPs to evaluate the performance of the w-YOLO. The calculation formulas for P, R, and mAP are shown in Equations (7)–(9).

$$\mathrm{P} = \frac{\mathrm{TP}}{\mathrm{TP} + \mathrm{FP}} \tag{7}$$

$$\mathrm{R} = \frac{\mathrm{TP}}{\mathrm{TP} + \mathrm{FN}} \tag{8}$$

$$\mathrm{mAP} = \frac{1}{\mathrm{n}} \sum_{\mathrm{i}=1}^{\mathrm{n}} \mathrm{AP_i} \tag{9}$$

where TP represents true positives, FP represents false positives, and FN represents false negatives. mAP denotes the average precision across multiple classes and depends on both precision and recall. mAP0.5 represents the average precision at an IOU threshold of 0.5 for all classes, while mAP0.5:0.95 represents the average precision for all classes at IOU thresholds ranging from 0.5 to 0.95 with a step size of 0.05. In object detection tasks, a higher mAP value indicates better detection performance and is a commonly used and authoritative evaluation metric. F1-score ($0 \leq \mathrm{F1} \leq 1$) is used to measure the balance between precision and recall, as shown in Equation (10). It represents the harmonic mean of precision and recall, and a higher F1-score indicates better results.

$$\text{F1-Score} = \frac{2 \times \mathrm{P} \times \mathrm{R}}{(\mathrm{P} + \mathrm{R})} \tag{10}$$

The parameter count serves as a metric for evaluating the complexity and resource consumption of a model. Generally, a higher parameter count indicates a more complex model, requiring more computational resources and memory space for training and execution. GFLOPs represent the number of floating-point operations executed by the model per second during inference, and can be used to assess the model's complexity.

### 3.3. Experimental Results

#### 3.3.1. Experimental Results of w-YOLO on the Walnut Dataset

We trained YOLOv8s and the w-YOLO model, and the training curves are shown in Figure 10. The training results of both YOLOv8s and w-YOLO are the best results obtained when the training converges. From the localization loss curve, it can be observed that the curve of w-YOLO converges faster, indicating that it learns better in the task of walnut localization. This suggests that w-YOLO has a better predictive ability for the location information of walnuts. Additionally, in terms of the detection accuracy metric mAP0.5, w-YOLO achieved 0.970, which is 0.004 higher than YOLOv8s (0.966). This indicates that the feature extraction network, feature fusion network, and detection layer of YOLOv8s were improved, enhancing the detection effectiveness for walnut objects in the w-YOLO model. In actual walnut target detection tasks, although there is only a slight improvement in the mAP0.5 metric with w-YOLO, it signifies that we can ensure overall detection performance and reliability while reducing the model size. Especially for small objects like walnuts, even a small improvement can have a significant impact. The ability of the walnut detection system to perform detection tasks accurately and reliably is crucial for practical deployment and application.

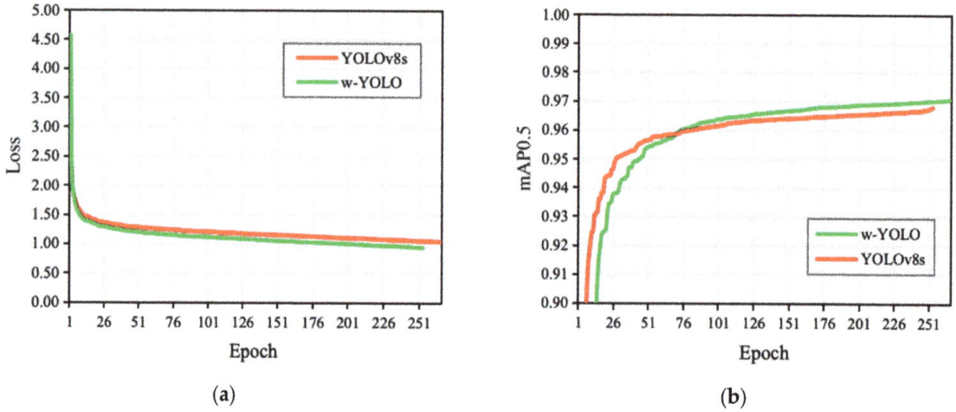


**Figure 10.** Comparison of the training curves of YOLOv8s and the w-YOLO model. (**a**) Loss curves of YOLOv8s and the w-YOLO model; (**b**) mAP0.5 curves of YOLOv8s and the w-YOLO model.

We tested YOLOv8s and w-YOLO using 165 walnut test images. Detailed information about these walnut test images is presented in Table 2. We subdivided the complex lighting conditions into facing light, side light, and backlight, and examples of walnuts in the test dataset under facing light, side light, and backlight are shown in (a–c) of Figure 11, respectively. Figure 12 illustrates the visual comparison of the detection results between YOLOv8s and w-YOLO. In the first row, it can be observed that YOLOv8s missed detections when walnuts were in backlighting conditions, whereas w-YOLO, after improvements, maintained stable accuracy even in such extreme lighting environments. In the second row, the circled walnuts are partially occluded and illuminated from the front. Due to the enhanced feature-capturing capability of w-YOLO, its detection performance is notably better than that of YOLOv8s. When the lighting on the walnut surface is uneven (third row), YOLOv8s struggles to distinguish between the features of leaves and walnuts, while w-YOLO still ensures accurate detection under such complex lighting conditions.

**Table 2.** The total number of targets in the test images and the number of targets under different lighting conditions.

| Target Number (pcs) | Facing Light (pcs) | Side Light (pcs) | Backlight (pcs) |
|---|---|---|---|
| 391 | 137 | 118 | 136 |

Note: pcs is the abbreviation for pieces.

Taking into account the lightweight design and detection performance of the model, we applied some lightweight optimizations to w-YOLO, significantly reducing the model's parameter count while improving detection accuracy. The comparison results between YOLOv8s and w-YOLO on metrics including P, R, F1-Score, Weighted file size, and Parameters are presented in Table 3. Thanks to the lightweight backbone network and C2f-Faster in the Neck, the Weighted file size of w-YOLO (11.1 MB) was reduced by 50.7%, and Parameters (5.31 M) decreased by 52.3%. w-YOLO achieves a good balance between model size and detection performance, which holds scientific value for future research in edge computing.

(a) (b) (c)

**Figure 11.** Walnuts under complex lighting conditions. The walnuts in the red circles represent examples of walnuts under different lighting conditions. (**a**) Facing light. (**b**) Side light. (**c**) Backlight.

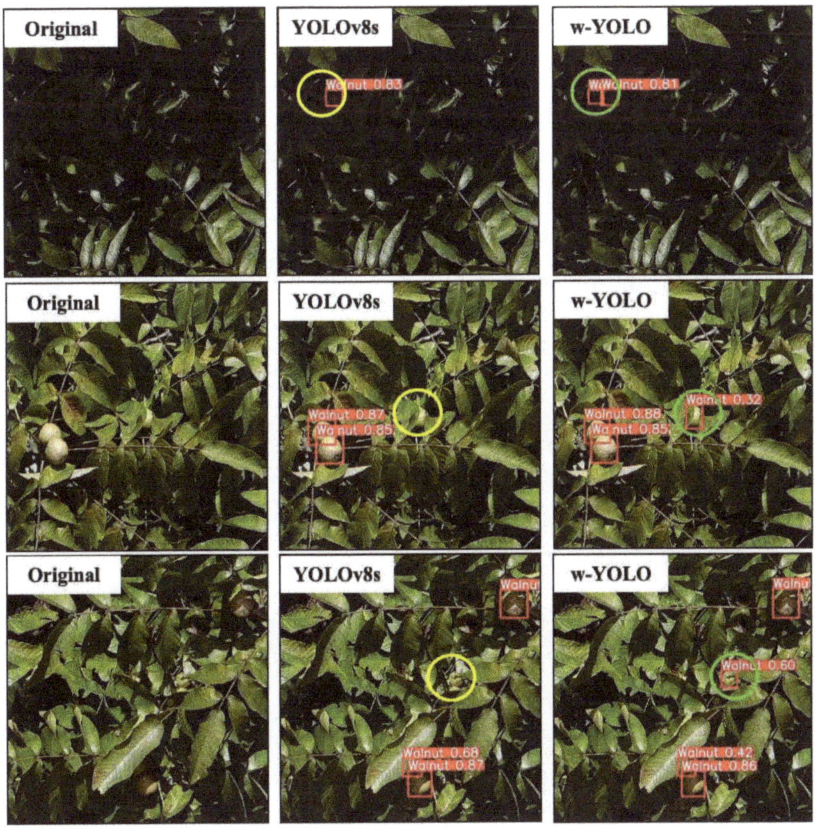


(a) (b) (c)

**Figure 12.** Visualization comparison of detection results between YOLOv8s and w-YOLO. The walnuts circled in yellow represent the walnuts missed by YOLOv8s, while those circled in green represent the detection results of w-YOLO for the walnuts missed by YOLOv8s. (**a**) Original image; (**b**) Detection result of YOLOv8s; (**c**) Detection result of w-YOLO.

Table 3. Comparison Results between YOLOv8s and w-YOLO.

| Model | P | R | F1-Score | Weighted File Size (MB) | Parameters (M) |
|---|---|---|---|---|---|
| YOLOv8 | 0.886 | 0.99 | 0.91 | 22.5 | 11.13 |
| w-YOLO | 0.900 | 0.99 | 0.92 | $11.1_{(-50.7\%)}$ | $5.31_{(-52.3\%)}$ |

### 3.3.2. Comparison with Other Popular Models

To demonstrate the superiority of w-YOLO in walnut detection under complex lighting conditions, we compared it with many other mainstream object detection models, including YOLOv3, YOLOv3-spp, YOLOv5s, YOLOv5m, YOLOv7, YOLOv7-Tiny, YOLOv8s, and YOLOv8m. The results of the comparative experiments are shown in Table 4. Compared to other models, w-YOLO exhibits the best detection performance and also has significant advantages in terms of model size.

Table 4. The comparative experimental results between w-YOLO and other mainstream models.

| Model | F1-Score | mAP0.5 | Parameters (M) | Weighted File Size (MB) |
|---|---|---|---|---|
| YOLOv3 | 0.92 | 0.956 | 61.50 | 123.5 |
| YOLOv3-spp | 0.92 | 0.960 | 62.55 | 125.6 |
| YOLOv5s | 0.91 | 0.959 | 7.01 | 14.4 |
| YOLOv5m | 0.92 | 0.964 | 20.85 | 42.2 |
| YOLOv7 | 0.91 | 0.962 | 37.20 | 71.3 |
| YOLOv7-Tiny | 0.88 | 0.936 | 6.01 | 11.7 |
| YOLOv8s | 0.91 | 0.966 | 11.13 | 22.5 |
| YOLOv8m | 0.91 | 0.969 | 25.84 | 52.0 |
| w-YOLO | 0.92 | 0.970 | 5.31 | 11.1 |

In terms of the F1-Score metric, w-YOLO belongs to the top tier, achieving a score of 0.92, similar to larger models such as YOLOv3, YOLOv3-spp, and YOLO5m, indicating that w-YOLO exhibits stronger generalization ability. Combined with mAP0.5, it can be observed that w-YOLO effectively balances recall and precision. Compared to the baseline model YOLOv8s and models with larger parameter counts, w-YOLO demonstrates higher detection accuracy and recall for walnut target detection tasks. In terms of parameters and weighted file size, w-YOLO has only 5.31 M parameters and a weight file size of only 11.1 MB. Compared to models such as YOLOv3, YOLOv3-spp, YOLOv5s, YOLOv5m, YOLOv7, YOLOv7-Tiny, YOLOv8s, and YOLOv8m, w-YOLO also has a significant advantage in model lightweightness, making it more suitable for deployment on edge computing devices.

### 3.3.3. Comparison of Detection Visualization with Other Models

In Section 3.3.2, we conducted a quantitative analysis. In this section, qualitative analysis is performed to visually demonstrate the detection capability of w-YOLO (Figure 13). In the images to be detected, there are a total of 10 walnut targets, with 5 in backlight, 3 in facing light, and 2 receiving only partial illumination.

From the perspective of walnut target detection under different lighting conditions, all models except YOLOv5m identified three walnut fruits illuminated by facing light and two under partial illumination. When detecting walnut targets in backlight conditions, YOLOv5s, YOLOv7, and YOLOv7-Tiny missed three targets, YOLOv3, YOLOv3-spp, YOLOv5m, and YOLOv8s missed two targets, while YOLOv8m and w-YOLO only missed one target.

### 3.3.4. Ablation Experiment

In this section, we discuss in detail the role of each module in the YOLOv8s model. A series of ablation experiments were conducted using YOLOv8s as the baseline model. The results of the ablation experiments are shown in Table 5. During the experiments, we sequentially introduced C2f-Faster, BiFPN, FasterNet, DyHead, and S2 to the model. Here, S2 refers to the operation of adding a 160 × 160 detection head and removing the 20 × 20 detection head.

**Figure 13.** The visual detection results of w-YOLO compared to other mainstream models. The green boxes in the Ground Truth represent all walnut targets in this test image.

**Table 5.** Ablation experiment.

| Group | C2f-Faster | BiFPN | FasterNet | DyHead | S2 | mAP0.5 | Parameters (M) | Weighted File Size (MB) |
|---|---|---|---|---|---|---|---|---|
| A | | | | | | 0.966 | 11.13 | 22.5 |
| B | √ | | | | | $0.967_{(+0.001)}$ | $9.75_{(-1.38)}$ | $19.8_{(-2.7)}$ |
| C | √ | √ | | | | $0.966_{(-0.001)}$ | $7.29_{(-2.46)}$ | $14.9_{(-4.9)}$ |
| D | √ | √ | √ | | | $0.963_{(-0.003)}$ | $4.76_{(-2.53)}$ | $9.8_{(-5.1)}$ |
| E | √ | √ | √ | √ | √ | $0.970_{(+0.007)}$ | $5.31_{(+0.55)}$ | $11.1_{(+1.3)}$ |

Group A represents the baseline model YOLOv8s. After improving the Neck with C2f (Group B), the model's Parameters and Weighted file size slightly decreased, but it did not negatively affect the walnut object detection accuracy. In experimental Group C, the BiFPN feature fusion structure further reduced the size of the model. Although the mAP0.5 decreased by 0.001 compared to Group B, it still maintained the initial performance. Building on Group C, we replaced the model's backbone network with the FasterNet structure (Group D). At this point, the model improved significantly in terms of parameters and weighted file size, decreasing by 2.53 M and 5.1 MB, respectively. To mitigate the negative impact of lightweighting, we enhanced the model's detection head. In experimental Group E, with the effect of DyHead and S2, the model's mAP0.5 increased by 0.007, while there was only a slight increase in model size.

## 4. Discussion

### *4.1. The Impact of Data Augmentation Parameters on the Model*

After the images are inputted into the w-YOLO model, they first pass through the data augmentation module. Geometry-based data augmentation is equivalent to introducing variations in viewpoint and spatial position within the dataset, thereby enhancing the model's robustness in these aspects and improving testing accuracy [47]. Therefore, To delve deeper into the impact of the data augmentation module on the detection performance of w-YOLO, we analyzed the following parameters: image rotation (Degree), image translation (Translate), image scale (Scale), image perspective (Perspective), image flip up–down (Flipud), image flip left–right (Fliplr), image mosaic (Mosaic), image mixup (Mixup), and segment copy–paste (Copy_paste). The corresponding experimental results are shown in Figure 14a–i. From the figures, it can be observed that when Degree, Translate, Scale, Flipud, Fliplr, and Mosaic are set to −5, 0.45, 0.7, 0.7, 0.5, and 0.7, respectively, they have a beneficial effect on the model. However, using Perspective, Mixup, and Copy_paste may have a negative impact on the model.

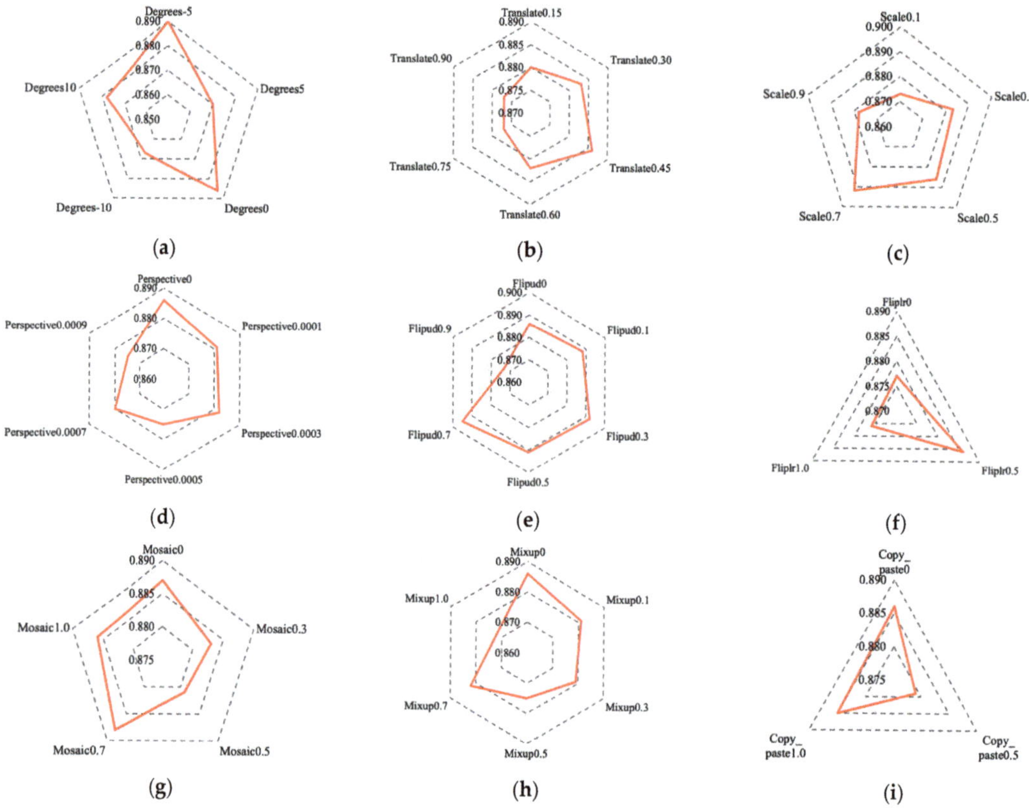


**Figure 14.** mAP0.5 values of w-YOLO under different parameters of various data augmentation methods. (**a–i**) represent the experimental results of image rotation, image translation, image scale, image perspective, image flip up–down, image flip left–right, image mosaic, image mixup, and segment copy–paste, respectively.

## 4.2. Detection Layer Analysis

One of the drawbacks of the detection head in the YOLO algorithm is that, since the detection head typically operates at the final layer of the network, it may miss some low-level detailed information. This can result in lower detection accuracy, particularly for small objects or in complex scenes [48,49]. Therefore, we redesigned the detection head of YOLO and thoroughly analyzed the performance of the detection head in w-YOLO.

### 4.2.1. Effect of the Number of Dyhead Blocks on Model Performance

To explore the impact of the number of DyHead blocks (Block_num) on the detection performance of w-YOLO, experiments were conducted by adding 1, 2, 3, 4, and 5 DyHead blocks. From Table 6, it can be observed that as the number of DyHead blocks increases, the model's complexity also increases. When the number is 1, the model achieves the optimal mean Average Precision (mAP0.5) value (0.970), with the lowest values observed for Layer, Parameters, GFLOPs, and Weighted file size indicators. This indicates that adding more

DyHead blocks does not necessarily imply stronger feature-capturing capability for the model. Therefore, increasing the depth of the model could have a negative impact on walnut detection.

**Table 6.** Comparison Experiment of Different Numbers of DyHead Blocks.

| Block_Num | mAP0.5 | Layer | Parameters (M) | GFLOPs | Weighted File Size (MB) |
|---|---|---|---|---|---|
| 1 | 0.970 | 349 | 5.31 | 41.3 | 11.1 |
| 2 | 0.968 | 373 | 5.80 | 43.5 | 12.1 |
| 3 | 0.968 | 397 | 6.30 | 45.6 | 13.1 |
| 4 | 0.967 | 421 | 6.79 | 47.7 | 14.1 |
| 5 | 0.965 | 445 | 7.29 | 49.8 | 15.1 |

#### 4.2.2. Effect of S2 Structure on Model Performance

The walnut target detection in UAV remote sensing images falls under the small object detection category. Therefore, in the design of w-YOLO, an additional detection layer specifically tailored for small objects with dimensions of 160 × 160 was added to address the detection of smaller walnut targets. To further compress the size of w-YOLO, the detection layer with dimensions of 20 × 20, suitable for larger targets, was removed.

In this section, we provide a detailed comparison of the experimental results regarding different combinations of detection head sizes to investigate their impact on the model's detection performance and parameter count. Using YOLOv8s as the baseline, the experimental results are presented in Table 7. From the table, it is evident that the metrics P (0.928) and mAP0.5 (0.968) in Group C outperform those in Groups A and B, while the values of layers and GFLOPs increased by only 13 and 5.7, respectively. Overall, the size configuration of Group C effectively balances the model's detection performance and complexity.

**Table 7.** Ablation experiment of different detection layers.

| Group | 20 × 20 | 40 × 40 | 80 × 80 | 160 × 160 | P | mAP0.5 | Layer | GFLOPs | Parameters (M) |
|---|---|---|---|---|---|---|---|---|---|
| A | √ | √ | √ | | 0.886 | 0.966 | 168 | 28.4 | 11.13 |
| B | √ | √ | √ | √ | 0.917 | 0.967 | 207 | 36.6 | 10.63 |
| C | | √ | √ | √ | 0.928 | 0.968 | 181 | 34.1 | 7.40 |

### 4.3. *Analysis of Other Model Parameters*

Optimizers and learning rates play crucial roles in the model training process. Choosing different optimizers can have varied effects on the model's performance. Utilizing appropriate optimizers can facilitate faster and more stable model convergence during training. Similarly, learning rates play a significant role, and setting them scientifically (Table 8) enables the model to converge to the optimal solution more quickly, accelerating the training process. Figure 15 depicts the training loss curves obtained with different optimizers. From the zoomed-in plots, it can be observed that SGD converges notably faster, followed by Adamax. At the convergence point towards the end of the curves, w-YOLO trained with the SGD optimizer stops training early around 200 epochs, while the curve for Adamax continues to descend. The training loss for w-YOLO using Adamax is lower, which is advantageous for detecting walnut objects.

**Table 8.** Results of mAP0.5 under different final learning rate values.

| Value | mAP0.5 |
|---|---|
| 0.03 | 0.967 |
| 0.05 | 0.968 |
| 0.07 | 0.970 |
| 0.10 | 0.967 |

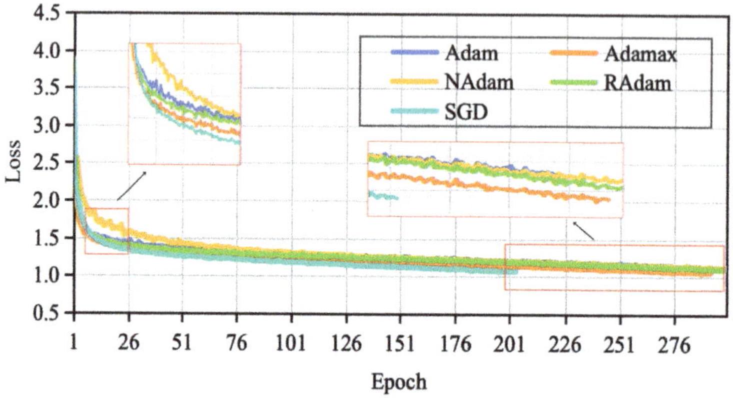


**Figure 15.** Training loss curves of different optimizers.

### *4.4. Model Performance Advantages and Limitations*

Although w-YOLO achieves a certain degree of lightweight design without sacrificing detection accuracy, our research still has the following limitations: (1) The GFLOPs of the detection head in the original size are only 28.4, and in the design of S2 (34.1), we focus more on parameter count and mAP0.5, but this increases the computational load of the model. (2) Despite w-YOLO outperforming YOLOv3, YOLOv3-spp, YOLOv5s, YOLOv5m, YOLOv7, YOLOv7-Tiny, and YOLOv8s in detecting walnut targets under backlight conditions, there are still cases of missed detections. (3) Although we made some progress in walnut object detection tasks under facing light, side light, and backlight conditions, we did not further analyze the detection of occluded walnuts. In future research, we will strive to investigate the impact of walnut occlusion on object detection more deeply and continuously optimize the detection performance of w-YOLO to make it applicable to a wider range of walnut detection tasks.

## 5. Conclusions

In walnut agriculture production, yield prediction is a crucial step, and traditional manual counting methods face significant challenges in hilly areas. Given the advantages of deep learning models and low-altitude remote sensing technology in agricultural production, in this study, we constructed a walnut small object dataset using high-resolution aerial images captured by UAVs, addressing the problem of data scarcity in this research field. The dataset consists of 2490 images, totaling 12,138 walnut targets. In hilly areas, the complex lighting conditions experienced by walnut fruits during UAV data collection to some extent affect the accuracy of the model. Therefore, based on the YOLOv8s model, we made a series of improvements to obtain w-YOLO, including the utilization of FasterNet, C2f-Faster, and BiFPN to simplify the model's feature extraction and fusion networks, reducing parameters by 6.37 M and shrinking the weight file size to 9.8 MB. Additionally, we employed a DyHead detection layer with attention mechanisms and redesigned a detection head combination more suitable for walnut object identification. In the walnut recognition task in complex lighting conditions of UAV remote sensing images, w-YOLO achieved a mAP0.5 of 97%, an increase of 0.4% compared to YOLOv8s, with parameters and weight file size reduced by 52.3% and 50.7%, respectively. It is worth noting that our study focuses on model lightweight and enabling w-YOLO to adapt to walnut fruit detection under different lighting conditions. The detection performance of w-YOLO under backlighting was significantly improved compared to the original model, but there are still instances of missed detections, making walnut identification under backlighting conditions challenging. Furthermore, w-YOLO has shown excellent detection results under facing and side lighting. We believe that the lightweight w-YOLO can provide valuable assistance for walnut production management and support the development of edge hardware devices for walnut detection.

However, we recognize that there is still significant room for improvement in the robustness of walnut recognition models. Therefore, our walnut dataset still needs to be further expanded, such as adding walnut data in different occlusion scenarios and multispectral walnut image data. In future research, we will also conduct radar-based three-dimensional modeling of walnut forests and calculate vegetation indices to provide more valuable resources for walnut agriculture production research.

**Author Contributions:** Conceptualization, M.W. and L.Y.; methodology, M.W.; software, M.W.; validation, M.W., Y.X. and Z.C.; formal analysis, C.X.; investigation, C.X.; resources, M.W. and L.Y.; data curation, M.W. and C.X.; writing—original draft preparation, M.W.; writing—review and editing, M.W.; visualization, M.W.; supervision, L.Y.; funding acquisition, L.Y. All authors have read and agreed to the published version of the manuscript.

**Funding:** This study was supported by Key Project of Yunnan Basic Research Program, grant number 202401AS070034.

**Institutional Review Board Statement:** Not applicable.

**Data Availability Statement:** The data used in this study is available by contacting the corresponding author.

**Conflicts of Interest:** The authors declare no conflicts of interest.

## References

1. Regueiro, J.; Sánchez-González, C.; Vallverdú-Queralt, A.; Simal-Gándara, J.; Lamuela-Raventós, R.; Izquierdo-Pulido, M. Comprehensive identification of walnut polyphenols by liquid chromatography coupled to linear ion trap–Orbitrap mass spectrometry. *Food Chem.* **2014**, *152*, 340–348. [CrossRef] [PubMed]
2. Aradhya, M.K.; Potter, D.; Gao, F.; Simon, C.J. Molecular phylogeny of Juglans (Juglandaceae): A biogeographic perspective. *Tree Genet. Genomes* **2007**, *3*, 363–378. [CrossRef]
3. Li, H.; Han, J.; Zhao, Z.; Tian, J.; Fu, X.; Zhao, Y.; Wei, C.; Liu, W. Roasting treatments affect oil extraction rate, fatty acids, oxidative stability, antioxidant activity, and flavor of walnut oil. *Front. Nutr.* **2023**, *9*, 1077081. [CrossRef] [PubMed]
4. Tang, Y.; Chen, M. The impact of agricultural digitization on the high-quality development of agriculture: An empirical test based on provincial panel data. *Land* **2022**, *11*, 2152. [CrossRef]
5. Zhang, X.; Fan, D. Can agricultural digital transformation help farmers increase income? An empirical study based on thousands of farmers in Hubei Province. *Environ. Dev. Sustain.* **2023**, 1–27. [CrossRef] [PubMed]
6. Momoh, O.Y.; Akpoko, J.G.; Akinola, M.O. Impact of agricultural services and training centre project on tomato farmers' livelihood in Plateau State, Nigeria. *J. Agric. Ext.* **2018**, *22*, 35–43. [CrossRef]
7. Zheng, Y.; Sarigul, E.; Panicker, G.; Stott, D. Vineyard LAI and canopy coverage estimation with convolutional neural network models and drone pictures. In Proceedings of the Sensing for Agriculture and Food Quality and Safety XIV. SPIE, Orlando, FL, USA, 3–7 April 2022; Volume 12120, pp. 29–38.
8. Li, X.; Ba, Y.; Zhang, M.; Nong, M.; Yang, C.; Zhang, S. Sugarcane nitrogen concentration and irrigation level prediction based on UAV multispectral imagery. *Sensors* **2022**, *22*, 2711. [CrossRef] [PubMed]
9. Weyler, J.; Quakernack, J.; Lottes, P.; Behley, J.; Stachniss, C. Joint plant and leaf instance segmentation on field-scale UAV imagery. *IEEE Robot. Autom. Lett.* **2022**, *7*, 3787–3794. [CrossRef]
10. Reddy, S.R.G.; Varma, G.P.S.; Davuluri, R.L. Optimized convolutional neural network model for plant species identification from leaf images using computer vision. *Int. J. Speech Technol.* **2023**, *26*, 23–50. [CrossRef]
11. Prasad, A.; Mehta, N.; Horak, M.; Bae, W.D. A two-step machine learning approach for crop disease detection using GAN and UAV technology. *Remote Sens.* **2022**, *14*, 4765. [CrossRef]
12. Reedha, R.; Dericquebourg, E.; Canals, R.; Hafiane, A. Transformer neural network for weed and crop classification of high resolution UAV images. *Remote Sens.* **2022**, *14*, 592. [CrossRef]
13. Shorewala, S.; Ashfaque, A.; Sidharth, R.; Verma, U. Weed density and distribution estimation for precision agriculture using semi-supervised learning. *IEEE Access* **2021**, *9*, 27971–27986. [CrossRef]
14. Cheng, X.; Zhang, Y.; Chen, Y.; Wu, Y.; Yue, Y. Pest identification via deep residual learning in complex background. *Comput. Electron. Agric.* **2017**, *141*, 351–356. [CrossRef]
15. Behroozi-Khazaei, N.; Maleki, M.R. A robust algorithm based on color features for grape cluster segmentation. *Comput. Electron. Agric.* **2017**, *142*, 41–49. [CrossRef]
16. Arribas, J.I.; Sánchez-Ferrero, G.V.; Ruiz-Ruiz, G.; Gómez-Gil, J. Leaf classification in sunflower crops by computer vision and neural networks. *Comput. Electron. Agric.* **2011**, *78*, 9–18. [CrossRef]

17. Zhao, W.; Ding, H.; Na, J.; Li, M. Height estimation from single aerial imagery using contrastive learning based multi-scale refinement network. *Int. J. Digit. Earth* **2023**, *16*, 2346–2364. [CrossRef]
18. Song, W.; Yan, D.; Yan, J.; Hu, C.; Wu, W.; Wang, X. Ship detection and identification in SDGSAT-1 glimmer images based on the glimmer YOLO model. *Int. J. Digit. Earth* **2023**, *16*, 4687–4706. [CrossRef]
19. Hartmann, M.C.; Purves, R.S. Seeing through a new lens: Exploring the potential of city walking tour videos for urban analytics. *Int. J. Digit. Earth* **2023**, *16*, 2555–2573. [CrossRef]
20. Yang, K.; Ye, Z.; Liu, H.; Su, X.; Yu, C.; Zhang, H.; Lai, R. A new framework for GEOBIA: Accurate individual plant extraction and detection using high-resolution RGB data from UAVs. *Int. J. Digit. Earth* **2023**, *16*, 2599–2622. [CrossRef]
21. Huang, F.; Chen, S.; Wang, Q.; Chen, Y. Using deep learning in an embedded system for real-time target detection based on images from an unmanned aerial vehicle: Vehicle detection as a case study. *Int. J. Digit. Earth* **2023**, *16*, 910–936. [CrossRef]
22. Tu, X.; Yuan, Z.; Liu, B.; Liu, J.; Hu, Y.; Hua, H.; Wei, L. An improved YOLOv5 for object detection in visible and thermal infrared images based on contrastive learning. *Front. Phys.* **2023**, *11*, 354. [CrossRef]
23. Girshick, R.; Donahue, J.; Darrell, T.; Malik, J. Rich feature hierarchies for accurate object detection and semantic segmentation. In Proceedings of the IEEE Conference on Computer Vision and Pattern Recognition, Columbus, OH, USA, 23–28 June 2014; pp. 580–587.
24. Girshick, R. Fast r-cnn. In Proceedings of the IEEE International Conference on Computer Vision, Santiago, Chile, 7–13 December 2015; pp. 1440–1448.
25. Ren, S.; He, K.; Girshick, R.; Sun, J. Faster R-CNN: Towards real-time object detection with region proposal networks. *IEEE Trans. Pattern Anal. Mach. Intell.* **2016**, *39*, 1137–1149. [CrossRef]
26. He, K.; Gkioxari, G.; Dollár, P.; Girshick, R. Mask r-cnn. In Proceedings of the IEEE International Conference on Computer Vision, Venice, Italy, 22–29 October 2017; pp. 2961–2969.
27. Redmon, J.; Divvala, S.; Girshick, R.; Farhadi, A. You only look once: Unified, real-time object detection. In Proceedings of the IEEE Conference on Computer Vision and Pattern Recognition, Las Vegas, NV, USA, 27–30 June 2016; pp. 779–788.
28. Liu, W.; Anguelov, D.; Erhan, D.; Szegedy, C.; Reed, S.E.; Fu, C.-Y.; Berg, A. Ssd: Single shot multibox detector. In Proceedings of the Computer Vision–ECCV 2016: 14th European Conference, Amsterdam, The Netherlands, 11–14 October 2016; Proceedings, Part I 14. Springer International Publishing: Cham, Switzerland, 2016; pp. 21–37.
29. Duan, K.; Bai, S.; Xie, L.; Qi, H.; Huang, Q.; Tian, Q. Centernet: Keypoint triplets for object detection. In Proceedings of the IEEE/CVF International Conference on Computer Vision, Seoul, Republic of Korea, 27 October–2 November 2019; pp. 6569–6578.
30. Redmon, J.; Farhadi, A. Yolov3: An incremental improvement. *arXiv* **2018**, arXiv:1804.02767.
31. Wang, C.Y.; Bochkovskiy, A.; Liao, H.Y.M. YOLOv7: Trainable bag-of-freebies sets new state-of-the-art for real-time object detectors. In Proceedings of the IEEE/CVF Conference on Computer Vision and Pattern Recognition, Vancouver, BC, Canada, 17–24 June 2023; pp. 7464–7475.
32. Chen, Y.; Xu, H.; Zhang, X.; Gao, P.; Xu, Z.; Huang, X. An object detection method for bayberry trees based on an improved YOLO algorithm. *Int. J. Digit. Earth* **2023**, *16*, 781–805. [CrossRef]
33. Hao, J.; Bing, Z.; Yang, S.; Yang, J.; Sun, L. Detection of green walnut by improved YOLOv3. *Trans. Chin. Soc. Agric. Eng.* **2022**, *38*, 183–190. (In Chinese)
34. Zhong, Z.; Yun, L.; Yang, X.; Chen, Z. Research on Walnut Recognition Algorithm in Natural Environment Based on Improved YOLOX. *J. Henan Agric. Sci.* **2024**, *53*, 152–161. (In Chinese)
35. Li, Q.; Luo, Z.; He, X.; Chen, H. LA_YOLOx: Effective Model to Detect the Surface Defects of Insulative Baffles. *Electronics* **2023**, *12*, 2035. [CrossRef]
36. Tan, M.; Pang, R.; Le, Q.V. Efficientdet: Scalable and efficient object detection. In Proceedings of the IEEE/CVF Conference on Computer Vision and Pattern Recognition, Seattle, WA, USA, 13–19 June 2020; pp. 10781–10790.
37. Jocher, G. Ultralytics YOLOv8. 2023. Available online: https://github.com/ultralytics/ultralytics (accessed on 17 January 2024).
38. Lin, T.Y.; Dollár, P.; Girshick, R.; He, K.; Hariharan, B.; Belongie, S. Feature pyramid networks for object detection. In Proceedings of the IEEE Conference on Computer Vision and Pattern Recognition, Honolulu, HI, USA, 21–26 July 2017; pp. 2117–2125.
39. Liu, S.; Qi, L.; Qin, H.; Shi, J.; Jia, J. Path aggregation network for instance segmentation. In Proceedings of the IEEE Conference on Computer Vision and Pattern Recognition, Salt Lake City, UT, USA, 18–23 June 2018; pp. 8759–8768.
40. Dai, X.; Chen, Y.; Xiao, B.; Chen, D.; Liu, M.; Yuan, L.; Zhang, L. Dynamic head: Unifying object detection heads with attentions. In Proceedings of the IEEE/CVF Conference on Computer Vision and Pattern Recognition, Nashville, TN, USA, 20–25 June 2021; pp. 7373–7382.
41. Sandler, M.; Howard, A.; Zhu, M.; Zhmoginov, A.; Chen, L.-C. Mobilenetv2: Inverted residuals and linear bottlenecks. In Proceedings of the IEEE Conference on Computer Vision and Pattern Recognition, Salt Lake City, UT, USA, 18–23 June 2018; pp. 4510–4520.
42. Zhang, X.; Zhou, X.; Lin, M.; Sun, J. Shufflenet: An extremely efficient convolutional neural network for mobile devices. In Proceedings of the IEEE Conference on Computer Vision and Pattern Recognition, Salt Lake City, UT, USA, 18–23 June 2018; pp. 6848–6856.
43. Han, K.; Wang, Y.; Tian, Q.; Guo, J.; Xu, C.; Xu, C. Ghostnet: More features from cheap operations. In Proceedings of the IEEE/CVF Conference on Computer Vision and Pattern Recognition, Seattle, WA, USA, 13–19 June 2020; pp. 1580–1589.

44. Mehta, S.; Rastegari, M. Mobilevit: Light-weight, general-purpose, and mobile-friendly vision transformer. *arXiv* **2021**, arXiv:2110.02178.
45. Chen, Y.; Dai, X.; Chen, D.; Liu, M.; Dong, X.; Yuan, L.; Liu, Z. Mobile-former: Bridging mobilenet and transformer. In Proceedings of the IEEE/CVF Conference on Computer Vision and Pattern Recognition, New Orleans, LA, USA, 18–24 June 2022; pp. 5270–5279.
46. Chen, J.; Kao, S.; He, H.; Zhuo, W.; Wen, S.; Lee, C.-H.; Chan, G. Run, Don't Walk: Chasing Higher FLOPS for Faster Neural Networks. In Proceedings of the IEEE/CVF Conference on Computer Vision and Pattern Recognition, Vancouver, BC, Canada, 17–24 June 2023; pp. 12021–12031.
47. Ma, D.; Tang, P.; Zhao, L.; Zhang, Z. Review of data augmentation for image in deep learning. *J. Image Graph.* **2021**, *26*, 0487–0502. (In Chinese)
48. Ge, Z.; Liu, S.; Wang, F.; Li, Z.; Sun, J. Yolox: Exceeding yolo series in 2021. *arXiv* **2021**, arXiv:2107.08430.
49. Wu, M.; Yun, L.; Wang, Y.; Chen, Z.; Cheng, F. Detection algorithm for dense small objects in high altitude image. *Digit. Signal Process.* **2024**, *146*, 104390. [CrossRef]

*agriculture*

MDPI

*Article*

# Optimizing the YOLOv7-Tiny Model with Multiple Strategies for Citrus Fruit Yield Estimation in Complex Scenarios

**Juanli Jing [1], Menglin Zhai [1], Shiqing Dou [1,*], Lin Wang [1], Binghai Lou [2], Jichi Yan [3] and Shixin Yuan [1]**

[1] College of Geomatics and Geoinformation, Guilin University of Technology, Guilin 541006, China; 2003080@glut.edu.cn (J.J.); 2120222042@glut.edu.cn (M.Z.); 2120211899@glut.edu.cn (L.W.); 1020232072@glut.edu.cn (S.Y.)

[2] Guangxi Academy of Specialty Crops, Guilin 541004, China; binghai.lou@hotmail.com

[3] College of Mechanical and Control Engineering, Guilin University of Technology, Guilin 541006, China; 2019178@glut.edu.cn

* Correspondence: doushiqing@glut.edu.cn; Tel.: +86-18278367609

**Abstract:** The accurate identification of citrus fruits is important for fruit yield estimation in complex citrus orchards. In this study, the YOLOv7-tiny-BVP network is constructed based on the YOLOv7-tiny network, with citrus fruits as the research object. This network introduces a BiFormer bilevel routing attention mechanism, which replaces regular convolution with GSConv, adds the VoVGSCSP module to the neck network, and replaces the simplified efficient layer aggregation network (ELAN) with partial convolution (PConv) in the backbone network. The improved model significantly reduces the number of model parameters and the model inference time, while maintaining the network's high recognition rate for citrus fruits. The results showed that the fruit recognition accuracy of the modified model was 97.9% on the test dataset. Compared with the YOLOv7-tiny, the number of parameters and the size of the improved network were reduced by 38.47% and 4.6 MB, respectively. Moreover, the recognition accuracy, frames per second (FPS), and F1 score improved by 0.9, 2.02, and 1%, respectively. The network model proposed in this paper has an accuracy of 97.9% even after the parameters are reduced by 38.47%, and the model size is only 7.7 MB, which provides a new idea for the development of a lightweight target detection model.

**Keywords:** citrus identification; computer vision; YOLOv7-tiny; deep learning; load estimation

## 1. Introduction

In the last decade, China has been a world leader in citrus fruit production [1], which has become an important factor in the economic growth of agricultural cultivation. In citrus fruit production, preharvest yield estimation is an essential tool for assessing fruit quality and planting techniques and for reflecting the market supply and demand; moreover, this method is crucial for guiding agricultural production [2]. Traditional methods for estimating production usually involve manual sampling, which is inefficient and labor intensive. Currently, deep learning technology has shown excellent advantages in image detection, and the fast and accurate detection ability of a target detection network can provide a more efficient means for citrus fruit yield estimation and information counting.

In recent years, object detection has played an important role in the field of agriculture. Sozzi et al. [3] used a detection network to determine grape yield estimation. Wang et al. [4] employed the YOLOv5 network to detect small apples for early yield estimation. Cardellicchio et al. [5] used the YOLOv5 detection network to study the phenotypic characteristics of tomato plants, so that tomato fruits and flowers could be accurately identified. In addition, the target detection networks have also demonstrated a good detection performance in fruit identification and counting [6–10]. Bi et al. [11] used migration learning to train the FastR-CNN model for citrus fruit recognition with an average accuracy of 86.6%. Chen et al. [12]

improved the YOLOv7 network for citrus fruit detection by introducing a small-target detection layer and lightweight convolution with a model size of 24.26 M. However, the above model suffers from an excessive number of parameters and computational effort. Therefore, the lightweight YOLO series of inspection networks is the preferred choice [13–15]. Lightweight networks can increase the inference speed and reduce the model size through cutting the depth and width of the model. Huang et al. [16] improved upon the YOLOv5 lightweight model via introducing the CBAM attention mechanism, which has an average recognition accuracy of 91.3% for citrus fruits. Wang et al. [17] used the YOLOv5 model by introducing migration learning to detect apples with a model size of 4.07 M, but the detection accuracy was only 83.1%. Ma et al. [18] introduced a BiFPN module based on YOLOv7-tiny for detecting apples under different weather conditions, with an accuracy of 80.1%. Wang et al. [19] replaced the regular convolution in the YOLOv7-tiny backbone network by using variability convolution and introducing an SE attention mechanism for detecting millet chili at different maturity levels, with a model accuracy of 90.3%. In the above studies, most of the networks focused on local area feature extraction and matching, and had a lightweight character while losing the semantic information of the target fruits. Thus, they are prone to both missing and misdetecting target fruits when performing a multiscale target detection task, resulting in a lower accuracy in complex scenarios or small-target contexts. Therefore, to further improve the network's ability to detect citrus plants in complex environments and to meet the current demand for lightweight edge devices, there is an urgent need to explore in-depth strategies and methods to further enhance the detection capability of lightweight YOLO networks.

The purpose of this study is to solve the problems of redundant parameters and the high computation of target detection networks in complex scenarios. We propose a lightweight and high-precision network, YOLOv7-tiny-BVP, through module replacement and parameter optimization of reconfigurable network layers. The network includes the BiFormer attention mechanism, which improves the capture of spatial and contextual semantic information about citrus plants via capturing bidirectional dependencies in the input sequence. Moreover, regular convolution is replaced with GSConv in the neck network, and the VoVGSCSP module is introduced to improve detection accuracy and to reduce the model inference time. In the backbone network, the simplified ELAN is replaced with PConv, which leads to a more efficient feature extraction and reduces the number of network computations. The network is lightweight while maintaining high performance, which opens up the possibility of future applications of lightweight networks.

## 2. Materials and Methods

### 2.1. Data Acquisition

The citrus fruit images were collected at the Pengyu Brothers Citrus Demonstration Base in Gongcheng County, Guilin City, Guangxi Zhuang Autonomous Region. The data were collected on 11 November 2022, at 2:00–5:00 p.m. In this study, to improve sample representativeness, model applicability, and device compatibility, images were acquired from the Huawei Mate40 Pro phone and Nova 7 phone. To ensure the representativeness of the sample data, we obtain citrus fruit images at different distances varying from 0.2 to 2.0 m away from the fruit trees, and various real growth conditions were taken into account, such as upward view, downward view, downlight, backlight, fruit sparseness, denseness, shading, overlapping, etc. Finally, a total of 525 high-definition citrus fruit images with a resolution of 4096 × 3072 were captured in JPG format.

### 2.2. Data Enhancement

In the natural environment, there are a variety of disturbing factors, such as plant stems, leaves, and light. To improve the robustness of the model and prevent any overfitting caused by too little training data [20], 182 representative images were selected for data augmentation in this study. First, the original images were manually labeled using the image annotation software LabelImg. Then, the mixup method, the mosaic method, image equalization, gray scaling, and gamma transformation were used for mixed data enhancement. After manual screening, 3182 data points were ultimately obtained. The effect after data enhancement is shown in Figure 1.

(a) (b) (c) (d) (e) (f)

**Figure 1.** Data enhancement chart. (**a**) Mosaic enhancement; (**b**,**c**) Mixup enhancement; (**d**) Mosaic enhancement and gray scaling; (**e**) Gray scaling; (**f**) Gamma transform.

### 2.3. Dataset Preparation

In this study, the data were stored according to the format of Microsoft's publicly available dataset MS COCO, and the dataset was randomly divided into training, validation, and testing sets at a ratio of 7:2:1. The dataset contained 3182 images, including 182 original citrus fruit images and 3000 enhanced images, and 39,483 citrus fruits were labeled. The training set consisted of 2290 images that labeled 28,279 citrus fruits. The validation set consisted of 573 images in which 7181 fruits were labeled. The remaining 319 images comprised the test set, which included 4023 citrus fruits. According to the COCO dataset standard, citrus fruit targets with a resolution of less than 32 pixels × 32 pixels in the image were defined as small targets in this experiment. Table 1 demonstrates the division of the dataset.

**Table 1.** Dataset segmentation.

| Dataset | Proportions | Images | Number of Captures |
|---|---|---|---|
| Training dataset | 70% | 2290 | 28,279 |
| Validation dataset | 20% | 573 | 7181 |
| Test dataset | 10% | 319 | 4023 |
| Full data | 100% | 3182 | 39,483 |

### 2.4. Test Platforms and Parameters

This experiment was based on Windows 10 with the following hardware configuration: an i7-9800X@3.80 GHz CPU, 32 GB of RAM, and an Nvidia GeForce RTX2060 graphics card. We used PyCharm as the IDE, Python version 3.9 as the compiler, and PyTorch 1.17 as the test framework. In addition, CUDA version 12.1 parallel computing was used in combination with the cuDNN version 11.7 deep neural network acceleration library, and OpenCV 4.3.5 was selected as the image processing library.

To ensure the effectiveness of the model, we set the parameters of the initialized model as follows: begin training from 0 without loading the official pretraining weights, fix the random seeds, and accelerate the model training process by means of adjusting the number of threads used by CPU during data loading. The network hyperparameters were configured as follows: the input image size was 640 pixels × 640 pixels, and the batch size was 32. For model optimization, the stochastic gradient descent (SGD) method was used by setting the initial learning rate, momentum factor, and weight decay factor to 0.001, 0.937, and 0.0005, respectively. The model was trained for a total of 300 epochs, the weights were saved every 10 epochs, and the weights with the highest accuracy were ultimately selected for validation and testing.

In this experiment, COCO dataset-related metrics were used to evaluate the model performance. The evaluation metrics included precision (P), recall (R), mean average precision (mAP), F1 score (F1), model parameters, model size, and FPS, and the coefficient of determination $R^2$ was selected as the evaluation indicator of the prediction model. The equations for the aforementioned assessment indicators are shown in (1)–(4):

$$\mathrm{P} = \frac{\mathrm{TP}}{\mathrm{TP} + \mathrm{FP}} \tag{1}$$

$$\mathrm{R} = \frac{\mathrm{TP}}{\mathrm{TP} + \mathrm{FN}} \tag{2}$$

$$\mathrm{mAP} = \frac{\Sigma_1^N \int_0^1 \mathrm{P(R)dR}}{\mathrm{N}} \tag{3}$$

$$\mathrm{F1} = \frac{2 \times \mathrm{P} \times \mathrm{R}}{\mathrm{P} + \mathrm{R}} \tag{4}$$

In the above equations, TP denotes the positive sample predicted by the model, referring to the number of citrus fruits correctly identified, FP denotes the negative sample predicted by the model, referring to the number of citrus fruits incorrectly identified, and FN denotes citrus fruits that are not correctly detected.

## 3. Multi-Strategy Construction of a New YOLOv7-Tiny-BVP Network

### 3.1. YOLOv7-Tiny Network Infrastructure

The YOLOv7 target detection network is a detector proposed by Alexey Bochkovskiy's team in July 2022. On the Microsoft COCO public dataset, YOLOv7 outperforms all currently known detectors in terms of both speed and accuracy. Compared to its predecessors, the authors innovatively proposed the composite model scaling method and the extended efficient layer aggregation network (E-ELAN) to increase the depth and width of the network, which improved the feature expressiveness and detection performance of the network. Moreover, methods such as the dynamic label assignment strategy are used to improve the inference speed and detection accuracy [21].

YOLOv7-tiny is a lightweight network of the YOLOv7 series that consists of three main parts: the backbone, the neck, and the head; its network structure is shown in Figure 2. To achieve the lightweight standard, YOLOv7-tiny reduces the number of convolutions in the ELAN module, MP module, and SPPCSPC module and uses regular convolution instead of the CBL module, but does not fully integrate the advantageous modules in the

YOLOv7 structure, which weakens the feature learning ability of the network to a certain extent [22].

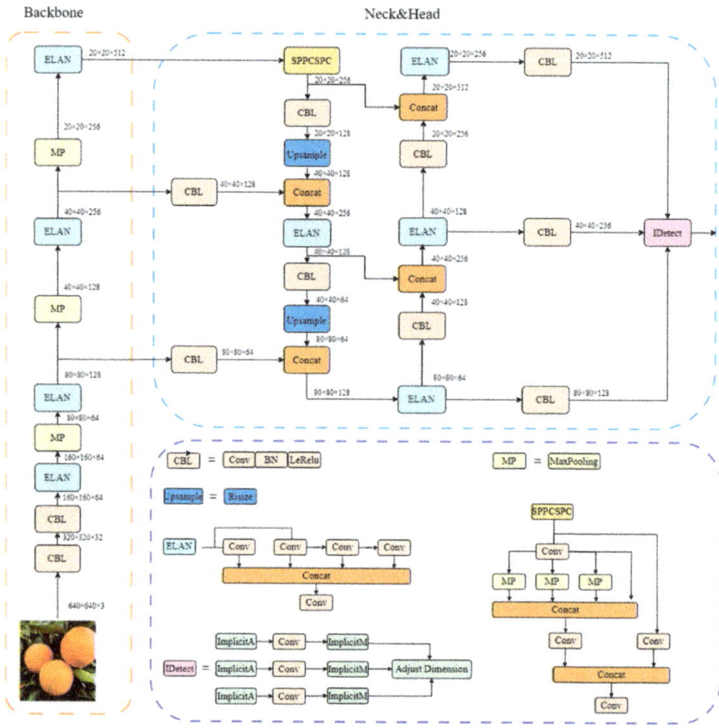


**Figure 2.** Diagram of the YOLOv7-tiny network.

### *3.2. Construction of a New YOLOv7-Tiny-BVP Network Using Multi-Strategy*

This study explored methods for improving the new network based on the YOLOv7-tiny model. That is, the BiFormer attention mechanism is used after the SPPCSPC module, which enhances the network's fruit feature extraction ability in complex scenes; second, in the neck and head part, the regular convolution is replaced by GSConv, and the VoVGSCSP module is introduced, improving the detection accuracy and reduces the model inference time, simultaneously. Finally, the PConv module is used to replace the simplified ELAN structure in the backbone, which can reduce the network parameters and redundant computations, while maintaining detection accuracy. The new network named YOLOv7-tiny-BVP is constructed through multipolicy fusion; the structure is shown in Figure 3.

#### 3.2.1. BiFormer Dynamic Sparse Attention Mechanism for Dual-Layer Routing

The BiFormer is an improvement and extension of the transformer model and consists of an encoder and a decoder. The encoder is used to encode the input sequence, and the decoder is used to generate the output sequence [23]. The BiFormer structure, which introduces a new bidirectional attention mechanism, is shown in Figure 4. The attention scores of each position are calculated via inputting the query vector (Q), key vector (K), and value vector (V) in the sequence, thereby establishing a global association within the sequence [24]. In citrus fruit recognition and detection tasks, contextual information is crucial for locating targets. The BiFormer can more comprehensively capture the background information and contextual relationships around citrus fruit plants, improving fruit detection accuracy.

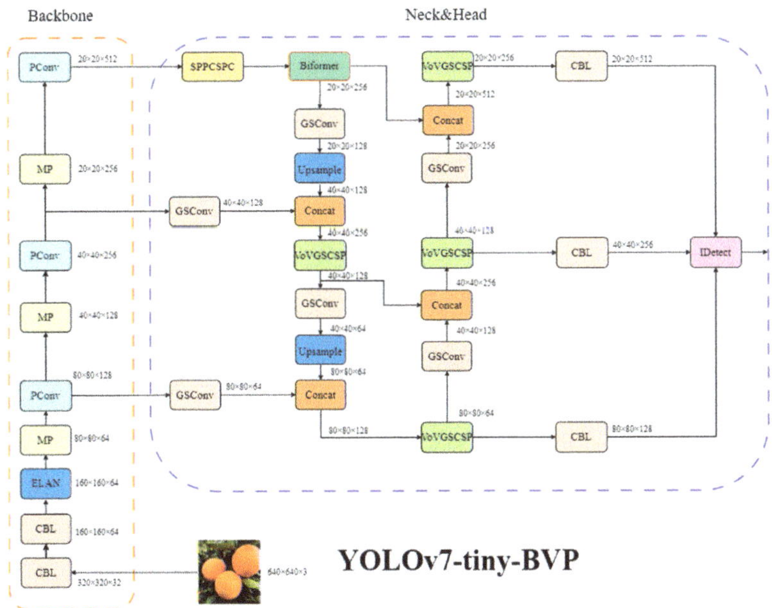


**Figure 3.** Diagram of the YOLOv7-tiny-BVP network.

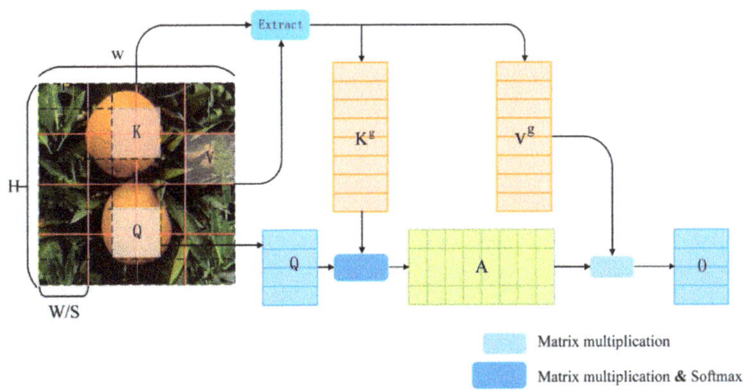


**Figure 4.** Diagram of the BiFormer attention mechanism.

The query vector Q is used to represent the content that the current position needs to focus on the attention mechanism. The weight in the attention allocation process is determined through calculating the similarity with the key vector K. The value vector V contains the actual feature representation, which is assigned different weights based on the similarity between Q and K to determine the importance of the different positions in the output. The relationships between them are shown in Equations (5)–(7):

$$K^g = \text{gather}(K, I^r) \tag{5}$$

$$V^g = \text{gather}(V, I^r) \tag{6}$$

$$O = \text{Attention}(Q, K^g, V^g) + \text{LCE}(V) \tag{7}$$

where $I^r$ in Equations (5) and (6) represents the index tensor, which is used to specify the position where the elements are collected from the input tensor. The gathering operation involves collecting the elements at corresponding positions from the input tensor based on the index; $K^g$ and $V^g$ are new key and value vectors obtained after the gathering operation. In Equation (7), attention represents the attention mechanism calculation, and LCE represents the cross-entropy loss function. O is the final output result that is calculated through the attention mechanism and combined with cross-entropy loss to obtain the final output, helping the model to better understand contextual information in detection tasks and to make accurate predictions.

### 3.2.2. VoVGSCSP Module Based on a Slim Neck

To further improve the model's ability to capture the positional information of citrus fruit features and reduce computational costs, the regular convolution in the neck network is replaced with GSConv, and the ELAN module is replaced with the VoVGSCSP module in this study to balance the model accuracy and speed [25].

In the convolutional neural network architecture, regular convolutional operations have a high accuracy, but the model is more complex and has a longer inference time. In contrast, although DWConv (depth-wise separable convolution) has a faster detection speed, it has a lower accuracy [26]. Therefore, the GSConv module was originally designed to reduce the model complexity while maintaining detection accuracy; the structure of the module is shown in Figure 5. GSConv infiltrates the semantic information generated by regular convolution into each part of DWConv, which fully utilizes the features of regular convolution and DWConv convolution, increasing both the speed of DWConv and the accuracy of regular convolution; thus, it achieves a better performance in citrus fruit detection tasks.

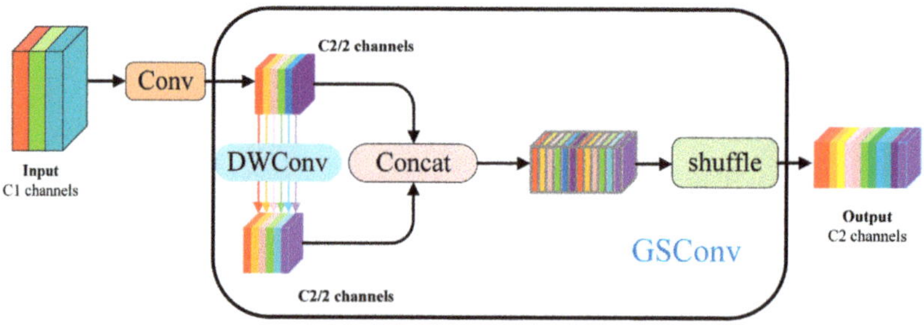


**Figure 5.** Diagram of the GSConv structure.

Although GSConv can significantly reduce the redundant information in the feature maps of citrus fruit detection models, it has limitations in further reducing inference time and maintaining accuracy. Therefore, the ELAN module is replaced with the VoVGSCSP module in the neck network section. The VoVGSCSP is formed by combining the modules in Figure 5 through a one-time aggregation method. The structure of VoVGSCSPC is shown in Figure 6. This approach can reduce the complexity of the computations and network structure and further minimize the memory footprint of the model, facilitating model deployment to edge devices with limited computational resources.

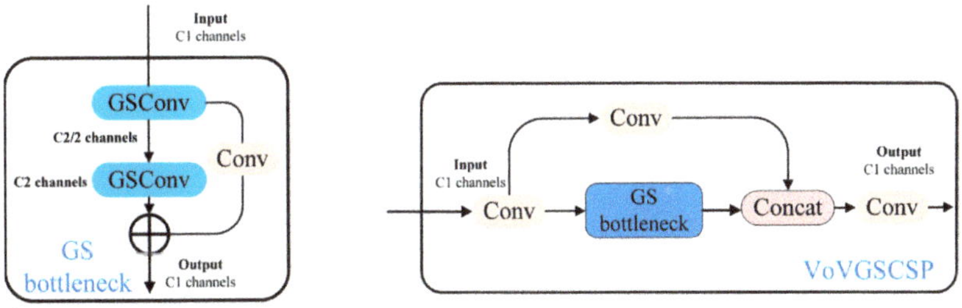


**Figure 6.** Structural diagram of the VoVGSCSPC.

#### 3.2.3. PConv Module Based on FasterNet

The floating-point operation per second (FLOPS) is the calculation speed; the larger the value, the better the network performance. The current mainstream conventional convolution, group convolution, and depth-wise separable convolution (DWConv) methods all suffer from low FLOPS problems, which is mainly due to frequent access memory [27]. For floating-point of operations (FLOPs), the greater the number of computations, the smaller the value, which is generally used to measure the model complexity. The original YOLOv7-tiny backbone network uses large numbers of regular convolutions for feature extraction, with many parameters and a large computational effort. Although convolution kernels can be utilized to reduce the number of parameters and FLOPs, memory access increases as the network width increases, to compensate for the accuracy degradation [28]. Partial convolution (PConv) uses partial channels for spatial feature extraction, which can effectively reduce redundant computations and memory access. Therefore, to reduce the network complexity while maintaining feature extraction ability, this experiment replaces the regular convolution in the backbone network with PConv. The working principle of PConv is shown in Figure 7.

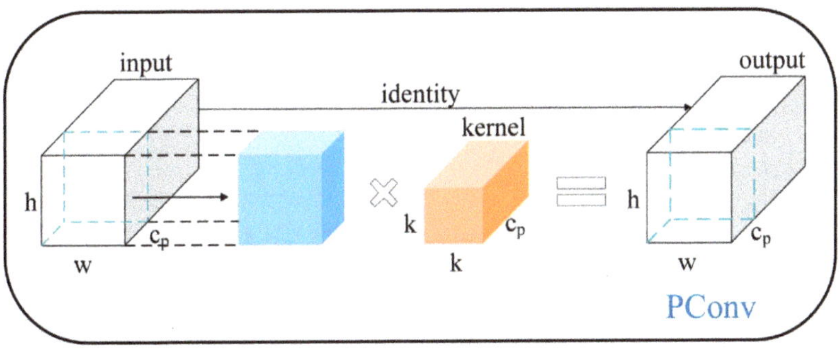


**Figure 7.** Diagram of the PConv structure.

In Figure 7, h and w represent the height and width of the feature map, respectively. c represents the total number of channels, cp represents the number of channels used, and the convolutional kernel size is k × k. In contrast to regular convolution, PConv requires only a portion of the input channels for spatial feature extraction, while keeping the remaining channels constant. For consecutive or regular memory access, the first or last consecutive channel is considered representative of the entire feature map for computation, and the input and feature maps are considered to have the same number of channels.

Without a loss of generality, the formula for calculating the FLOPs and memory access of PConv is as follows:

$$h \times w \times k^2 \times c_p^2 \tag{8}$$

$$h \times w \times 2c_p + k^2 \times c_p^2 \approx h \times w \times 2c_p \quad (9)$$

According to Equations (8) and (9), the FLOPs of PConv are only 1/16 of regular Conv, and the memory access is only 1/4 of the regular Conv.

## 4. Results and Discussion

### 4.1. Comparative Testing of Different Detection Models

To verify the ability of the new YOLOv7-tiny-BVP model to recognize citrus fruits, the current mainstream target detection networks in the YOLOv5 [29] series (YOLOv5s, YOLOv5x, YOLOv5n, YOLOv5l, and YOLOv5m) and YOLOv7 [22] series (YOLOv7, YOLOv7x, and YOLOv7-tiny) were selected for a comprehensive comparison in this experiment. During training, all the networks used the same dataset without loading the official default weights, and all the hyperparameters remained consistent. After training was completed, the weight of the network with the highest accuracy was selected. With the same test set, a total of 319 images were comprehensively evaluated. In this evaluation, the FPS metric was calculated as follows: when batch size = 1 is maintained, the result is calculated by dividing 1000 ms by the sum of image reasoning time, preprocessing time, and post-processing time. The results of the above target detection model after the completion of the test are shown in Table 2.

**Table 2.** Comparison of the results of different network models.

| Models | Layers | Para | Size/mb | P/% | R/% | mAP@.5/% | FPS | F1 |
|---|---|---|---|---|---|---|---|---|
| YOLOv5s | 157 | 7,012,822 | 14.4 | 96.8% | 88.5% | 95.2% | 76.80 | 0.93 |
| YOLOv5x | 322 | 86,173,414 | 173.1 | 97.0% | 89.6% | 95.6% | 16.00 | 0.93 |
| YOLOv5n | 157 | 1,760,518 | 3.9 | 95.1% | 87.4% | 94.2% | 91.90 | 0.91 |
| YOLOv5l | 267 | 46,108,278 | 92.8 | 96.6% | 90.5% | 95.6% | 26.10 | 0.93 |
| YOLOv5m | 212 | 20,852,934 | 42.2 | 96.7% | 89.0% | 95.7% | 42.57 | 0.93 |
| YOLOv7 | 314 | 36,481,772 | 74.8 | 97.5% | 91.6% | 97.1% | 34.96 | 0.95 |
| YOLOv7x | 362 | 70,782,444 | 142.1 | 97.6% | 92.0% | 96.8% | 21.27 | 0.95 |
| YOLOv7-tiny | 172 | 6,007,596 | 12.3 | 97.0% | 95.0% | 98.0% | 87.26 | 0.96 |
| YOLOv7-tiny-BVP | 257 | **3,696,396** | **7.7** | **97.9%** | 94.6% | 97.8% | **89.28** | **0.97** |

Note: Bold font is the optimal value for the model detection indicators; mAP@.5 average accuracy at IoU = 0.5.

Table 2 shows that the current YOLOv5 series network performs well in citrus fruit detection, but the overall performance is weaker than that of the YOLOv7 series network. Although YOLOv5n has a small number of parameters and a smaller proportion of models, it is still weaker than YOLOv7 series networks in terms of the other indicators. Although the P, R, and mAP indices of the YOLOv7 and YOLOv7x networks are higher than those of the YOLOv5 series, the overall model parameters are large, the model proportion is too large, and the FPS frame rate is low, making it difficult to deploy edge computing devices with insufficient GPU resources for detection.

Compared with YOLOv7-tiny, the overall parameters decreased by 38.47%, the model scale was reduced by 4.6 Mb, and P increased by 0.9%. Meanwhile, the FPS and F1 scores increased by 2.02 and 1%, respectively. Obviously, YOLOv7-tiny-BVP effectively reduces the computational load on the device, and still maintains a high performance.

### 4.2. Comparative Experiments in Complex Scenarios

Under natural conditions, the orchard environment is more complex. The growth characteristics of fruit trees result in severe damage between the branches and leaves of the fruit. Moreover, fruits of different sizes and shapes commonly overlap each other. Additionally, there are many disturbing factors in the field environment, such as light and darkness, which make fruit identification and counting difficult [30]. Therefore, for the above two real-world scenarios, the YOLOv5s model, which has the most balanced performance among the YOLOv5 series; the YOLOv7-tiny; and the YOLOv7-tiny-BVP proposed in this paper were selected for comparative analysis. In Figures 8–11, the red and green circles indicate the network failing to detect under two different scenarios.

YOLOv5s YOLOv7-tiny YOLOv7-tiny-BVP

**Figure 8.** Mild occlusion recognition results.

YOLOv5s YOLOv7-tiny YOLOv7-tiny-BVP

**Figure 9.** Heavy occlusion recognition results.

YOLOv5s YOLOv7-tiny YOLOv7-tiny-BVP

**Figure 10.** Recognition effect in a dark scene.

YOLOv5s YOLOv7-tiny YOLOv7-tiny-BVP

**Figure 11.** Recognition results in bright scenes.

4.2.1. Comparison of Recognition Results under Different Occlusion Scenes

Figure 8 shows that both YOLOv5s and YOLOv7-tiny can recognize more obvious targets in the slightly occluded scene, but the network is not sensitive to small targets at the edge of the image due to foliage occlusion, and both suffer from missed detections. YOLOv7-tiny-BVP can accurately detect the location of citrus fruit plants by introducing an attention mechanism, which increases the sensitivity of the network to the feature information around the citrus plants.

Figure 9 shows that in the case of heavy occlusion, overlapping fruits and leafy branch misalignment result in target feature loss. Although the improved network can still accurately recognize the target in the case of heavy occlusion, YOLOv7-tiny suffers from the omission of detection, and YOLOv5s can accurately recognize the fruits but suffers from misdiagnosis, incorrectly recognizing the two adhered fruits as one. In summary, the YOLOv7-tiny-BVP network proposed in this paper achieves a better performance in different occlusion scenarios.

4.2.2. Comparison of Recognition Results in Bright and Dark Scenes

Brightness also poses a considerable challenge for fruit identification and counting during actual field inspections. Figure 10 shows the actual detection results of the three network models for different light and dark scenes. As Figure 10 shows, in the case of dim light, YOLOv5s has a poor recognition performance compared to the other two networks, and three fruits were missed. YOLOv7 has two unrecognized fruits, while the recognition effect of YOLOv7-tiny-BVP is better overall than that of the other two networks. It is concluded that the color features of citrus fruit plants are masked in a dimly lit environment, the network cannot extract effective textural features, and there is a complication associated with overlapping leaves and fruits in the figure, leading to individual fruits not being detected.

In addition, in a well-lit environment, under the conditions of severe leaf and fruit overlap, the recognition ability of YOLOv7-tiny-BVP and YOLOv7-tiny proposed in this study significantly improved compared to that of the YOLOv5s network. Figure 11 shows that the recognition effect of YOLOv7-tiny is more accurate than that of dim light scenarios. It is concluded that under sufficient light conditions, the network is more likely to obtain semantic color information about fruits, resulting in more accurate fruit recognition than in dim light scenarios. Therefore, the YOLOv7-tiny-BVP network proposed in this paper is more suitable for fruit detection in complex environments, has a high accuracy, and fewer missed detections under different scenes of light and darkness.

### *4.3. Ablation Experiment*

To validate the effectiveness of the three improvement strategies proposed in this article for the YOLOv7-tiny-BVP network, ablation experiments were conducted to observe the changes in various indicators on the same training set after 300 epochs of training. The training results are shown in Figure 12, where the model gradually starts to converge after 250 iterations, and there is no significant change between the three independent improvement schemes and the original network model YOLOv7-tiny in terms of precision, recall, and mAP@.5 in the training set. However, in terms of the mAP@.5:.95 indicator, the YOLOv7-tiny-BVP achieves better results when three improvement strategies are introduced simultaneously, rather than when the BiFormer attention mechanism, VoVGSCSP, or PConv are introduced each time alone. In addition, YOLOv7-tiny-BVP also has a lower localization (boss_loss) and confidence loss (obj_loss) than the other individual improvement strategies, indicating that the model overall performs better after introducing the three improvement strategies.

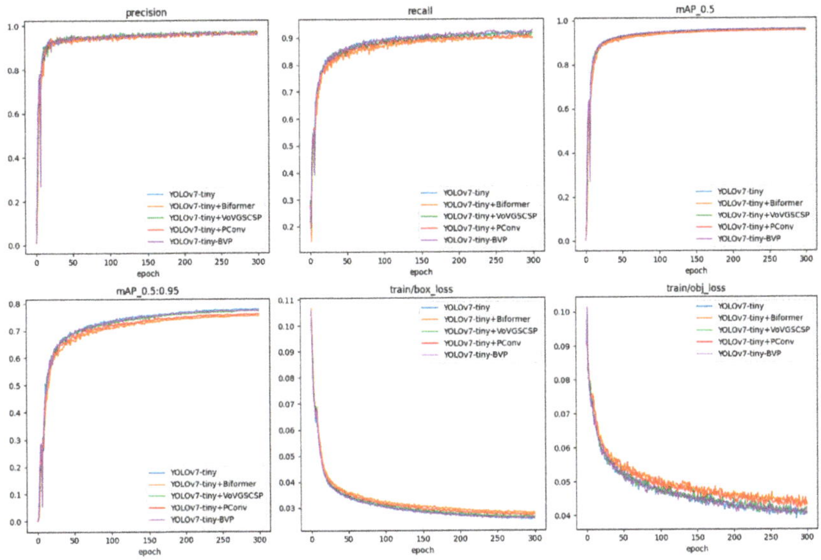


**Figure 12.** Ablation experiment.

After the training was completed, the weights with optimal accuracy were selected for a comprehensive model evaluation in the same validation set of 573 images; the evaluation results are shown in Table 3. In Table 3, time is the sum of the image preprocessing time, inference speed, and the NMS non extremely large suppression time; a smaller value indicates a better model performance. The analysis of various indicators in the table shows that the accuracy of the model improves by 1% after introducing the BiFormer attention mechanism. The analysis suggests that introducing an attention mechanism enhances the network's ability to extract features and improve detection accuracy. However, at the same time, the proportion of the model is relatively increased, which does not meet the need for lightweight materials. In addition, after introducing the VoVGSCSP module, the overall parameter quantity and the model size decrease slightly, while the accuracy improves by 1%. Moreover, the model's image processing time increases by 1.2 ms, resulting in a relatively poor real-time detection efficiency.

Moreover, after the ELAN module was replaced with the PConv module in the YOLOv7-tiny backbone network, the overall parameter quantity and the proportion of the model significantly decreases, while the accuracy decreases by 2% compared to those of the original network. Additionally, the analysis shows that the number of convolutions and the depth of the network decrease after the network is replaced with PConv, after which the number of parameters is reduced. Therefore, the network's feature extraction ability is weakened. In addition, after introducing the three improvement schemes simultaneously, the recall and mAP@.5 metrics of the model decrease by 1%, and the overall parameters of the model decrease significantly. The improved model has the highest accuracy and shortest image processing time, and the model is lighter, more accurate, more real-time, and easier to deploy.

**Table 3.** Comparison of ablation experiment indices.

| Model | Layers | Para | Size/mb | P/% | R/% | mAP@.5/% | Time |
|---|---|---|---|---|---|---|---|
| YOLOv7-tiny | 172 | 6,007,596 | 12.3 | 96.4 | 91.4 | 96.1 | 8.4 ms |
| YOLOv7-tiny + BiFormer | 183 | 6,273,324 | 12.8 | 97.5 | 91.3 | 96.2 | 8.2 ms |
| YOLOv7-tiny + VoVGSCSP | 267 | 5,620,492 | 11.6 | 97.1 | 88.7 | 95.5 | 9.6 ms |
| YOLOv7-tiny + PConv | 151 | 3,817,772 | 7.8 | 94.3 | 91.7 | 95.5 | 8.2 ms |
| YOLOv7-tiny-BVP | 257 | **3,696,396** | **7.7** | **97.4** | 89.7 | 95.5 | **7.9 ms** |

Note: Bold font is the optimal value for the model detection indicators; mAP@.5 average accuracy at IoU = 0.5.

### 4.4. Comparison of the Latest Methods for Fruit Testing

Table 4 shows the latest research results. Lai et al. [10] used the YOLOv7 network to detect pineapple fruits and classify pineapple fruits according to their ripeness. Ma et al. [18] detect apples under complex weather conditions by using the YOLOv7-tiny model and a public apple dataset. Compared to their model, our model has a significant improvement in P, F1 score, and mAP. Chen et al. [12] used the YOLOv7 network to detect the ripeness of citrus fruits; though the mAP of the model reached 97.29%, the model has too much computation and too many parameters. In addition, Zhang et al. [31] used UAV to collect citrus fruit images and employed the YOLOv7-tiny network to recognize the fruits; though the size of the model is only 3.96 MB, our model has a great improvement in both P and F1 scores. In summary, the model proposed in this study has a more comprehensive performance than the models compared above and is more suitable for citrus fruit detection.

**Table 4.** Latest research on fruit detection.

| Authors | Crops | Model | P/% | mAP@.5/% | F1/% | Size/mb | Time/ms |
|---|---|---|---|---|---|---|---|
| Lai et al. [10] | Pineapple | YOLOv7 | 94.17 | 95.82 | 91.95 | / | 23.81 |
| Ma et al. [18] | Apple | YOLOv7-tiny | 80.1 | 80.4 | 76.8 | 5.06 | / |
| Chen et al. [12] | Citrus | YOLOv7 | 94.25 | 97.29 | 93.81 | 24.26 | 69.38 |
| Zhang et al. [31] | Citrus | YOLOv7-tiny | 89.02 | 90.34 | 86.0 | 3.96 | / |
| Ours | Citrus | YOLOv7-tiny | 97.9 | 97.8 | 97.0 | 7.7 | 7.9 |

### 4.5. Discussion on Lightweight Networks

With the rapid development of mobile and embedded devices nowadays, more and more algorithms are deployed to mobile devices for people's convenience, but the aforementioned devices suffer from the problems of small memory, insufficient storage space, and limited computational power [32,33]. The lightweight network proposed in this study can both accurately identify citrus fruits and significantly reduce the number of network participants. As shown in Figure 13 the image is the side view of a citrus fruit tree, and the improved network maintains a high recognition rate for citrus fruits, which provides a new way to deploy the aforementioned mobile devices for fruit identification and yield estimation.

Although the network model proposed in this study is small and has a high accuracy, it still has some limitations. (1) The date and location of image collection is relatively homogeneous. (2) There is a lack of green or yellow-green citrus fruits in the images. (3) The proposed network only demonstrates the promise of a lightweight network and was not deployed to mobile devices for testing. In future research, we will focus more on image acquisition, including date, location, and type, to optimize the proposed lightweight network for the real-time detection of citrus fruits during the full-growth period.

**Figure 13.** Fruit identification results of citrus fruit trees from a side view. (**a**–**e**) Results of different fruit tree tests.

## 5. Conclusions

Based on YOLOv7-tiny, a new network model, YOLOv7-tiny-BVP, is constructed in this study by introducing the BiFormer attention mechanism, the VoVGSCSPC module, and the PConv convolution. In this study, a total of eight different networks from the YOLOv5 and YOLOv7 series were selected for comparative testing with the same dataset. The results show that compared with YOLOv7-tiny, the total number of parameters and model scale of YOLOv7-tiny-BVP are reduced by 38.47% and 4.6 MB, respectively.

Meanwhile, the accuracy, FPS, and F1 were improved to different degrees, respectively, and the detection accuracy was no less than that of YOLOv7-tiny. Therefore, the proposed YOLOv7-tiny-BVP network model is not only lightweight, but also has the advantage of a high detection accuracy of the YOLOv7 model, which provides a new idea for the direction of the future research of lightweight models, and also provides the possibility of applying the lightweight target detection model in the field of agricultural yield estimation.

**Author Contributions:** Conceptualization, J.J., M.Z. and S.D.; Data curation, L.W., S.Y. and J.Y.; Formal analysis, J.J. and S.D.; Funding acquisition, J.J. and S.D.; Investigation, B.L., J.Y. and M.Z.; Methodology, S.D., J.J. and L.W.; Project administration, S.D.; Resources, M.Z. and J.J.; Software, M.Z., S.Y. and B.L.; Validation, M.Z. and J.Y.; Visualization, S.D. and J.J.; Writing—original draft, J.J., M.Z. and L.W.; Writing—Review and editing, J.J. and S.D. All authors have read and agreed to the published version of the manuscript.

**Funding:** This research was funded by the National Natural Science Foundation of China (Nos. 42161028, 42061059), the Natural Science Foundation of Guangxi (No. 2023JJA150097), the Guilin Science and Technology Bureau Development Program (No. 2020010701, 20210226-2), and the "BaGui Scholars" program of the provincial government of Guangxi, the Guilin University of Technology Foundation (No. GUTQDJJ2019046).

**Institutional Review Board Statement:** Not applicable.

**Data Availability Statement:** Data will be made available on request.

**Conflicts of Interest:** The authors declare no conflicts of interest.

## Abbreviations

The following abbreviations are used in this manuscript:

| | |
|---|---|
| ELAN | efficient layer aggregation network |
| PConv | partial convolution |
| FPS | frames per second |
| CBAM | convolutional block attention module |
| BiFPN | bidirectional feature pyramid network |
| SE | squeeze-and-excitation networks |
| DWConv | Depth-wise separable convolution |

## References

1. Zhang, F. Statistical Analysis of Fruit Production in China in 2020. *China Fruit News* **2021**, *38*, 29–39.
2. Wang, L.; Zhao, Y.; Xiong, Z.; Wang, S.; Li, Y.; Lan, Y. Fast and precise detection of litchi fruits for yield estimation based on the improved YOLOv5 model. *Front. Plant Sci.* **2022**, *13*, 965425. [CrossRef]
3. Sozzi, M.; Cantalamessa, S.; Cogato, A.; Kayad, A.; Marinello, F. Automatic Bunch Detection in White Grape Varieties Using YOLOv3, YOLOv4, and YOLOv5 Deep Learning Algorithms. *Agronomy* **2022**, *12*, 319. [CrossRef]
4. Wang, D.; He, D. Channel pruned YOLO V5s-based deep learning approach for rapid and accurate apple fruitlet detection before fruit thinning. *Biosyst. Eng.* **2021**, *210*, 271–281. [CrossRef]
5. Cardellicchio, A.; Solimani, F.; Dimauro, G.; Petrozza, A.; Summerer, S.; Cellini, F. Detection of tomato plant phenotyping traits using YOLOv5-based single stage detectors. *Comput. Electron. Agric.* **2023**, *207*, 107757. [CrossRef]
6. Liu, Y.; Ren, H.; Zhang, Z.; Men, F.; Zhang, P.; Wu, D.; Feng, R. Research on multi-cluster green persimmon detection method based on improved Faster RCNN. *Front. Plant Sci.* **2023**, *14*, 1177114. [CrossRef]
7. Wang, C.; Wang, Y.; Liu, S.; Lin, G.; He, P.; Zhang, Z.; Zhou, Y. Study on Pear Flowers Detection Performance of YOLO-PEFL Model Trained with Synthetic Target Images. *Front. Plant Sci.* **2022**, *13*, 911473. [CrossRef]
8. Zhou, J.; Zhang, Y.; Wang, J. RDE-YOLOv7: An Improved Model Based on YOLOv7 for Better Performance in Detecting Dragon Fruits. *Agronomy* **2023**, *13*, 1042. [CrossRef]
9. Zhou, J.; Zhang, Y.; Wang, J. A Dragon Fruit Picking Detection Method Based on YOLOv7 and PSP-Ellipse. *Sensors* **2023**, *23*, 3803. [CrossRef]
10. Lai, Y.; Ma, R.; Chen, Y.; Wan, T.; Jiao, R.; He, H. A Pineapple Target Detection Method in a Field Environment Based on Improved YOLOv7. *Appl. Sci.* **2023**, *13*, 2691. [CrossRef]
11. Bi, S.; Gao, F.; Chen, J.; Zhang, L. Detection Method of Citrus Based on Deep Convolution Neural Network. *Trans. Chin. Soc. Agric. Mach.* **2019**, *50*, 181–186.
12. Chen, J.; Liu, H.; Zhang, Y.; Zhang, D.; Ouyang, H.; Chen, X. A Multiscale Lightweight and Efficient Model Based on YOLOv7: Applied to Citrus Orchard. *Plants* **2022**, *11*, 3260. [CrossRef]
13. Zhao, C.; Liang, X.; Yu, H.; Wang, H.; Fan, S.; Li, B. Automatic Identification and Counting Method of Caged Hens and Eggs Based on Improved YOLOv7. *Trans. Chin. Soc. Agric. Mach.* **2023**, *54*, 300–312.
14. Xiong, J.; Zheng, Z.; Liang, J.; Zhong, Z.; Liu, B.; Sun, B. Citrus Detection Method in Night Environment Based on Improved YOLO v3 Network. *Trans. Chin. Soc. Agric. Mach.* **2020**, *51*, 199–206.
15. Huang, H.; Hu, J.; Li, Z.; Wei, Z.; Liu, S. Design of citrus fruit intelligent recognition system based on edge computing. *J. Hunan Agric. Univ. (Nat. Sci.)* **2021**, *47*, 727–732.
16. Huang, T.; Huang, H.; Li, Z.; Lv, S.; Xue, X.; Dai, q.; Wen, W. Citrus fruit recognition method based on the improved model of YOLOv5. *J. Huazhong Agric. Univ.* **2022**, *41*, 170–177. [CrossRef]
17. Wang, Z.; Jin, L.; Wang, S.; Xu, H. Apple stem/calyx real-time recognition using YOLO-v5 algorithm for fruit automatic loading system. *Postharvest Biol. Technol.* **2022**, *185*, 111808. [CrossRef]
18. Ma, L.; Zhao, L.; Wang, Z.; Zhang, J.; Chen, G. Detection and Counting of Small Target Apples under Complicated Environments by Using Improved YOLOv7-tiny. *Agronomy* **2023**, *13*, 1419. [CrossRef]

19. Wang, F.; Jiang, J.; Chen, Y.; Sun, Z.; Tang, Y.; Lai, Q.; Zhu, H. Rapid detection of Yunnan Xiaomila based on lightweight YOLOv7 algorithm. *Front. Plant Sci.* **2023**, *14*, 1200144. [CrossRef]
20. Mekhalfi, M.L.; Nicolo, C.; Ianniello, I.; Calamita, F.; Goller, R.; Barazzuol, M.; Melgani, F. Vision System for Automatic On-Tree Kiwifruit Counting and Yield Estimation. *Sensors* **2020**, *20*, 4214. [CrossRef]
21. Wang, C.Y.; Bochkovskiy, A.; Liao, H.Y.M. YOLOv7: Trainable bag-of-freebies sets new state-of-the-art for real-time object detectors. *arXiv* **2022**, arXiv:2207.02696.
22. Wu, C.; Ye, M.; Zhang, J.; Ma, Y. YOLO-LWNet: A Lightweight Road Damage Object Detection Network for Mobile Terminal Devices. *Sensors* **2023**, *23*, 3268. [CrossRef] [PubMed]
23. Yang, Z.; Feng, H.; Ruan, Y.; Weng, X. Tea Tree Pest Detection Algorithm Based on Improved Yolov7-Tiny. *Agriculture* **2023**, *13*, 1031. [CrossRef]
24. Zhu, L.; Wang, X.; Ke, Z.; Zhang, W.; Lau, R.W. BiFormer: Vision Transformer with Bi-Level Routing Attention. *arXiv* **2023**, arXiv:2303.08810. [CrossRef]
25. Li, H.; Li, J.; Wei, H.; Liu, Z.; Zhan, Z.; Ren, Q. Slim-neck by GSConv: A better design paradigm of detector architectures for autonomous vehicles. *arXiv* **2022**, arXiv:2206.02424. [CrossRef]
26. Dai, Y.; Li, C.; Su, X.; Liu, H.; Li, J. Multi-Scale Depthwise Separable Convolution for Semantic Segmentation in Street-Road Scenes. *Remote Sens.* **2023**, *15*, 2649. [CrossRef]
27. Liu, C.; Wang, T.; Dong, S.; Zhang, Q.; Yang, Z.; Gao, F. Hybrid Convolutional Network Combining 3D Depthwise Separable Convolution and Receptive Field Control for Hyperspectral Image Classification. *Electronics* **2022**, *11*, 3992. [CrossRef]
28. Chen, J.; Kao, S.; He, H.; Zhuo, W.; Wen, S.; Lee, C.H.; Chan, S.H.G. Run, Don't Walk: Chasing Higher FLOPS for Faster Neural Networks. *arXiv* **2023**, arXiv:2303.03667. [CrossRef]
29. Zhu, X.; Liu, S.; Wang, X.; Zhao, Q. TPH-YOLOv5: Improved YOLOv5 Based on Transformer Prediction Head for Object Detection on Drone-captured Scenarios. *arXiv* **2021**, arXiv:2108.11539. [CrossRef]
30. Dorj, U.O.; Lee, M.; Yun, S. A yield estimation in citrus orchards via fruit detection and counting using image processing. *Comput. Electron. Agric.* **2017**, *140*, 103–112. [CrossRef]
31. Zhang, Y.; Fang, X.; Guo, J.; Wang, L.; Tian, H.; Yan, K.; Lan, Y. CURI-YOLOv7: A Lightweight YOLOv7tiny Target Detector for Citrus Trees from UAV Remote Sensing Imagery Based on Embedded Device. *Remote Sens.* **2023**, *15*, 4647. [CrossRef]
32. Yin, H. Research on Fall Detection Algorithm and Algorithm Deployment in Embedded Platform. Master's Thesis, University of Electronic Science and Technology of China, Chengdu, China, 2023.
33. Zhu, H. Research on Lightweight and Mobile Deployment Method of Road Target Detection Algorithm Based on Deep Learning. Master's Thesis, Inner Mongolia Agricultural University, Hohhot, China, 2022.

*agriculture*

MDPI

*Article*

# AG-YOLO: A Rapid Citrus Fruit Detection Algorithm with Global Context Fusion

**Yishen Lin, Zifan Huang, Yun Liang *, Yunfan Liu and Weipeng Jiang**

College of Mathematics and Informatics, South China Agricultural University, Guangzhou 510642, China; linyishen@scau.edu.cn (Y.L.); hzfan@stu.scau.edu.cn (Z.H.); lyf316@stu.scau.edu.cn (Y.L.); 20223170038@stu.scau.edu.cn (W.J.)

* Correspondence: yliang@scau.edu.cn

**Abstract:** Citrus fruits hold pivotal positions within the agricultural sector. Accurate yield estimation for citrus fruits is crucial in orchard management, especially when facing challenges of fruit occlusion due to dense foliage or overlapping fruits. This study addresses the issues of low detection accuracy and the significant instances of missed detections in citrus fruit detection algorithms, particularly in scenarios of occlusion. It introduces AG-YOLO, an attention-based network designed to fuse contextual information. Leveraging NextViT as its primary architecture, AG-YOLO harnesses its ability to capture holistic contextual information within nearby scenes. Additionally, it introduces a Global Context Fusion Module (GCFM), facilitating the interaction and fusion of local and global features through self-attention mechanisms, significantly improving the model's occluded target detection capabilities. An independent dataset comprising over 8000 outdoor images was collected for the purpose of evaluating AG-YOLO's performance. After a meticulous selection process, a subset of 957 images meeting the criteria for occlusion scenarios of citrus fruits was obtained. This dataset includes instances of occlusion, severe occlusion, overlap, and severe overlap, covering a range of complex scenarios. AG-YOLO demonstrated exceptional performance on this dataset, achieving a precision (P) of 90.6%, a mean average precision (mAP)@50 of 83.2%, and an mAP@50:95 of 60.3%. These metrics surpass existing mainstream object detection methods, confirming AG-YOLO's efficacy. AG-YOLO effectively addresses the challenge of occlusion detection, achieving a speed of 34.22 frames per second (FPS) while maintaining a high level of detection accuracy. This speed of 34.22 FPS showcases a relatively faster performance, particularly evident in handling the complexities posed by occlusion challenges, while maintaining a commendable balance between speed and accuracy. AG-YOLO, compared to existing models, demonstrates advantages in high localization accuracy, minimal missed detection rates, and swift detection speed, particularly evident in effectively addressing the challenges posed by severe occlusions in object detection. This highlights its role as an efficient and reliable solution for handling severe occlusions in the field of object detection.

**Keywords:** occlusion detection; NextViT; attention mechanism

## 1. Introduction

In China, citrus, being a prevalent type of citrus fruit, holds significant importance within the agricultural sector. Accurate yield estimation during citrus cultivation is paramount, as precise estimations aid fruit growers in making informed decisions for orchard management, including fertilization, irrigation, and harvesting plans. Furthermore, it contributes to production planning and supply chain management for fruit growers. However, traditional yield estimation methods may lead to significant deviations, especially when fruits are obstructed by dense foliage or mutual fruit occlusion, posing substantial challenges in accurately determining fruit yield. The issue of severe fruit occlusion poses a significant challenge in accurately estimating the yield at present.

The current methods for object detection in the agricultural domain can be broadly categorized into two types: non-deep learning methods and deep learning methods.

Non-deep learning methods, exemplified by Hamuda et al. (2018) [1], employ Kalman filtering and the Hungarian algorithm for crop detection in agricultural fields. However, due to the simplicity of the experimental background that lacked overlapping crops, these methods are unsuitable for detecting densely distributed fruits with occlusions. Lu et al. (2015) [2] introduced a citrus fruit recognition approach based on color and contour information under varying canopy lighting conditions. Although this method can adapt to complex natural environments with diverse lighting and backgrounds, its detection performance diminishes when dealing with smaller citrus fruits. Lin et al. (2013) [3] proposed an improved method based on the circular random Hough transform (CRHT) to detect occluded camellia fruit. The algorithm incorporates modules such as edge pre-detection and rapid center-point localization to enhance the recognition rate. Song et al. (2012) [4] proposed a method for recognizing occluded apple targets based on the convex hull theory. This method converts images from RGB to Lab color space and employs K-means clustering to categorize images into three classes: leaves, branches, and fruits. Subsequently, through morphology and convex hull techniques, the method extracts contour curves of fruit edges and estimates parameters of smooth curves, achieving occluded fruit localization. Feng et al. (2015) [5] used the 2R-G-B color difference model to extract color features of ripe red tomatoes and employed a dynamic threshold segmentation method for fruit identification. However, this approach suffers from prolonged identification times and does not adequately address accurate identification in complex environments with leaf occlusions. Sun et al. (2017) [6] utilized the K-means clustering algorithm based on the Lab color space to segment apple targets occluded by branches. The mathematical morphology method was employed to extract the contour of the apples. Subsequently, employing the minimum bounding rectangle method removed false contours, and the curvature features of the contour facilitated the reconstruction of the apple targets. This study's approach accurately identifies, locates, and reconstructs occluded apple targets by branches, effectively reducing reconstruction time. Sun et al. (2019) [7] proposed an object extraction algorithm based on geometric morphology and iterative random circles. This algorithm effectively segments and identifies adhered fruits in images. It addresses the issue of segmenting fruits in complex environments where multiple fruits are adhered or slightly occluded to some extent. The aforementioned machine vision-based fruit recognition algorithms are constrained by strong environmental dependencies and poor robustness in complex occlusion scenarios, often resulting in slow recognition speeds and low identification accuracy (Kuznetsova et al. 2020 [8], Sozzi et al. 2022 [9], Zhang et al. 2021 [10]).

The powerful processing capabilities of deep learning and its adaptability to complex scenarios make it an ideal choice for addressing occlusion issues in fruit detection. In the realm of deep learning for fruit detection, the prevailing frameworks are categorized into single-stage detectors and two-stage detectors. YOLO (Redmon et al. 2016 [11], Redmon and Farhadi 2017 [12], Redmon and Farhadi 2018 [13], Bochkovskiy et al. 2020 [14]) is a one-stage object detector known for its outstanding performance in both detection accuracy and speed. Koirala et al. (2019) [15] conducted a comparative analysis of six different deep learning architectures. They combined the distinctive features of YOLOv2 (Tiny) to devise a novel architecture termed 'MangoYOLO'. This new model was specifically designed for mango detection in canopy images. MangoYOLO integrates the strengths of both YOLOv2 and YOLOv3, offering both high detection speed and high accuracy in fruit detection. Tian et al. (2019) [16] proposed an improved YOLOv3 model designed for fruit orchard scenarios characterized by fluctuations in lighting, complex backgrounds, fruit overlap, and foliage obstruction. They leveraged DenseNet (Huang et al. 2017) [17] to process the low-resolution feature maps of the YOLOv3 model, effectively enhancing feature propagation and reuse. Yang et al. (2022) [18] utilized an enhanced YOLOv4-tiny model for tomato fruit recognition. They improved the identification accuracy of small tomatoes by incorporating a detection layer with dimensions of 76 pixels by 76 pixels. Additionally, they introduced a convolutional attention module to enhance the recognition accuracy of occluded tomatoes and employed a densely connected convolutional network to improve global feature integration. Huang et al. (2022) [19] proposed an optimized YOLOv5 model for citrus fruit recognition. By integrating the CBAM (convolutional block attention module) attention mechanism and the $\alpha$-IoU (alpha-intersection over union) loss function, they significantly enhanced the detection accuracy of occluded targets and improved the precision of bounding box localization. The two-stage detectors are primarily based on the R-CNN algorithm (Girshick et al. 2014 [20]). Mu et al. (2020) [21] utilized deep learning methods, integrating regional convolutional networks (R-CNN) and Resnet-101 (Kaiming et al. 2016 [22]), to detect unripe tomatoes in heavily shaded conditions. This was applied for both ripeness detection and predicting tomato yield. Afonso et al. (2020) [23] applied Mask R-CNN (Kaiming et al. 2017) [24] to tomato datasets for detection, utilizing various neural networks as backbones to extract feature maps. Bi et al. (2019) [25] enhanced the multi-scale image detection capability and real-time performance of citrus fruit recognition models using a multiple segmentation approach. Testing the citrus fruit target dataset in natural environments containing various interference factors, the results indicated the robustness and real-time performance of the citrus recognition model against common interference factors and occlusions in natural picking environments. The average recognition accuracy was 86.6%, with an average single-frame image detection time of 80 ms.

Although the aforementioned methods have shown improvements in both the accuracy and speed of detection models and have proposed feasible solutions, they struggle to effectively address severe occlusion issues in close-range object detection. This paper adopts the NextViT (Li et al. 2022) [26] structure, a new generation Vision Transformer (Dosovitskiy et al. 2020) [27], as the backbone network. Additionally, it integrates the Global Context Fusion Module (GCFM) at the end of the neck network, presenting a model with enhanced contextual awareness and feature representation capabilities, named AG-YOLO (Attention-Guided YOLO).

The main contributions of this paper are as follows:

(1) The study introduces the Global Context Fusion Module (GCFM) to effectively merge local and global contexts. In contrast to conventional methodologies, the GCFM employs self-attention mechanisms to selectively emphasize essential features of occluded targets. Moreover, it enables interaction and fusion between occluded regions and other image areas, facilitating a more comprehensive understanding of the image's semantic content and structural composition. This utilization of global context information notably enhances the model's capacity to discern the relationship between occluded fruit targets and their background, thereby improving the detection capability for such occluded targets.

(2) The utilization of NextViT as the backbone network highlights its superior global perception and representation capabilities compared to traditional convolutional neural networks, attributed to its Vision Transformer-based architecture. The Transformer (Vaswani et al. 2017) [28] architecture of NextViT provides the network with extensive global awareness, allowing it to process information from all areas across the input image simultaneously, rather than focusing solely on localized regions.This capability proves particularly advantageous for addressing challenges in object detection tasks involving issues like occlusion and scale variations.

(3) To meet the training requirements of the proposed model in this paper, a dataset specifically for occluded citrus fruits was collected, generated, and selected. The dataset contains ripe yellow fruit data and unripe green fruit data, segmented into subsets based on occlusion types, including branch-occluded, densely occluded, severely branch-occluded, and severely densely occluded within each color category. This rigorous classification method ensures the diversity and comprehensiveness of the dataset.

## 2. Datasets

### 2.1. Image Data Collection

Comprehensive image data collection was carried out to encompass a diverse range of natural lighting conditions and detection environments within the dataset. The image data collection work was carried out in the citrus orchard located in Wuming District, Nanning City, Guangxi Zhuang Autonomous Region, during the summer (to collect unripe fruits) and autumn (to collect ripe fruits) seasons. To adapt to various detection environments under natural light, the citrus trees were shot at distances of 0.3 m to 1 m from the citrus target, fully capturing images under diverse scenes such as sunny days, cloudy days, front lighting, and back lighting. Over 8000 images were collected in total, with the resolution preprocessed to 1280 × 720. After manual screening, 957 images meeting the task requirements were obtained to compose the 'Citrus Occlusion Dataset'. Sample images of the dataset are shown in Figure 1.

**Figure 1.** Images of citrus fruit under various conditions of oranges.

### 2.2. Dataset Creation

The processed dataset was manually annotated using the LabelImg (LabelImg is an open-source annotation tool developed by Tzutalin and contributors released on GitHub) data annotation tool. Subsequently, the dataset was divided into training, validation, and test sets in a ratio of 7:2:1. The training set comprises 670 images, the validation set includes 190 images, and the test set contains 97 images. Further subdivision of the dataset was conducted to delve deeper into the occlusion challenges in citrus fruit detection. The dataset was categorized into two distinct subsets: 666 images for branch and leaf occlusion and 291 images for fruit overlap occlusion. Among the branch and leaf occlusion dataset, images with fruit occlusion exceeding 60% and constituting more than 30% of occluded samples were categorized as the 'severe occlusion' dataset, totaling 200 images. Within the fruit overlap occlusion dataset, images with overlaps exceeding 60% were categorized as the 'severe overlap' dataset, comprising 89 images. The delineation of the citrus fruit occlusion dataset is presented in Table 1.

**Table 1.** Citrus fruit occlusion dataset.

| Dataset | Occlusion | Severe Occlusion | Overlap | Severe Overlap |
|---|---|---|---|---|
| Unripe fruit | 129 | 58 | 100 | 20 |
| Ripe fruit | 337 | 142 | 102 | 69 |

## 3. Methods

### 3.1. The Architecture of AG-YOLO

To mitigate the effects of occlusion and overlap in citrus fruit detection, a detection network model with enhanced global perception and stronger feature extraction capabilities is proposed in this study, as depicted in Figure 2. Initially, the NextViT backbone network extracts global features from citrus fruit images, capturing long-range dependencies and semantic information within the images. This empowers the model to exhibit robust feature representation when dealing with large-scale images and complex scenes. Subsequently, the Feature Pyramid Network (FPN) and the Path Aggregation Network (PANet) extract multi-scale features to strengthen feature fusion. Next, the Global Context Fusion Module (GCFM) effectively integrates local features with global context information, significantly enhancing the model's capability to understand and analyze occlusions, scale variations, and object relationships. This is particularly important for citrus fruit detection tasks, as citrus fruits might be obscured by dense foliage or other citrus fruits, while leveraging global context information helps the model better understand the image. Finally, AG-YOLO completes the detection task through prediction layers of different scales. These prediction layers handle target category predictions, confidence estimation in target presence, and regression predictions for object positions. Through this hierarchical processing, the model accurately generates detection boxes for citrus fruits.

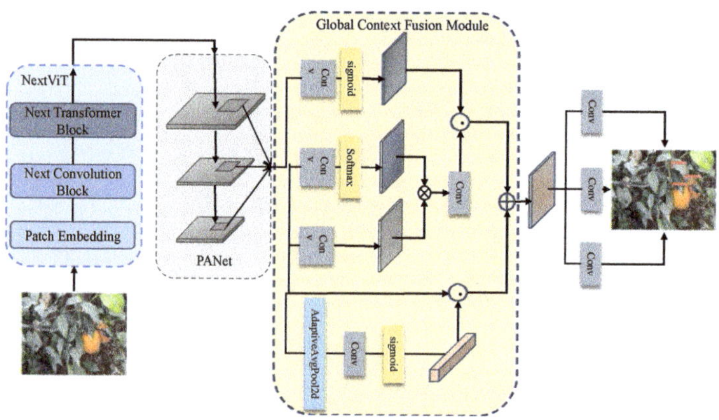


**Figure 2.** Network architecture of AG-YOLO.

### 3.2. Backbone Network NextViT

In the past, most object detection models aimed at addressing severe occlusion issues have relied on convolutional neural networks for feature extraction, such as VGG (Simonyan et al. 2014) [29] , ResNet, MobileNet (Howard et al. 2017) [30], and EfficientNet (Tan et al. 2019) [31], among others. These conventional CNN architectures primarily focus on capturing local information within images, leading to certain limitations when handling occlusion challenges. On the contrary, NextViT, based on the Vision Transformer architecture, leverages self-attention mechanisms to capture the global relationships within input images. This approach effectively captures long-range dependencies and semantic information within the images. NextViT demonstrates stronger representational capabilities when dealing with large-scale images and complex scenes, enabling the extraction of richer and more accurate features. NextViT follows a hierarchical pyramid structure, where each stage comprises a patch embedding layer along with a series of next-generation convolutional modules (Next Convolution Block) and next-generation Transformer modules (Next Transformer Block). As depicted in Figure 3, the next-generation convolutional modules play a pivotal role within the model. They leverage convolutional operations and multi-head attention mechanisms for feature extraction. By incorporating a path dropout mechanism, these next-generation convolutional modules effectively reduce redundant information within feature maps, thereby enhancing the model's robustness and generalization capability. Its design enables the model to better adapt to occluded object detection and localization tasks, resulting in a significant performance improvement, particularly in scenarios involving severe occlusion.

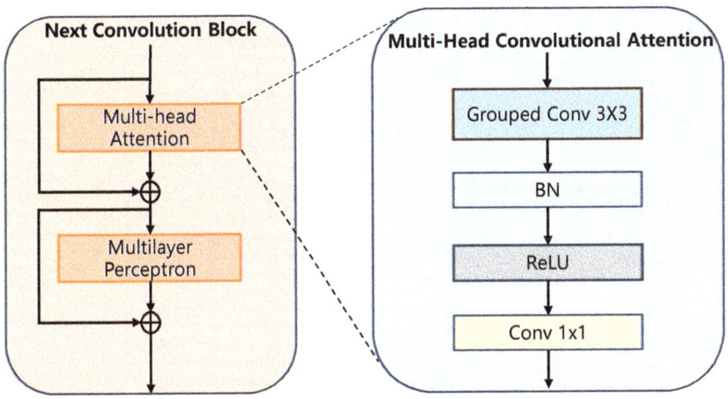


**Figure 3.** Next-generation convolutional module diagram.

As shown in Figure 4, the next-generation Transformer module combines the self-attention mechanism of Transformers with convolutional operations, specifically designed to handle severe occlusion in object instances. It achieves global information interaction and feature fusion on feature maps through the self-attention mechanism, thereby better capturing the details and context of occluded targets. Simultaneously, it performs local feature extraction through convolutional operations, enhancing the recognition capability of occluded targets. Leveraging self-attention and convolutional operations at different scales effectively integrates multi-scale features. This strategy allows the model to comprehensively perceive occluded targets, thereby improving fruit detection and localization capabilities in scenarios involving severe occlusion.

### *3.3. Global Context Fusion Module GCFM*

The Global Context Fusion Module (GCFM) takes into full consideration the specificity of occlusion scenarios, enhancing the recognition capability of occluded targets through a strategy that fuses global contextual information. As shown in Figure 5, the Global Context Fusion Module primarily comprises two components: the Local Attention Module and the Global Attention Module. Within the Local Attention Module, the module conducts convolution operations and a Softmax activation function on the feature map, resulting in a positional weight map. Subsequently, this positional weight map undergoes similarity computations with the feature map. The results of these similarity computations then undergo a convolutional operation and are weighted fused with a local weighting map obtained from convolutional layers and a sigmoid activation function. The self-attention mechanism facilitates the capture of interdependencies and feature expressions within local regions, thereby reinforcing the feature representation of the target area.

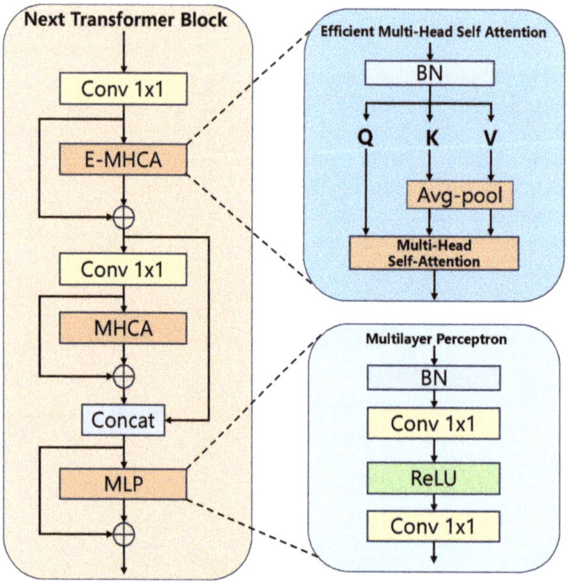


**Figure 4.** Next-generation Transformer module diagram.

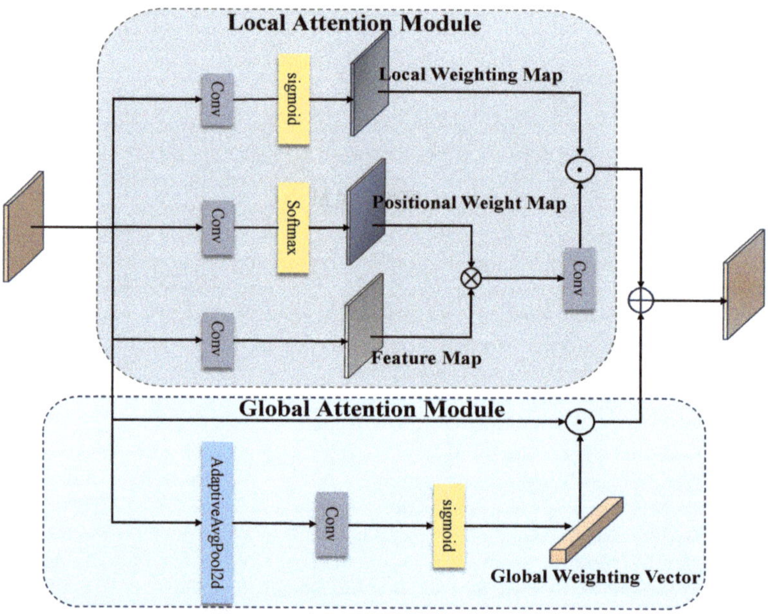


**Figure 5.** Global Context Fusion Module.

In the Global Attention Module, the input feature map undergoes a series of operations, including adaptive average pooling, convolutional layers, and sigmoid activation functions, resulting in a global weighting vector. Subsequently, the weighted adjusted global features are combined with the self-attention strengthened local features through addition fusion. Even when the target is partially occluded, the module is capable of extracting effective features through global correlations, thereby enhancing recognition accuracy. Whereas the local features encompass detailed contextual information, the global features encapsulate holistic semantic information. Their combination enables the model to gain a more comprehensive understanding of the target, thereby enhancing the detection performance of occluded targets.

The Global Context Fusion Module, combined with the NextViT backbone network, maximizes the advantage of effectively capturing contextual global information in complex scenes. This integration contributes to enhancing the detection capability of occluded targets. Furthermore, the Global Context Fusion Module serves as an independent module that can be easily embedded into various object detection architectures without altering the overall network structure. Such modular design allows for more flexibility and convenience in its usage.

## 4. Experiments

### *4.1. Experimental Environment and Parameter Settings*

The experiment is based on the Ubuntu 22.04.1 operating system, utilizing an Intel Xeon Silver 4214R CPU, Nvidia RTX 3090 GPU, PyTorch version 1.13 for deep learning framework, and CUDA version 11.7. The model undergoes 600 iterations with an initial learning rate set to 0.01. The bounding box loss function selected is the CIoU loss. The batch size is set to 16, utilizing the SGD optimizer. During training, input images are uniformly resized to 640 × 640 pixels.

### *4.2. Model Evaluation Metrics*

To validate the model's effectiveness, evaluations were conducted qualitatively and quantitatively. For qualitative assessment, the model's performance was evaluated by comparing detection images between the AG-YOLO model and other models, specifically focusing on object localization precision, detection count, and the presence of false negatives or false positives. For quantitative evaluation, the selected metrics include: precision (P), recall (R), and mean average precision (mAP) for ripe and unripe fruits. Within mAP, mAP50 and mAP50:95 are chosen: mAP50 represents the average precision at an IoU threshold of 0.5, and mAP50:95 signifies the mean detection accuracy across 10 IoU thresholds from 0.5 to 0.95, with a step size of 0.05. The specific mathematical formulas are shown as Equations (1)–(4).

$$P = \frac{T_P}{T_P + F_P} \tag{1}$$

$$R = \frac{T_P}{T_P + F_N} \tag{2}$$

In Equation (1), $T_P$ represents the number of true positive samples detected, and $F_P$ represents the number of false positive samples predicted; in Equation (2), $F_N$ represents the number of positive samples not detected.

$$AP = \int_0^1 P \cdot RdR \tag{3}$$

$$mAP = \frac{\sum_{i=0}^{n} AP_i}{n} \tag{4}$$

In Equation (4), $n$ represents the number of detected target categories.

### *4.3. Quantitative Analysis Evaluation*

To validate the superiority of the proposed algorithm, comparisons were conducted with various state-of-the-art algorithms from recent years. The algorithms compared in this study include CNN-based object detectors such as YOLOv8 and Faster R-CNN, and Transformer-based object detectors like DAB-DETR (Liu et al. 2022) [32], DN-DETR (Li et al. 2022) [33], SAP-DETR (Liu et al. 2022) [34], Anchor-DETR (Wang et al. 2021) [35], and Conditional-DETR (Meng et al. 2021) [36]. Through these comparisons, the aim is to demonstrate the algorithm's advantages in object detection over both traditional CNN-based detectors and more recent Transformer-based detectors, particularly in addressing occlusion challenges, as illustrated in Tables 2 and 3.

**Table 2.** Comparative experiments with CNN-based detectors.

| Network | P/% | R/% | mAP50/% | mAP50:95/% |
|---|---|---|---|---|
| AG-YOLO (ours) | 90.6 | 73.4 | 83.2 | 60.3 |
| YOLOv5s | 86.7 | 67.7 | 79.0 | 57.1 |
| YOLOv7 | 85.4 | 69.6 | 79.9 | 53.2 |
| YOLOv8s | 85.1 | 67.5 | 77.6 | 57.0 |
| Faster RCNN | 50.6 | 81.1 | 77.3 | 47.2 |
| SSD | 87.8 | 61.5 | 70.9 | 42.8 |

**Table 3.** Comparative experiments with Transformer-based detectors.

| Network | mAP50/% | mAP50:95/% | $AP_s$/% | $AP_m$/% | $AP_l$/% |
|---|---|---|---|---|---|
| AG-YOLO (ours) | 83.2 | 60.3 | 18.8 | 47.2 | 69.2 |
| DAB-DETR | 77.5 | 48.5 | 12.0 | 39.0 | 65.0 |
| DN-DETR | 74.0 | 49.1 | 15.8 | 38.6 | 65.5 |
| SAP-DETR | 78.4 | 50.9 | 15.5 | 41.4 | 67.0 |
| Anchor-DETR | 76.3 | 47.7 | 12.9 | 38.2 | 63.9 |
| Conditional-DETR | 78.0 | 49.6 | 13.8 | 40.0 | 66.1 |

Based on the data from Tables 2 and 3, in the comparative experiments between CNN-based detectors and Transformer-based detectors, AG-YOLO achieved the best performance across metrics, including precision, recall, mAP50, mAP50:95, $AP_s$ (AP for small objects), $AP_m$ (AP for medium objects), and $AP_l$ (AP for large objects). This achievement can be attributed to the comprehensive utilization of Transformer model's contextual modeling ability and the global context fusion mechanism. This enables the model to better understand the contextual information within the images, effectively addressing severe occlusion issues and improving the accuracy of object detection.

### 4.4. Ablation Study

To validate the effectiveness of individual modules and analyze their impact on the precision of the AG-YOLO algorithm, this study conducted ablation experiments. YOLOv5s with the CSPDarknet53 backbone network was employed as the baseline for comparison. The results of the ablation experiments conducted in this research are presented in Table 4. According to the data in Table 4, the ablation experiments illustrate that using NextViT as the backbone network and introducing the Global Context Fusion Module enhances both the precision and recall of the model, consequently reducing misidentification and omission rates. Compared to the baseline model, employing the NextViT backbone network results in an improvement of 4.0 and 2.7 points in mAP50 and mAP50:95, respectively. This signifies that utilizing NextViT as the backbone network helps the network focus on obscured target features, thus enhancing downstream task accuracy. Notably, integrating the Global Context Fusion Module with the NextViT backbone network optimizes model performance. In comparison to the baseline model, metrics such as P, R, mAP50, and mAP50:95, respectively, improve by 3.9, 5.7, 4.2, and 3.2 points, highlighting the effectiveness of NextViT combined with the Global Context Fusion Module.

**Table 4.** Ablation experiments of NextViT backbone and Global Context Fusion Module.

| Network | Backbone | P/% | R/% | mAP50/% | mAP50:95/% |
|---|---|---|---|---|---|
| AG-YOLO w/o GCFM | CSPDarknet53 | 86.7 | 67.7 | 79.0 | 57.1 |
| AG-YOLO w/o GCFM | NextViT | 92.3 | 72.1 | 83.0 | 59.8 |
| AG-YOLO w/ GCFM | CSPDarknet53 | 86.7 | 67.0 | 78.4 | 57.3 |
| AG-YOLO w/ GCFM | NextViT | 90.6 | 73.4 | 83.2 | 60.3 |

## 4.5. Qualitative Analysis Evaluation

To more intuitively demonstrate the superior detection performance of AG-YOLO proposed in this study, comparisons were conducted on the test set of the citrus dataset. The evaluation specifically focused on both one-stage and two-stage detection models: YOLOv5s, YOLOv7, YOLOv8s, Faster R-CNN, and SSD. These models were employed to evaluate AG-YOLO's effectiveness in detecting severely occluded fruit targets. Based on Figures 6 and 7, in scenarios where immature and mature fruits are heavily obscured by leaves and branches, YOLOv5s, YOLOv7, YOLOv8s, Faster R-CNN, and SSD exhibit instances of missed detections. However, AG-YOLO still detects obscured fruit targets. In the GradCAM heatmaps depicted in Figures 6 and 7, color intensity represents the importance of each pixel or region concerning the predicted target class. Darker colors (such as red or blue) indicate more significant influence on the prediction outcome. The YOLOv5s, YOLOv7, YOLOv8s, Faster R-CNN, and SSD models demonstrate insufficient attention to obscured regions, failing to learn the feature connections between obscured and non-obscured areas, thereby failing to detect obscured target fruits. AG-YOLO exhibits higher responsiveness to obscured regions, indicating that the AG-YOLO model can effectively focus on and localize targets even in severely occluded situations. It displays robustness, extracting crucial information from complex scenes to identify and locate obscured targets.

Based on the confusion matrix results in Figure 8, it suggests that AG-YOLO can more accurately identify samples within the citrus category compared to other models. Moreover, it tends to misclassify fewer samples of other categories as citrus compared to these other models. This indicates the superiority of AG-YOLO in recognizing and reducing misclassifications in the citrus category.

**Figure 6.** Comparisons of image detection results and GradCAM thermograms for localized shading conditions in unripe fruits.

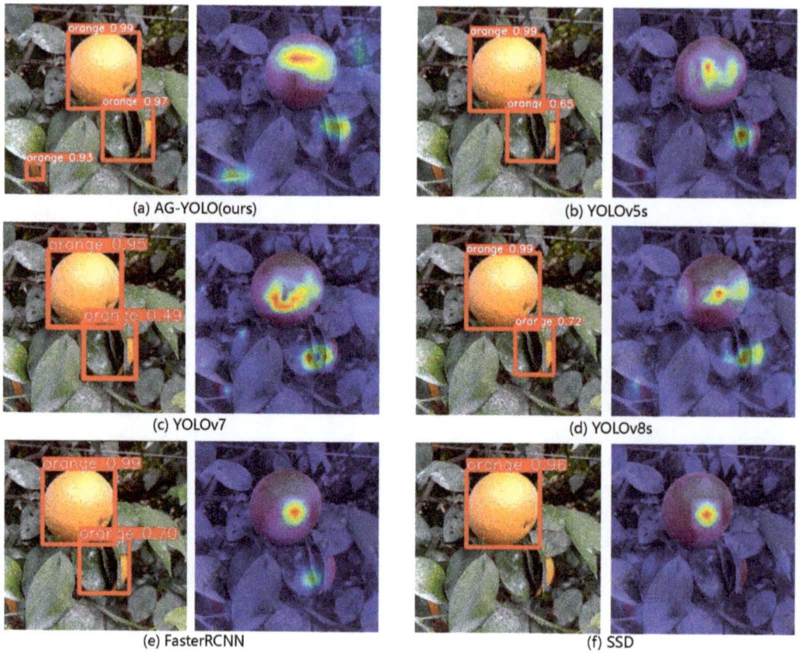


**Figure 7.** Comparisons of image detection results and GradCAM thermograms for localized shading conditions in ripe fruits.

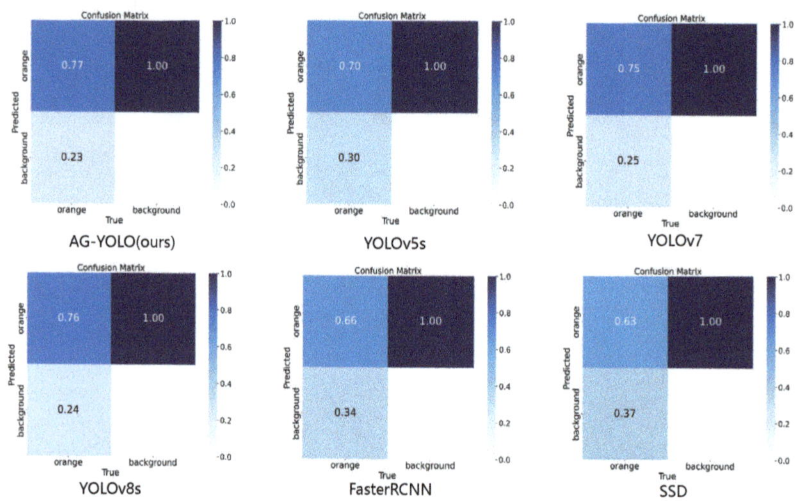


**Figure 8.** Confusion matrix diagrams for the models.

### *4.6. The Detection Performance of AG-YOLO in Different Weather Conditions and PR Curves*

The detection performance of AG-YOLO in cloudy weather is illustrated in Figure 9, and its performance in sunny weather is depicted in Figure 10. Experimental results indicate that AG-YOLO demonstrates stable performance in occluded scenes under various lighting conditions, exhibiting strong robustness. Figure 11 illustrates the PR curves of AG-YOLO.

Figure 9. AG-YOLO's performance in cloudy weather.

Figure 10. AG-YOLO's performance in sunny weather.

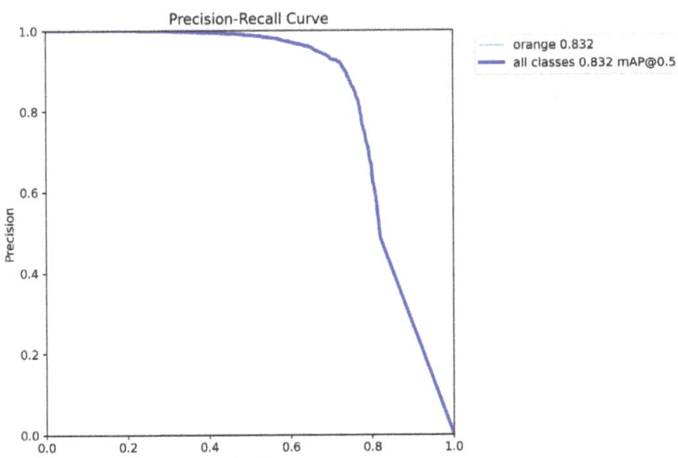


**Figure 11.** PR curves of AG-YOLO.

## 5. Conclusions

This study aims to address the issue of missed detections and low detection rates in citrus target detection due to severe occlusions. To address this challenge, a citrus occlusion dataset was constructed. An innovative network model named AG-YOLO was proposed, incorporating the NextViT backbone along with the Global Context Fusion Module (GCFM). This design led to a significant enhancement in the model's performance and robustness.

NextViT, serving as the backbone of AG-YOLO, surpasses the limitations of traditional convolutional neural networks by introducing a Transformer-based feature extraction method. This allows the model to better capture the global context information within images. NextViT accurately captures the correlation between target regions and their surrounding environment, enhancing target localization accuracy. This robustly addresses the issue of missed detections caused by severe occlusions in citrus detection.

Simultaneously, the Global Context Fusion Module plays a critical role in AG-YOLO. It utilizes self-attention mechanisms to facilitate the interaction and fusion of local and global features. This is crucial in addressing the low detection rates of occluded targets in complex scenes, as occluded targets are often influenced by factors such as complex environments and indistinct image details. However, the Global Context Fusion Module can capture global associative information through contextual information fusion mechanisms, improving the detection effectiveness of occluded targets.

In the experiment, a comprehensive evaluation of the AG-YOLO model was conducted on the citrus occlusion dataset, comparing it against classic CNN-based detection methods as well as the latest Transformer-based detection methods.The results indicate a significant improvement in AG-YOLO, with its precision (P), recall (R), mAP at IoU 50% (mAP50), mAP at IoU 50-95% (mAP50:95), AP for small objects ($AP_s$), AP for medium objects ($AP_m$), and AP for large objects ($AP_l$) reaching 90.6%, 73.4%, 83.2%, 60.3%, 18.8%, 47.2%, and 69.2%, respectively. Such results confirm the effectiveness and practicality of the approach, illustrating AG-YOLO's superiority in addressing severe occlusion issues in fruit detection.

In summary, the proposed AG-YOLO model in this study successfully addressed the challenge of severe occlusion in citrus detection by effectively leveraging the global perceptual capabilities of the NextViT backbone network and the feature interaction and fusion abilities of the Global Context Fusion Module.This research contributes to the advancement of object detection techniques and offers valuable insights and explorations for research and practical applications in agricultural intelligence.

**Author Contributions:** Conceptualization, Y.L. (Yishen Lin) and Y.L. (Yun Liang); methodology, Y.L. (Yishen Lin) and Z.H.; software, Y.L. (Yishen Lin) and Z.H.; validation, Z.H.; investigation, Z.H.; resources, Y.L. (Yun Liang); data curation, Z.H., Y.L. (Yunfan Liu), and W.J.; writing—original draft preparation, Y.L. (Yishen Lin); writing—review and editing, Y.L. (Yishen Lin); supervision, Y.L. (Yun Liang); project administration, Y.L. (Yun Liang). All authors have read and agreed to the published version of the manuscript.

**Funding:** This research received no external funding.

**Data Availability Statement:** The raw data supporting the conclusions of this article will be made available by the authors on request.

**Conflicts of Interest:** The authors declare no conflicts of interest.

## References

1. Hamuda, E.; Mc Ginley, B.; Glavin, M.; Jones, E. Improved image processing-based crop detection using Kalman filtering and the Hungarian algorithm. *Comput. Electron. Agric.* **2018**, *148*, 37–44. [CrossRef]
2. Lu, J.; Sang, N. Detecting citrus fruits and occlusion recovery under natural illumination conditions. *Comput. Electron. Agric.* **2015**, *110*, 121–130. [CrossRef]
3. Lin, X.; Li, L.; Gao, Z.; Yi, C.; Li, Q. Revised quasi-circular randomized Hough transform and its application in camellia-fruit recognition. *Trans. Chin. Soc. Agric. Eng.* **2013**, *29*, 164–170.
4. Song, H.; He, D.; Pan, J. Recognition and localization methods of occluded apples based on convex hull theory. *Trans. Chin. Soc. Agric. Eng.* **2012**, *28*, 174–180.

5. Feng, Q.; Chen, W.; Yang, Q. Identification and localization of overlapping tomatoes based on linear structured vision system. *J. China Agric. Univ.* **2015**, *20*, 100–106.
6. Sun, Sashuang.; Wu, Q.; Tan, J.; Long, Y.; Song, H. Recognition and reconstruction of single apple occluded by branches. *J. Northwest A&F Univ. (Nat. Sci. Ed.)* **2017**, *45*, 138–146.
7. Sun, J.; Sun, Y.; Zhao, R.; Ji, Y.; Zhang, M.; Li, H. Tomato Recognition Method Based on Iterative Random Circle and Geometric Morphology. *Trans. Chin. Soc. Agric. Mach.* **2019**, *50*, 22–26.
8. Kuznetsova, A.; Maleva, T.; Soloviev, V. Using YOLOv3 algorithm with pre-and post-processing for apple detection in fruit-harvesting robot. *Agronomy* **2020**, *10*, 1016. [CrossRef]
9. Sozzi, M.; Cantalamessa, S.; Cogato, A.; Kayad, A.; Marinello, F. Automatic bunch detection in white grape varieties using YOLOv3, YOLOv4, and YOLOv5 deep learning algorithms. *Agronomy* **2022**, *12*, 319. [CrossRef]
10. Zhang, C.; Li, T.; Zhang, W. The detection of impurity content in machine-picked seed cotton based on image processing and improved YOLOV4. *Agronomy* **2021**, *12*, 66. [CrossRef]
11. Redmon, J.; Divvala, S.; Girshick, R.; Farhadi, A. You only look once: Unified, real-time object detection. In Proceedings of the IEEE Conference on Computer Vision and Pattern Recognition, Las Vegas, NV, USA, 26 June–1 July 2016; pp. 779–788.
12. Redmon, J.; Farhadi, A. YOLO9000: better, faster, stronger. In Proceedings of the IEEE Conference on Computer Vision and Pattern Recognition, Honolulu, HI, USA 21–26 July 2017; pp. 7263–7271.
13. Redmon, J.; Farhadi, A. Yolov3: An incremental improvement. *arXiv* **2018**, arXiv:1804.02767.
14. Bochkovskiy, A.; Wang, C.Y.; Liao, H.Y.M. Yolov4: Optimal speed and accuracy of object detection. *arXiv* **2020**, arXiv:2004.10934.
15. Koirala, A.; Walsh, K.; Wang, Z.; McCarthy, C. Deep learning for real-time fruit detection and orchard fruit load estimation: Benchmarking of 'MangoYOLO'. *Precis. Agric.* **2019**, *20*, 1107–1135. [CrossRef]
16. Tian, Y.; Yang, G.; Wang, Z.; Wang, H.; Li, E.; Liang, Z. Apple detection during different growth stages in orchards using the improved YOLO-V3 model. *Comput. Electron. Agric.* **2019**, *157*, 417–426. [CrossRef]
17. Huang, G.; Liu, Z.; Van Der Maaten, L.; Weinberger, K.Q. Densely connected convolutional networks. In Proceedings of the IEEE Conference on Computer Vision and Pattern Recognition, Honolulu, HI, USA, 21–26 July 2017; pp. 4700–4708.
18. Yang, J.; Qian, Z.; Zhang, Y.; Qin, Y.; Miao, H. Real-time recognition of tomatoes in complex environments based on improved YOLOv4-tiny. *Trans. Chin. Soc. Agric. Eng* **2022**, *9*, 215–221.
19. Tongbin, H. Citrus fruit recognition method based on the improved model of YOLOv5. *J. Huazhong Agric. Univ.* **2022**, *41*, 170–177.
20. Girshick, R.; Donahue, J.; Darrell, T.; Malik, J. Rich feature hierarchies for accurate object detection and semantic segmentation. In Proceedings of the IEEE Conference on Computer Vision and Pattern Recognition, Columbus, OH, USA, 23–28 June 2014; pp. 580–587.
21. Mu, Y.; Chen, T.S.; Ninomiya, S.; Guo, W. Intact detection of highly occluded immature tomatoes on plants using deep learning techniques. *Sensors* **2020**, *20*, 2984. [CrossRef]
22. He, K.; Zhang, X.; Ren, S.; Sun, J. Deep residual learning for image recognition. In Proceedings of the IEEE Conference on Computer Vision and Pattern Recognition, Las Vegas, NV, USA, 26 June–1 July 2016; pp. 770–778.
23. Afonso, M.; Fonteijn, H.; Fiorentin, F.S.; Lensink, D.; Mooij, M.; Faber, N.; Polder, G.; Wehrens, R. Tomato fruit detection and counting in greenhouses using deep learning. *Front. Plant Sci.* **2020**, *11*, 571299. [CrossRef]
24. He, K.; Gkioxari, G.; Dollár, P.; Girshick, R. Mask r-cnn. In Proceedings of the IEEE International Conference on Computer Vision, Venice, Italy, 22–29 October 2017; pp. 2961–2969.
25. Bi, S.; Gao, F.; Chen, J.; Zhang, L. Detection Method of Citrus Based on Deep Convolution Neural Network. *Trans. Chin. Soc. Agric. Mach.* **2019**, *50*, 181–186.
26. Li, J.; Xia, X.; Li, W.; Li, H.; Wang, X.; Xiao, X.; Wang, R.; Zheng, M.; Pan, X. Next-vit: Next generation vision transformer for efficient deployment in realistic industrial scenarios. *arXiv* **2022**, arXiv:2207.05501.
27. Dosovitskiy, A.; Beyer, L.; Kolesnikov, A.; Weissenborn, D.; Zhai, X.; Unterthiner, T.; Dehghani, M.; Minderer, M.; Heigold, G.; Gelly, S.; et al. An image is worth 16x16 words: Transformers for image recognition at scale. *arXiv* **2020**, arXiv:2010.11929.
28. Vaswani, A.; Shazeer, N.; Parmar, N.; Uszkoreit, J.; Jones, L.; Gomez, A.N.; Kaiser, Ł.; Polosukhin, I. Attention is all you need. *Adv. Neural Inf. Process. Syst.* **2017**, *30*, 5998–6008.
29. Simonyan, K.; Zisserman, A. Very deep convolutional networks for large-scale image recognition. *arXiv* **2014**, arXiv:1409.1556.
30. Howard, A.G.; Zhu, M.; Chen, B.; Kalenichenko, D.; Wang, W.; Weyand, T.; Andreetto, M.; Adam, H. Mobilenets: Efficient convolutional neural networks for mobile vision applications. *arXiv* **2017**, arXiv:1704.04861.
31. Tan, M.; Le, Q. Efficientnet: Rethinking model scaling for convolutional neural networks. In Proceedings of the International Conference on Machine Learning, PMLR, Long Beach, CA, USA, 9–15 June 2019; pp. 6105–6114.
32. Liu, S.; Li, F.; Zhang, H.; Yang, X.B.; Qi, X.; Su, H.; Zhu, J.; Zhang, L. DAB-DETR: Dynamic Anchor Boxes are Better Queries for DETR. *arXiv* **2022**, arXiv:2201.12329v4.
33. Li, F.; Zhang, H.; Liu, S.; Guo, J.; Ni, L.M.; Zhang, L. DN-DETR: Accelerate DETR Training by Introducing Query DeNoising. In Proceedings of the 2022 IEEE/CVF Conference on Computer Vision and Pattern Recognition (CVPR), New Orleans, LA, USA, 18–24 June 2022; pp. 13609–13617.
34. Liu, Y.; Zhang, Y.; Wang, Y.; Zhang, Y.; Tian, J.; Shi, Z.; Fan, J.; He, Z. SAP-DETR: Bridging the Gap Between Salient Points and Queries-Based Transformer Detector for Fast Model Convergency. In Proceedings of the 2023 IEEE/CVF Conference on Computer Vision and Pattern Recognition (CVPR), Vancouver, BC, Canada 18–22 June 2023; pp. 15539–15547.

35. Wang, Y.; Zhang, X.; Yang, T.; Sun, J. Anchor DETR: Query Design for Transformer-Based Object Detection. *arXiv* **2021**, arXiv:2109.07107.
36. Meng, D.; Chen, X.; Fan, Z.; Zeng, G.; Li, H.; Yuan, Y.; Sun, L.; Wang, J. Conditional DETR for Fast Training Convergence. In Proceedings of the 2021 IEEE/CVF International Conference on Computer Vision (ICCV), Montreal, QC, Canada, 10–17 October 2021; pp. 3631–3640.

MDPI AG
Grosspeteranlage 5
4052 Basel
Switzerland
Tel.: +41 61 683 77 34

*Agriculture* Editorial Office
E-mail: agriculture@mdpi.com
www.mdpi.com/journal/agriculture

www.ingramcontent.com/pod-product-compliance
Lightning Source LLC
Chambersburg PA
CBHW061759210926
51850CB00006B/596
* 9 7 8 3 0 3 9 2 8 5 9 7 6 *